Series in Pure Mathematics

COMPLEX ANALYSIS

SERIES IN PURE MATHEMATICS

Editor: C C Hsiung
Associate Editors: S S Chern, S. Kobayashi, I Satake, Y-T Siu, W-T Wu
and M Yamaguti

Part I.   Monographs and Textbooks

Part II. Lecture Notes

Series in Pure Mathematics – Volume 9

COMPLEX ANALYSIS

Theral O. Moore
Department of Mathematics
University of Florida, USA

and

Edwin H. Hadlock
Late Prof. Emeritus

World Scientific
Singapore • New Jersey • London • Hong Kong

Published by

World Scientific Publishing Co. Pte. Ltd.
P O Box 128, Farrer Road, Singapore 9128
USA office: 687 Hartwell Street, Teaneck, NJ 07666
UK office: 73 Lynton Mead, Totteridge, London N20 8DH

COMPLEX ANALYSIS

ISBN 981-02-0246-6
      981-02-0247-4 (pbk)

Printed in Singapore by Loi Printing Pte. Ltd.

To Janet Hadlock
and
Nancy Moore

Preface

This book was written to be used as a textbook for an introductory course in complex functions. Such a course is expected to serve both undergraduate and graduate students in engineering, physics, and mathematics. Thus our goals are manifold.

Our first objective was to make the material accessible to students whose backgrounds include little or no mathematics beyond the calculus. Our second goal was to be rigorous in proofs and clear and precise in defining concepts and stating theorems—without sacrificing ease of material for students with little experience in proofs. To help these students, we have written explicit, detailed proofs, citing the justification for almost every step in each proof. We have also included solutions to many exercises. In a short course, students may not have enough time to digest all proofs in the part of the book covered in that course. Nevertheless, these proofs are recorded for those who do have sufficient time and inclination, since this book may be used in a three, four, or five hour course, or even a two semester course. Each instructor may choose selected proofs which will give the student adequate experience and growth.

Another goal of the authors has been to enhance the students appreciation of the nature of mathematics. We hope that our preciseness and concepts and exactness in proofs will help to meet this goal. If we can meet this goal and at the same time give the student a deeper understanding of the subject, then why not do so? To the student who does not find time to explore the appendix in this book while taking the first course in complex functions, we cordially extend an invitation to browse through this appendix at a later date as time permits. We believe the non-mathematics major who uses higher mathematics will be delighted to reflect on the ideas in the appendix. We think that such a professional will enjoy the realization that the system of real numbers is not something which we should fully understand just because we may feel we grew up with it throughout our school days. We hope that the first section in the appendix will convey the understanding that the system of real numbers is an abstract system which satisfies all of the conditions of a complete ordered field as indicated in A.1.7. We also believe it will be a joy to many students as they read or review the material on countable and uncountable sets, which shows that the set of rational numbers is countable but the set of irrational numbers is uncountable.

In a brief course (unless it is an advanced course), the material on compactness will surely need to be omitted. It will then be emphasized to the student that throughout the book, compactness in the complex plane is equivalent to closed and boundedness.

Also in a short course, it can be accepted without proof (as in the usual elementary textbook) that a continuous real-valued function on a closed and bounded set D in the complex plane will assume a maximum value and a minimum value in D. Even though many classes omit the material on compactness, the student can see what is involved here.

We do hope to suggest to the student that some of the most basic concepts in real and complex analysis are topological concepts.

The theory on Laurent series, except for the uniqueness theorem, is developed in Chapter Six without the use of uniform convergence. Thus Sections 4.4 and 4.5 and part of Chapter Three may be omitted.

The formal style of Definition—Theorem—Proof is used to help the inexperienced student keep track of and locate results already established. This also helps the student put together the pieces in a proof and know when the proof is complete. To the experienced person, at a glance, this formal approach may seem more difficult than the informal relaxed exposition. However, one must find proofs of the same result in the two styles and compare these proofs line by line in order to decide which is easier. Often the inexperienced student glides over some points in the informal argument, never being aware that these steps must be justified. We mention that in Section 1.2, we use the polar form to prove the associative law of multiplication. In polar form, this proof is trivial, but in informal presentations, the associative law is usually discussed in cartesian form. In this form, the details are extremely tedious if indeed they are verified by the reader.

It is not possible for the undersigned author to express adequately on these pages his heartfelt praise and appreciation of his wife, Nancy, who spent countless hours typesetting this entire book. I also express my appreciation to my sons, Steven and David, for their assistance in the preparation of the figures. We are especially grateful to George Fischer and his very capable assistant, Brian Bartholomew, for the invaluable technical assistance which they have given in the Mathematics Computer Laboratory of the Department of Mathematics at the University of Florida. This author is extremely grateful to Jerome Schwartz and Jose Melgarejo for their valuable help and suggestions in the preparation of this book. To Pamela Ryan and Joel Mathews, I express my thanks for many pleasant hours of valuable assistance.

Theral O. Moore
Gainesville, Florida

Contents

Chapter 1

Complex Numbers

1.1 Complex Numbers

In the calculus, it became clear that there is a very important structure in the real number system other than the algebraic operations of addition and multiplication. The concepts of limit and continuity depend upon this structure so that, indeed, **it is basic to all of calculus**. (Derivatives, definite integrals and sums of infinite series may be defined as limits.) These limits may be defined in terms of the "less than" relation ($<$), called an **order relation**.

We use R to denote the **set** of all real numbers and to emphasize both the algebraic and "limit" structures in R, we use

$$(R, +, \cdot, <)$$

to denote the **system of real numbers** .

The student has met complex numbers as roots of quadratic equations with negative discriminants. He probably wrote a complex number z in the form $z = a + ib$ where a and b are in R. The complex number z was determined by the ordered pair (a, b) of real numbers, and conversely. This suggests that we may use the notation (a, b) for the complex number $(a + ib)$, if we like. In practice, the notation $a + ib$ is desirable, and we shall return to it. But for smoothness in our definitions, we shall use the ordered pair notation. (Also, the student should be introduced to this scheme, since it is often used.)

We shall use R^2 to denote the set of all **ordered pairs** of real numbers where it is understood that if (x, y) and (u, v) are members of R^2, then

$$(x, y) = (u, v) \quad \text{iff} \quad x = u \text{ and } y = v. \tag{1.1}$$

We use "iff" as an abbreviation for "if and only if" throughout this book.

1

Definition 1.1.1 The **complex number system** is the system $(R^2, +, \cdot, |\ |)$ where for each **pair** $z = (x, y)$ and $w = (u, v)$ of members of R^2,

$$z + w = (x + u,\ y + v), \tag{1.2}$$

$$zw = z \cdot w = (xu - yv,\ xv + yu), \tag{1.3}$$

and

$$|z| = \sqrt{x^2 + y^2} \quad \text{(the non-negative square root)}. \tag{1.4}$$

In this system, each member z of R^2 is called a **complex number**, and $|z|$ is called the **absolute value of z** or the **modulus of z**. If $z = (x, y)$, then x and y are called the **real** and **imaginary parts** of z, respectively, and we write

$$x = \mathcal{R}(z) \quad \text{and} \quad y = \mathcal{I}(z).$$

The **commutative law** of addition in R^2

$$z + w = w + z$$

follows immediately from (1.2) and the commutativity of addition in R as follows

$$z + w = (x, y) + (u, v) = (x + u,\ y + v) = (u + x,\ v + y) = w + z.$$

Similarly, we use the **associative law** of addition in R and (1.2) to prove the associative law of addition in R^2. If $z = (x, y)$, $w = (u, v)$, and $p = (a, b)$ are any complex numbers, then

$$
\begin{aligned}
(z + w) + p &= ([x + u] + a,\ [y + v] + b) \quad \text{by (1.2)} \tag{1.5} \\
&= (x + [u + a],\ y + [v + b]) \quad \text{by an associative law in } R \\
&= z + (w + p) \quad \text{by (1.2)}.
\end{aligned}
$$

To prove the **distributive law** of multiplication over addition in R^2, we observe that

$$
\begin{aligned}
p(z + w) &= (a, b)(x + u,\ y + v) & \text{by (1.2)} \\
&= (a[x + u] - b[y + v],\ a[y + v] + b[x + u]) & \text{by (1.3)} \\
&= (ax - by,\ ay + bx) + (au - bv,\ av + bu) & \text{by (1.2) and laws in } R \\
&= pz + pw & \text{by (1.3)}.
\end{aligned}
$$

Proofs of the **commutative and associative laws** of multiplication in R^2 are postponed, since these laws (and other properties of multiplication) follow much more readily in polar form.

We now observe that by $(1.2), (1.3)$ and (1.4), the following equations hold.

$$
\begin{aligned}
(x, 0) + (u, 0) &= (x + u, 0) \\
(x, 0)(u, 0) &= (xu, 0) \\
|(x, 0)| &= \sqrt{x^2} = |x|
\end{aligned} \tag{1.6}
$$

Equations (1.6) suggest that it is convenient and natural to identify each complex number of the form $(x, 0)$ with the real number x. Indeed, we make this identification and we write

$$(x, 0) = x \quad \text{for each real number } x.$$

Thus we replace each pair $(x, 0)$ by the real number x. Now each real number is a special complex number; and in view of (1.6), $(R, +, \cdot)$ is a subsystem of $(R^2, +, \cdot)$. To use a real number x as a complex number, we write $(x, 0)$ for x and apply our rules for ordered pairs. In particular, by 1.3,

$$xw = (x, 0)(u, v) = (xu, xv). \tag{1.7}$$

Taking $x = 1$ in (1.7), it follows that $1w = w$. By (1.7),

$$y(0, 1) = (0, y) \quad \text{for each real number } y. \tag{1.8}$$

It is convenient and customary to denote $(0, 1)$ by i. Thus we write

$$i = (0, 1). \tag{1.9}$$

Then for each complex number $z = (x, y)$, we have

$$z = (x, y) = (x, 0) + (0, y) = x + yi \qquad \text{by (1.8) and (1.9)}.$$

Henceforth, we may use (x, y) or $x + yi$ to denote z.

We next observe that for each $z = (x, y)$ in R^2,

$$0 + z = z + 0 = (x, y) + (0, 0) = z \tag{1.10}$$

and

$$z + (-x, -y) = 0 = (-x, -y) + z. \tag{1.11}$$

Because of (1.11), we use $-z$ to denote $(-x, -y)$ (since clearly no other member of R^2 can be added to z to give 0).

Theorem. If z and w are complex numbers, then there is a unique p in R^2 such that

$$p + w = z;$$

namely, $p = z + (-w)$.

Proof. First, if $p = z + (-w)$, then $p + w = [z + (-w)] + w = z$ by $(1.5), (1.11)$ and (1.10).

Conversely, if

$$p + w = z, \tag{1.12}$$

then

$$\begin{aligned} p &= (p + w) + (-w) & \text{by (1.5), (1.11), and (1.10)} \\ &= z + (-w) & \text{by (1.12).} \end{aligned}$$

Because of this theorem, it is natural to use $z - w$ to denote $z + (-w)$. Thus if $z = (x, y)$ and $w = (u, v)$, then

$$z - w = z + (-w) = (x - u, y - v). \tag{1.13}$$

Exercises 1.1.2 Answers to many exercises in this book appear at the right margin by the respective exercises.

1. Prove each of the following results where z is an arbitrary complex number and b is an arbitrary real number.

    (a) $i^2 = -1$ where i^2 means $i \cdot i$

    (b) $0z = 0$

    (c) $(-1)z = -z$

    (d) $(-b)i = -(bi)$

2. If x, y, u and v are real, use (1.3) to write the product $(x + yi)(u + vi)$ in the form $a + bi$ where a and b are real.

3. (a) Conclude from the answer to Ex. 2 that we may compute $(x + yi)(u + vi)$ as if each factor were a real binomial and replace i^2 by -1.

    (b) Conclude that,

$$\begin{aligned} x - yi &= x + (-(yi)) \quad \text{by (1.13)} \\ &= x + (-y)i \quad \text{by Ex. 1(d)} \end{aligned}$$

    so that $(x - yi)(u \pm vi)$ may be computed just as stated in Part (a).

4. Write $(2 + 3i)(1 - 5i)$ in the form $a + bi$ and also in the form $a - ci$ where a, b and c are real.   Ans. $17 + (-7)i = 17 - 7i$

Proof of Ex. 1(c). $\begin{aligned}[t] (-1)z &= (-1)(x, y) = (-x, -y) \quad \text{by (1.7)} \\ &= -z \qquad\qquad\qquad\quad \text{by definition of } -z \end{aligned}$

Proof of Ex. 1(d). $(-b)i = (-b)(0, 1) = (0, -b) = -(0, b) = -(bi)$

Solution of Ex. 2. $(x + yi)(u + vi) = (x, y)(u, v) = (xu - yv, \ xv + yu) = xu - yv + (xv + yu)i$ by (1.3).

Geometric Representations. It is natural to represent the complex number $z = x + yi = (x, y)$ by the point in the xy-plane whose rectangular cartesian coordinates are (x, y) relative to a given set of axes and a given unit of distance in the plane. The origin represents $z = 0$. The x-axis is called the **real** axis, since points on it represent the real complex numbers $x = (x, 0)$. The y-axis is called the **imaginary** axis since its points represent the so-called **pure imaginary numbers** — those of the form $(0, y) = yi$.

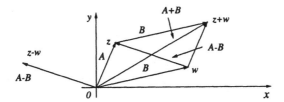

Figure 1.1: Vector representation of complex numbers

If $z = (x, y)$, then $|z| = \sqrt{x^2 + y^2}$, which is the distance from the origin to the point (x, y). If $w = (u, v)$, then by (1.13),

$$|z - w| = \sqrt{(x - u)^2 + (y - v)^2}, \qquad (1.14)$$

the distance from (x, y) to (u, v). When the plane is used to represent complex numbers, it is called the **complex plane**.

Sometimes it is useful to represent the complex number $z = (x, y)$ by the **vector** determined by the directed line segment from the origin to the point (x, y). Thus x and y are the respective components of the vector representing z. We recall that to add (or subtract) vectors, we add (or subtract) componentwise; (1.2) and (1.13) show that we add (or subtract) complex numbers in exactly the same way. Thus if A and B are the vectors representing z and w respectively, then $A \pm B$ represent $z \pm w$, respectively. Figure 1.1 shows that the vector representing $z - w$ is determined by the directed segment from point w to point z, and (1.14) confirms the fact that $|z - w|$ is the length of the vector with end points w and z.

Definition 1.1.3 If $z = (x, y)$ in R^2, then $(x, -y)$ is called the **complex conjugate of z** and is denoted by $\overline{z}$.

When $P(z)$ is a statement about z, the notation $\{z : \ P(z)\}$ is used to denote the set of all z's (and only those z's) for which $P(z)$ is true. Thus $\{z : \ \mathcal{I}(z) = 0\}$ is the set of all real complex numbers or the set of all points on the real axis. The set of all points on the line $y = x$ in the complex plane is $\{z : \ \mathcal{I}(z) = \mathcal{R}(z)\}$. Letters other than z may be used in this notation. Thus $\{x : \ x \text{ is real and } x < 2\}$ is the set of all real numbers less than 2.

Exercises 1.1.4

1. Let $z = (x, y)$. Locate the points $z, \overline{z}, -z$ and $-\overline{z}$ in the complex plane. They are the vertices of what kind of figure if $xy \neq 0$?

2. Prove that $z + \overline{z} = 2\mathcal{R}(z)$. State a formula for $z - \overline{z}$.

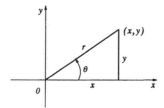

Figure 1.2: Polar form of a complex number

3. Prove each of the following results.

    (a) $z\bar{z} = |z|^2$ (b) $\overline{z \pm w} = \bar{z} \pm \bar{w}$

    (c) $\bar{\bar{z}} = z$ (d) $|\bar{z}| = |z| = |-z|$

4. Describe geometrically each of the following sets and draw a figure for each.

    (a) $\{z : \ |z| = 3\}$ (b) $\{z : \ |z| < 3\}$ (c) $\{z : \ |z| \leq 3\}$

    (d) $\{z : \ z\bar{z} < 9\}$ (e) $\{z : \ \mathcal{R}(z) = 3\}$

    (f) $\{z : \ |z - w| < 3\}$ where $w = (2, 4)$ [See (1.14) for (f).]

1.2 Polar Form of Complex Numbers

Let $z = (x, y) = x + yi$, and let (r, θ) be polar coordinates of the point (x, y) with $r \geq 0$ where the polar axis is the non-negative real axis. See Figure 1.2. (Unless otherwise stated, the number θ is determined by using the radian measure of angles.) Then

$$r = \sqrt{x^2 + y^2} = |z|, \quad x = r\cos\theta \quad \text{and} \quad y = r\sin\theta.$$

Hence

$$z = (x, y) = (r\cos\theta, \ r\sin\theta) = r(\cos\theta, \ \sin\theta) \quad \text{by (1.7).} \quad (1.15)$$

For convenience, we use $\text{cis}\,\theta$ to denote $(\cos\theta, \ \sin\theta) = \cos\theta + (\sin\theta)i$. Then by (1.15), the **polar form** of z can be written as

$$z = r\,\text{cis}\,\theta \quad \text{where } r = |z|$$

and where θ is called **an argument of z**. Now $r = |z|$ is uniquely determined by z, but θ is not unique. For $(r, \theta + 2k\pi)$ are also polar coordinates of the point $z = (x, y)$ for each k in J where J denotes the set $\{0, \pm 1, \pm 2, \cdots\}$ of all integers. Since the original choice of θ could have been any one of the values $\theta + 2k\pi$ where k is in J, each of these values is an argument of z. However, we have the following results.

Remarks 1.2.1 (a) If $w = s \operatorname{cis} \phi$ where $s \geq 0$, then $s = |w|$.

(b) If $w = |w| \operatorname{cis} \theta = |w| \operatorname{cis} \phi \neq 0$, then $\theta - \phi = 2k\pi$ for some integer k.

(c) If $z \neq 0$, then there is a unique argument of z, called the **principal argument** of z and denoted by $\operatorname{Arg} z$, satisfying the condition

$$-\pi < \operatorname{Arg} z \leq \pi.$$

We use $\arg z$ to mean any argument of z.

(d) If $w = r \operatorname{cis} \theta$, then $\overline{w} = r \operatorname{cis}(-\theta)$. For $\overline{w} = \overline{r(\cos\theta, \sin\theta)} = r(\cos\theta, -\sin\theta) = r(\cos[-\theta], \sin[-\theta]) = r \operatorname{cis}(-\theta)$.

Theorem 1.2.2 If $z = r \operatorname{cis} \theta$ and $w = s \operatorname{cis} \phi$, then

$$zw = rs \operatorname{cis}(\theta + \phi). \tag{1.16}$$

(Thus the modulus of the product zw of two complex numbers is the product of their moduli, and an $\arg zw = \arg z + \arg w$.)

Proof. By (1.3), we have

$$
\begin{aligned}
zw &= (r\cos\theta, \; r\sin\theta)(s\cos\phi, \; s\sin\phi) \\
&= (rs[\cos\theta\cos\phi - \sin\theta\sin\phi], \; rs[\cos\theta\sin\phi + \sin\theta\cos\phi]) \\
&= rs(\cos[\theta + \phi], \; \sin[\theta + \phi]).
\end{aligned}
$$

We easily obtain many properties of multiplication (and division) from (1.16). First, we observe by (1.16) that

$$zw = wz \quad \textbf{(commutative law)}, \tag{1.17}$$

since

$$rs = sr \quad \text{and} \quad \theta + \phi = \phi + \theta.$$

Also, by (1.16), if $p = t \operatorname{cis} \alpha$, then

$$(zw)p = z(wp) \quad \textbf{(associative law)}, \tag{1.18}$$

since

$$(rs)t = r(st) \quad \text{and} \quad (\theta + \phi) + \alpha = \theta + (\phi + \alpha).$$

Again by (1.16) and 1.2.1(a),

$$|zw| = |z||w|, \tag{1.19}$$

that is, the absolute value of the product of two complex numbers equals the product of their absolute values. Also,

$$\overline{zw} = \overline{z}\,\overline{w} \tag{1.20}$$

since

$$
\begin{aligned}
\overline{rs\,\text{cis}\,(\theta + \phi)} &= rs\,\text{cis}\,[-(\theta + \phi)] &&\text{by 1.2.1 (d)} \\
&= [r\,\text{cis}\,(-\theta)][s\,\text{cis}\,(-\phi)] &&\text{by (1.16)} \\
&= \overline{z}\,\overline{w}.
\end{aligned}
$$

Theorem 1.2.3 If $z = r\,\text{cis}\,\theta$ and $w = s\,\text{cis}\,\phi \neq 0$, then there is a unique q in R^2 such that

$$wq = z;$$

namely, $q = \left(\frac{r}{s}\right)\text{cis}\,(\theta - \phi)$.

Proof. Clearly $wq = z$. For $wq = (s\,\text{cis}\,\phi)\frac{r}{s}\,\text{cis}\,(\theta - \phi) = r\,\text{cis}\,\theta = z$ by (1.16). To show uniqueness, let

$$wq' = z.$$

Multiplying each side by $\frac{1}{s}\,\text{cis}\,(-\phi)$, we have

$$\left[\frac{1}{s}\,\text{cis}\,(-\phi)\right](s\,\text{cis}\phi)\,q' = \left[\frac{1}{s}\,\text{cis}\,(-\phi)\right]r\,\text{cis}\,\theta \qquad \text{by (1.18)}.$$

Now simplifying both sides of the last equation, we have

$$q' = \left(\frac{r}{s}\right)\text{cis}\,(\theta - \phi) = q.$$

In view of 1.2.3, the **quotient** $\frac{z}{w}$ is defined by $\frac{z}{w} = \frac{r}{s}\,\text{cis}\,(\theta - \phi)$ where $z = r\,\text{cis}\,\theta$ and $w = s\,\text{cis}\,\phi \neq 0$.

From the definition of $\frac{z}{w}$, we observe that the modulus of the quotient of two complex numbers is equal to the quotient of their moduli, and $\arg\left(\frac{z}{w}\right) = \arg z - \arg w$.

From 1.2.2 and the definition of a quotient, we obtain the usual rule for multiplying fractions

$$\frac{z}{w}\frac{p}{q} = \frac{zp}{wq} \qquad \text{if } wq \neq 0. \tag{1.21}$$

(See Ex. 7 in 1.2.4.)

Now, if $w \neq 0$, then

$$
\begin{aligned}
\frac{z}{w} &= \left(\frac{z}{w}\right)\left(\frac{\overline{w}}{\overline{w}}\right) \\
&= \left(\frac{1}{|w|^2}\right)z\overline{w} \qquad \text{by 1.21 and Ex. 3(a) in 1.1.4.}
\end{aligned}
$$

The last equation gives the extremely useful rule for writing the quotient $\frac{z}{w}$ in the form $a + bi$.

> To write $\dfrac{z}{w}$ in the form $a + bi$, we multiply numerator and
>
> denominator by $\overline{w}$, the conjugate of the denominator. (1.22)

By Ex. 3(b) in 1.1.4, (1.20) and Ex. 6 in 1.2.4, the operation of taking conjugates is distributive over addition, subtraction, multiplication and division.

By (1.17),
$$\cos\theta + (\sin\theta)i = \cos\theta + i\sin\theta$$

which explains the notation $\text{cis}\,\theta$ (c for cos and s for sin).

Since $(zw)p = z(wp)$, we use zwp to denote their common value. Similiarly, $zwpq$ has only one meaning, etc.

We use N to denote the set of all **natural numbers** (positive integers) $1, 2, 3, \cdots$. The symbol $\in$ means "is a member (or element) of " and $\notin$ means "is **not** a member of." Now, if $z \in R^2$ and $n \in N$, then z^n denotes the product $z_1 z_2 \cdots z_n$ where $z_1 = z_2 = \cdots = z_n = z$.

Exercises 1.2.4 Answers to many exercises in this book appear at the right margin by the respective exercises.

1. Use 1.22 to write each quotient in the form $a \pm bi$.

   (a) $\dfrac{2 - i}{3i - 1}$ $\qquad\qquad$ $-\dfrac{1}{2} - \left(\dfrac{1}{2}\right)i$

   (b) $\dfrac{6 - 3i}{2i}$ $\qquad\qquad$ $-\dfrac{3}{2} - 3i$

   (c) $\dfrac{1 + i}{(2 - i)(3 + i)}$ $\qquad\qquad$ $\dfrac{3}{25} + \left(\dfrac{4}{25}\right)i$

   (d) $\dfrac{(\overline{1 + 2i})^2}{3i - 4}$ $\qquad\qquad$ i

2. Write each number in polar form.

   (a) 2 $\qquad\qquad$ $2\,\text{cis}\,0$

   (b) $3i$ $\qquad\qquad$ $3\,\text{cis}\,\dfrac{\pi}{2}$

   (c) -1 $\qquad\qquad$ $\text{cis}\,\pi$

   (d) $-2i$ $\qquad\qquad$ $2\,\text{cis}\,\dfrac{3\pi}{2}$

   (e) $1 - \sqrt{3}i$ $\qquad\qquad$ $2\left[\dfrac{1}{2} - \left(\dfrac{\sqrt{3}}{2}\right)i\right] = 2\,\text{cis}\,\dfrac{5\pi}{3}$

    (f) $-1-i$ $\qquad\qquad\qquad\qquad\qquad\qquad\qquad\qquad$ $\sqrt{2}\,\mathrm{cis}\,\dfrac{5\pi}{4}$

    (g) $2+3i$ $\qquad\qquad\qquad\qquad\qquad\qquad\qquad$ $\sqrt{13}\,\mathrm{cis}\left(\mathrm{Arctan}\,\dfrac{3}{2}\right)$

3. Let $z=r\,\mathrm{cis}\,\theta$ where $r>0$. Write each of the following in polar form.

    (a) z^2 $\qquad\qquad\qquad\qquad\qquad\qquad\qquad\qquad\qquad\qquad$ $r^2\,\mathrm{cis}\,2\theta$

    (b) z^n where $n\in N$ $\qquad\qquad\qquad\qquad\qquad\qquad$ $r^n\,\mathrm{cis}\,n\theta$

    (c) two values of w such that $w^2=z$ $\qquad\quad$ $\sqrt{r}\,\mathrm{cis}\,\dfrac{\theta+2k\pi}{2}$ for $k=0,1$

    (d) three values of p such that $p^3=z$ $\qquad$ $\sqrt[3]{r}\,\mathrm{cis}\,\dfrac{\theta+2k\pi}{3}$ for $k=0,1,2$

4. Prove each result.

    (a) If $z_1z_2=z_1z_3$ and $z_1\neq0$, then $z_2=z_3$. **Hint.** Multiply by $\frac{1}{z_1}$.

    (b) If $z_1z_2=0$, then $z_1=0$ or $z_2=0$. **Hint.** If $z_1\neq0$, then $z_1z_2=0=z_10$.

5. Prove each result.

    (a) $(z_1+z_2)(z_3+z_4)=z_1z_3+z_1z_4+z_2z_3+z_2z_4$

    (b) If $z\neq1$ and $n\in N$, then $1+z+z^2+\cdots+z^{n-1}=\dfrac{1-z^n}{1-z}$.

6. Let $z=r\,\mathrm{cis}\,\theta$ and $w=s\,\mathrm{cis}\,\phi$. Prove if $w\neq0$, then $\overline{\left(\dfrac{z}{w}\right)}=\dfrac{\bar{z}}{\bar{w}}$.

7. If $z_k=r_k\,\mathrm{cis}\,\theta_k$ for $k=1,2,3,4$ and $z_2z_4\neq0$, prove $\dfrac{z_1}{z_2}\dfrac{z_3}{z_4}=\dfrac{z_1z_3}{z_2z_4}$.

Definition 1.2.5 If $z\in R^2$, $z\neq0$ and $n\in N$, then

$$z^0=1 \quad\text{and}\quad z^{-n}=\frac{1}{z^n}.$$

Theorem 1.2.6 If $z=r\,\mathrm{cis}\,\theta$ and $r>0$, then for each n in J,

$$z^n=r^n\,cis\,n\theta. \tag{1.23}$$

When $r=1$, this is called **DeMoivre's Theorem**.

Proof. If $n\in N$, then (1.23) is Ex. 3(b) in 1.2.4. If $n=0$, then (1.23) is clear. Now suppose $n=-m$ where $m\in N$. Then

$$z^n=z^{-m}=\frac{1}{z^m}=\frac{1\,\mathrm{cis}\,0}{r^m\,\mathrm{cis}\,m\theta}=r^{-m}\,\mathrm{cis}\,(-m\theta)=r^n\,\mathrm{cis}\,n\theta.$$

Theorem 1.2.7 Let $n \in N$. The equation

$$z^n = a \quad \text{where } a = r \operatorname{cis} \theta \text{ and } r > 0 \tag{1.24}$$

has exactly n distinct **roots**, and they are given by

$$z = \sqrt[n]{r} \operatorname{cis} \frac{\theta + 2k\pi}{n} \quad \text{where } k = 0, 1, 2, \cdots, n-1. \tag{1.25}$$

Each of these roots is called an nth root of a.

Proof. By 1.2.6, each number in (1.25) is a root of (1.24). Now let $s \operatorname{cis} \phi$ be any root of (1.24) where $s \geq 0$. Then

$$(s \operatorname{cis} \phi)^n = s^n \operatorname{cis} n\phi = a = r \operatorname{cis} \theta.$$

Hence by 1.2.1(a) and 1.2.1(b), there is an integer m such that $s^n = r$ and $n\phi = \theta + 2m\pi$, that is, $s = \sqrt[n]{r}$ and $\phi = \frac{\theta + 2m\pi}{n}$. Thus all roots of (1.24) are of the form

$$\sqrt[n]{r} \operatorname{cis} \frac{\theta + 2m\pi}{n} \quad \text{where } m \in J.$$

But if $m < 0$ or if $m \geq n$, we let q and k be integers such that

$$m = qn + k \quad \text{and} \quad 0 \leq k < n.$$

Then

$$\operatorname{cis} \frac{\theta + 2m\pi}{n} = \operatorname{cis} \frac{\theta + 2qn\pi + 2k\pi}{n} = \operatorname{cis} \frac{\theta + 2k\pi}{n}.$$

Hence (1.25) gives all roots of (1.24).

Finally, if $0 \leq k_1 < k_2 \leq n-1$, then

$$\frac{\theta + 2k_2\pi}{n} - \frac{\theta + 2k_1\pi}{n} = \left(\frac{k_2 - k_1}{n} \right) 2\pi,$$

which is not an integral multiple of 2π. Hence by 1.2.1(b), the roots in (1.25) are distinct.

Remark 1.2.8 (a) Geometrically, the n distinct nth roots of a in 1.2.7 are the vertices of a regular polygon inscribed in the circle with center at $z = 0$ and radius $\sqrt[n]{|a|}$ [since arguments of consecutive roots in (1.25) differ by $\frac{2\pi}{n}$].

(b) We use ω_n to denote $\operatorname{cis} \frac{2\pi}{n}$ where $n \in N$. The n distinct nth roots of 1 are ω_n^k where $k = 0, 1, 2, \cdots, n-1$.

Exercises 1.2.9

1. Find all cube roots of each number.

   (a) -8 $-2,\; 1 \pm \sqrt{3}i$

   (b) i $-i,\; \dfrac{\pm\sqrt{3}+i}{2}$

2. (a) Find all 4th roots of $-2 + 2\sqrt{3}i$. $\pm\dfrac{\sqrt{2}}{2}\left(1 - \sqrt{3}i\right),\; \pm\dfrac{\sqrt{2}}{2}\left(\sqrt{3}+i\right)$

   (b) Solve the equation $z^4 + 16 = 0$ and
       use the result to write $z^4 + 16$ as the
       product of quadratic factors with
       real coefficients. $\pm\sqrt{2}(1 + i),\; \pm\sqrt{2}(1 - i)$

3. Prove $(-z)w = -(zw) = z(-w)$. **Hint.** $(-z)w = [(-1)z]w = (-1)(zw) = -(zw) = -(wz) = (-w)z = z(-w)$

4. (a) Prove if $n > 1$, then $1 + \omega_n + \omega_n^2 + \cdots + \omega_n^{n-1} = 0$. Use Ex. 5(b) in 1.2.4.
   (b) Show that the roots of $az^2 + bz + c = 0$ are given by the usual quadratic
       formula when $a, b,$ and c are in R^2 and $a \neq 0$.

5. Prove if z and w are complex numbers and m and n are in N, then

$$z^m z^n = z^{m+n}, \quad (z^m)^n = z^{mn}, \quad \text{and} \quad (zw)^n = z^n w^n.$$

   Hint. Use the definition of z^n or 1.2.6.

6. Prove i^n has one of the values ± 1 or $\pm i$, if n is an integer. **Hint.** $n = 4q + r$ where q and r are integers and $0 \leq r < 4$.

7. Prove the nth roots of a are $z\omega_n^k$ for $k = 0, 1, 2, \cdots, n - 1$ where z is any nth root of a.

1.3   Inequalities

From $|z| = \sqrt{[\mathcal{R}(z)]^2 + [\mathcal{I}(z)]^2}$, we have

$$|z| \geq |\mathcal{R}(z)| \geq \mathcal{R}(z) \quad \text{and} \quad |z| \geq |\mathcal{I}(z)| \geq \mathcal{I}(z). \tag{1.26}$$

Theorem 1.3.1 (Triangle Inequalities). If $z, w, p, Z,$ and W are complex numbers, then

(a) $|z + w| \leq |z| + |w|,$

(b) $|Z - W| \leq |Z - p| + |p - W|$, and

(c) $|Z - W| \geq ||Z| - |W||$.

Proof of (a).

$$
\begin{aligned}
|z + w|^2 &= (z + w)\overline{(z + w)} = (z + w)(\overline{z} + \overline{w}) && \text{by Ex. 3 in 1.1.4} \\
&= |z|^2 + (z\overline{w} + \overline{z}w) + |w|^2 \\
&= |z|^2 + 2\mathcal{R}(z\overline{w}) + |w|^2 && \text{by Ex. 2 in 1.1.4 since } \overline{z\overline{w}} = \overline{z}w \\
&\leq |z|^2 + 2|z\overline{w}| + |w|^2 && \text{by (1.26)} \\
&= |z|^2 + 2|z||w| + |w|^2 && \text{since } |\overline{w}| = |w| \\
&= (|z| + |w|)^2 .
\end{aligned}
$$

Thus $|z + w| \leq |z| + |w|$.

Proof of (b). Part (b) follows from Part (a) by taking $z = Z - p$ and $w = p - W$.

Proof of (c). Now $|Z| = |(Z - W) + W| \leq |Z - W| + |W|$ by Part (a). Hence

$$|Z - W| \geq |Z| - |W|. \tag{1.27}$$

Also,

$$
\begin{aligned}
|W| &= |-W| \\
&= |(Z - W) + (-Z)| \\
&\leq |Z - W| + |Z| && \text{by Part (a)}
\end{aligned}
$$

and hence

$$|Z - W| \geq |W| - |Z|. \tag{1.28}$$

Now Part (c) follows from (1.27) and (1.28). The reader should draw figures to illustrate Parts (a), (b), and (c). The figures explain the name "Triangle Inequalities."

Exercises 1.3.2

In Ex. $1-2$, prove each result.

1. (a) $|z + w + p| \leq |z| + |w| + |p|$

   (b) $\left| \sum_{k=1}^{n} z_k \right| \leq \sum_{k=1}^{n} |z_k|$

2. (a) $|z + w| \geq ||z| - |w||$

   (b) $\left| \dfrac{p}{z + w} \right| \leq \dfrac{|p|}{||z| - |w||}$ if $|z| \neq |w|$

3. Let $\theta = \arg z$ and $\phi = \arg w$.

   (a) Prove $z\overline{w} = |z||w| \operatorname{cis}(\theta - \phi)$.

(b) Suppose $zw \neq 0$. Prove $z\overline{w} = |z||w|$ iff $\theta - \phi = 2k\pi$ for some integer k.

(c) Suppose $zw \neq 0$. Prove $|z+w| = |z|+|w|$ iff $\theta - \phi = 2k\pi$ for some integer k.
   Hint. As in the proof of 1.3.1(a), we have $|z + w|^2 = |z|^2 + z\overline{w} + \overline{z}w + |w|^2$.
   Now use Part (b).

4. Prove $|\mathcal{R}(z)| + |\mathcal{I}(z)| \leq \sqrt{2}\,|z|$. **Hint.** Since $0 \leq (|x| - |y|)^2$, we have $2|x||y| \leq |x|^2 + |y|^2$. Add $|x|^2 + |y|^2$ to each side.

5. See (1.14) and describe $\{z : \; 2 < |z - w| < 5\}$ where $w = 3 + 4i$. Draw a figure.

1.4   Basic Topological Concepts in R^2

Let $p \in R^2$ and let $\epsilon > 0$. Then $\{z : \; |z - p| < \epsilon\}$ is called the ϵ-neighborhood of p and is denoted by $N_\epsilon(p)$ or by $N(p, \epsilon)$. Thus $N_\epsilon(p)$ is the circular **disk** with center at p and radius ϵ, not including the points of the circle $\{z : \; |z - p| = \epsilon\}$.

The set obtained from $N_\epsilon(p)$ by deleting the center p, is called the **deleted ϵ-neighborhood of p** and is denoted by $N'_\epsilon(p)$ or by $N'(p, \epsilon)$. Each ϵ-neighborhood is also called a **neighborhood**. We note that $N'_\epsilon(p) = \{z : \; 0 < |z - p| < \epsilon\}$.

If A and B are sets, then the set of all members which are common to both A and B is called the **intersection of A and B** and is denoted by $A \cap B$. Thus

$$A \cap B = \{z : \; z \in A \; \text{and} \; z \in B\}.$$

The **union of A and B**, denoted by $A \cup B$, is defined as

$$A \cup B = \{z : \; z \in A \; \text{or} \; z \in B\}.$$

We use "$A \subset B$" to mean that A is a **subset** of B, i.e., that each member of A is also a member of B. We use $\emptyset$ to denote the **empty set**, called the **null set**.

Definition 1.4.1 Let $S \subset R^2$ and let $p \in R^2$.

(a) The point p is a **limit point of S** iff each deleted neighborhood of p contains at least one point of S. This means that p is a limit point of S iff

$$\text{for each } \epsilon > 0, \text{ we have } S \cap N'_\epsilon(p) \neq \emptyset.$$

(b) The set of all limit points of S is denoted by S^* and is called the **derived set of S**. The set $S \cup S^*$ is called the **closure of S** and is denoted by S^- or by $\overline{S}$.

(c) The set S is **closed** iff $S^* \subset S$ (that is, iff each limit point of S belongs to S).

(d) The set S is **open** iff

$$\text{for each point } p \text{ in } S, \text{ there is a number } \epsilon > 0 \text{ such that } N_\epsilon(p) \subset S.$$

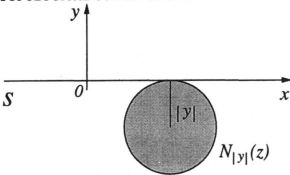

Figure 1.3: $S \cap N_{|y|}(z) = \emptyset$

(e) The set S is **bounded** iff there is a number $r > 0$ such that $S \subset N_r(0)$.

Example 1.4.2 (a) If S is the real axis, then S is closed. For if $z = (x, y) \notin S$, then $N_{|y|}(z)$ contains no point of S so that $z \notin S^*$. [See Figure 1.3 and recall the definition of $N_{|y|}(z)$.]

Thus z is not a limit point of S. This means that each limit point of S must be in S so that $S^* \subset S$. Hence S is closed by 1.4.1 (c). Clearly S is not bounded. Finally S is not open, for S contains no disk.

(b) Let $E = \{z : \mathcal{I}(z) \neq 0\}$. Then $E^* = R^2$. For let z be **any** point in R^2. Certainly each deleted neighborhood of z contains points of E (even if z is on the x-axis). Thus $z \in E^*$. Hence E is not closed. But E is open. For if $z = (x, y) \in E$, then $|y| > 0$ and $N_{|y|}(z) \subset E$. Clearly E is not bounded.

(c) The set $F = \{z : 1 < \mathcal{R}(z) < 4$ and $\mathcal{I}(z) = 0\}$ is not closed, since $1 \in F^*$ but $1 \notin F$. (Also $4 \in F^*$.) F is not open (since F contains no disk). Now $F \subset R$; and if we were discussing "openness in R," we would call F an open set — an open interval. Thus it is important to remember that (for sets in both R and R^2) "open in R" is different from "open in R^2."

It seems intuitively clear that the set of all points inside an ellipse (excluding the points on the ellipse) is open. We accept this conclusion in our later work as obvious — though a proof pointing to the precise mathematical properties leading to this conclusion would be awkward at this stage. Likewise, we accept the fact that the "interior of a simple closed curve" is open. (See Figure 1.4.)

We now prove that $N_r(z)$ is open. A triangle inequality is crucial to this proof.

Theorem 1.4.3 If $z \in R^2$ and $r > 0$, then $N_r(z)$ is open.

Figure 1.4: The interior of a simple closed curve is open.

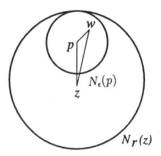

Figure 1.5: $N_r(z)$ is open.

Proof. Let $p \in N_r(z)$ and let $\epsilon = r - |p - z|$. Then $\epsilon > 0$. Now for each w in $N_\epsilon(p)$,

$$
\begin{aligned}
|w - z| &\leq |w - p| + |p - z| \qquad \text{by 1.3.1 (b)} \\
&< \epsilon + |p - z| \\
&= r - |p - z| + |p - z| \\
&= r.
\end{aligned}
$$

Hence $w \in N_r(z)$ so that $N_\epsilon(p) \subset N_r(z)$. Thus $N_r(z)$ is open by 1.4.1(d). See Figure 1.5.

Exercises 1.4.4

In Ex. 1–14 state whether the given subset of R^2 is (a) closed, (b) open, (c) bounded.

1. $\{z : \ |z| \leq 3\}$ 2. $\{z : \ 0 < |z| < 3\}$

3. $\{z : \ 1 < |z + i| \leq 2\}$ 4. $\{z : \ \mathcal{I}(z) = \mathcal{R}(z)\}$

5. $\{z : \ |\mathcal{R}(z)| > 2\}$ 6. $\{z : \ |z| \geq 2\}$

7. $\left\{z : \ z = \dfrac{1}{n} \text{ and } n \in N\right\}$ 8. N

9. $\{z : \ \mathcal{R}(z) \text{ is rational}\}$ 10. Any nonempty finite set

11. $\{z : \ \mathcal{R}(z)\mathcal{I}(z) = 0\}$ 12. $\{z : \ \mathcal{I}(z^2) < 0\}$

13. $\left\{z : \ \arg(z - 1 - i) = \dfrac{\pi}{3}\right\}$ 14. $\left\{z : \ \dfrac{\pi}{6} < \arg z < \dfrac{\pi}{3}\right\}$

Answers

	1.	2.	3.	4.	5.	6.	7.	8.	9.	10.	11.	12.	13.	14.
closed	Y	N	N	Y	N	Y	N	Y	N	Y	Y	N	Y	N
open	N	Y	N	N	Y	N	N	N	N	N	N	Y	N	N
bounded	Y	Y	Y	N	N	N	Y	N	N	Y	N	N	N	N

The set in Ex. 5 of 1.4.4 can be written as the union of A and B where

$$A = \{z: \ \mathcal{R}(z) < -2\} \quad \text{and} \quad B = \{z: \ \mathcal{R}(z) > 2\}.$$

The sets A and B are so well separated that we would say S is not connected (in one piece). The set E in 1.4.2 (b) is the union of U and L where

$$U = \{z: \ \mathcal{I}(z) > 0\} \quad \text{and} \quad L = \{z: \ \mathcal{I}(z) < 0\}.$$

Here U and L are extremely close together. But they are **not** so crowded together that either contains a limit point of the other. Thus we say that U and L are separated and that E is not connected.

Definition 1.4.5 Let E, A, and B be subsets of R^2.

(a) The pair (A, B) is a **separation of E** iff $E = A \cup B$, $A \neq \emptyset \neq B$ and $A^- \cap B = \emptyset = A \cap B^-$. (The last condition states that neither A nor B contains a point or a limit point of the other.)

(b) The set E is **connected** iff there exists no separation of E.

It is easy to see that the set

$$E = \{z: \ \mathcal{I}(z) = 0 \quad \text{and} \quad \mathcal{R}(z) \text{ is rational}\}$$

is not connected. For let A be the set of all rationals on the real axis less that $\sqrt{2}$ and let B be the set of all rationals on the real axis greater than $\sqrt{2}$. Then (A, B) is a separation of E.

Without experience, it usually is not easy to prove that a given connected set is, in fact, connected. The proof that any interval of real numbers is connected is left to real analysis and topology. For those who have had real analysis or topology, it is easy to show from 1.4.5(b) that we have the following result for a subset E of R^2.

If each pair of points in E can be joined by a
polygonal line segment (or even a "curve")
which is contained in E, then E is connected. (1.29)

See Figure 1.6. This criterion is "intuitively obvious," and we shall accept it without proof.

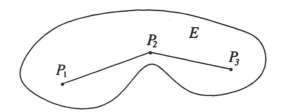

Figure 1.6: E is a connected set.

Figure 1.7: Some domains

Definition 1.4.6 Let $S \subset R^2$ and let $p \in R^2$.

(a) The set S is a **domain** iff S is open, connected, and nonempty.

(b) The point p is an **interior point of S** iff there is some $r > 0$ such that $N_r(p) \subset S$.

(c) The set of all interior points of S is called the **interior** of S and is denoted by S°.

(d) Let A and B be sets. Then $A - B$ denotes the set $\{z : \ z \in A \text{ and } z \notin B\}$.

(e) The point p is called a **boundary point of S** iff each neighborhood of p contains at least one point of S and at least one point of $R^2 - S$.

(f) The set of all boundary points of S is called the **boundary of S** and is denoted by $B(S)$.

(g) The set S is a **region** iff S is the union of a domain and a set consisting of some, none, or all of the boundary points of the domain.

Each set displayed in Figure 1.7 is a domain.

Exercises 1.4.7

1. Determine (a) which of the sets in 1.4.4 are connected and (b) which are domains. **Hint**. Use (1.29) if its hypothesis is satisfied; otherwise, use 1.4.5(b).

2. Let $D \subset R^2$. Prove if D is the union of nonempty open sets G and H where $G \cap H = \emptyset$, then D is not connected. **Hint**. Let $p \in G$ and show that $p \notin H^-$. See the solution in Ex. 2 of 9.1.7.

3. Prove that a subset S of R^2 is open iff $S = S^\circ$.

The structure in R^2 (and in other mathematical systems) which yields the theory of limits and continuity, is the topological structure. The collection of all open sets is called the **topology** for the given system. We shall see in the next chapter that neighborhoods (the basic open sets in terms of which all open sets are defined) are basic in our study of limits and continuity.

In a short course, the rest of this section (and its applications in the sequel) may be omitted. However, the concept of compactness is essential for a full appreciation of the basic theory in complex analysis.

Definition 1.4.8 Let $E \subset R^2$ and let $\mathcal{C}$ be a collection of subsets of R^2.

(a) The set $\{z : z \in A$ for at least one member A of $\mathcal{C}\}$ is called the **union of** $\mathcal{C}$ and is denoted by $\bigcup \mathcal{C}$.

(b) The collection $\mathcal{C}$ is a **cover for** E (or covers E) iff $E \subset \bigcup \mathcal{C}$.

(c) The collection $\mathcal{C}$ is an **open cover for** E iff $\mathcal{C}$ covers E and each member of $\mathcal{C}$ is open.

(d) A collection $\mathcal{D}$ is a **subcover of** $\mathcal{C}$ for E iff $\mathcal{D}$ covers E and $\mathcal{D} \subset \mathcal{C}$ (that is, each member of $\mathcal{D}$ is a member of $\mathcal{C}$).

(e) The set E is **compact** iff each open cover for E has a finite subcover for E.

Theorem 1.4.9 Let $\mathcal{C}$ be any collection of open subsets of R^2. Then $\bigcup \mathcal{C}$ is an open set.

Proof. Suppose $z \in \bigcup \mathcal{C}$. Then there is some E in $\mathcal{C}$ such that $z \in E$. But E is open. Thus by 1.4.1(d), there is some neighborhood $N(z)$ such that $N(z) \subset E \subset \bigcup \mathcal{C}$. Thus $\bigcup \mathcal{C}$ is open.

Exercises 1.4.10

1. Let $E = N_1(0) = \{z : z \in R^2$ and $|z| < 1\}$. Prove E is not compact by finding an open cover for E which has no finite subcover for E.

2. Let $F = \left\{z: \quad z = \dfrac{1}{n} \text{ and } n \in N\right\}$. Prove F is not compact.

3. Let $H = \left\{z: \quad z = 0 \text{ or } z = \dfrac{1}{n} \text{ and } n \in N\right\}$. Prove H is compact.

4. Prove if p is a limit point of a set $E \subset R^2$, then each neighborhood of p contains infinitely many points of E. **Hint.** Suppose there is some $r > 0$ such that $E \cap N'_r(p) = \{z_1, z_2, \cdots, z_k\}$. Let $\epsilon = \min\{|z_1 - p|, |z_2 - p|, \cdots, |z_k - p|\}$. Then $E \cap N'_\epsilon(p) = \emptyset$, which implies $p \notin E^*$.

Proof of Ex. 1. Let $\mathcal{C} = \{N_r(0): \quad 0 < r < 1\}$. Let $z \in E$. Then $|z| < 1$. By A.1.15, there is some r_0 in R such that $|z| < r_0 < 1$. Then $z \in N_{r_0}(0)$ and $N_{r_0}(0) \in \mathcal{C}$. Thus $\mathcal{C}$ is an open cover for E [by 1.4.3 and 1.4.8 (c)]. Clearly no finite subfamily of $\mathcal{C}$ covers E. [For let $\mathcal{D} = \{N_{r_1}(0), N_{r_2}(0), \cdots, N_{r_k}(0)\}$ be a nonempty finite subfamily of $\mathcal{C}$. Let $s = \max\{r_1, r_2, \cdots, r_k\}$. Then $s < 1$. Thus by A.1.15, there is some point z in E such that $|z| > s$. Hence $\mathcal{D}$ does not cover E.]

Proof of Ex. 2. For each n in N, let $U_n = N_{r_n}\left(\dfrac{1}{n}\right)$ where $r_n = \dfrac{1}{2n(n+1)}$. Let $\mathcal{C} = \{U_n: \quad n \in N\}$. Now

$$\left|\frac{1}{n} - \frac{1}{n+1}\right| = \frac{1}{n(n+1)} > r_n.$$

So $\dfrac{1}{n+1} \notin U_n$. Indeed U_k is the only member of $\mathcal{C}$ which contains $\dfrac{1}{k}$. Thus $\mathcal{C}$ is an open cover for F which certainly has no finite subcover for F. (For if a single member U_n of $\mathcal{C}$ is removed, the resulting subfamily fails to cover F since $\dfrac{1}{n}$ is in none of the remaining members of $\mathcal{C}$.)

Proof of Ex. 3. Let $\mathcal{C}$ be any open cover for H. There is some member A of $\mathcal{C}$ such that $0 \in A$. By 1.4.1 (d), let $r > 0$ such that $N_r(0) \subset A$. Now $\dfrac{1}{n} \in N_r(0) \subset A$ for each n in N such that $\dfrac{1}{n} < r$. But $\dfrac{1}{n} < r$ if $\dfrac{1}{r} < n$. So we let k be an integer $> \dfrac{1}{r}$. Then $\dfrac{1}{n} \in A$ for each $n > k$. For $n = 1, 2, 3, \cdots, k$, let $G_n \in \mathcal{C}$ such that $\dfrac{1}{n} \in G_n$. Then $\{A, G_1, G_2, \cdots, G_k\}$ is a finite subcover of $\mathcal{C}$ for H.

Remark 1.4.11 We notice that in Ex. 3 of 1.4.10, the set H is closed and bounded. Our next theorem (the Heine-Borel Theorem, which is of fundamental importance in analysis) shows that each closed and bounded set in R^2 is compact. Ex. 1 in 1.4.16 shows that the converse is also true. Thus in R^2 (and moreover, in any n-dimensional Euclidean space for each n in N) a set is compact iff it is closed and bounded.

We shall accept the following property in R (sometimes called the completeness property).

The Completeness Property 1.4.12 Each nonempty subset of R with an upper bound has a **least upper bound**; and each nonempty subset of R with a lower bound has a **greatest lower bound**.

We denote the **least upper bound** of a set E by lub E and the **greatest lower bound** of E by glb E.

Lemma 1.4.13 (Nested Interval Property). For each positive integer n, let $I_n = [a_n, b_n]$ be a closed interval in R such that $I_n \supset I_{n+1}$. Then there is some point p such that $p \in I_n$ for each n in N, [that is, there is a real number (or point) common to all these intervals I_n].

Proof. Since $I_n \supset I_{n+1}$, it follows that $a_n \leq a_{n+1} < b_{n+1} \leq b_n$ for each n in N. Thus each b_n is an upper bound of the set $E = \{a_n : n \in N\}$. By the Completeness Property 1.4.12, let $p = $ lub E. Thus $a_n \leq p \leq b_n$ for each n in N, which means $p \in I_n$ for each n in N.

Lemma 1.4.14 For each n in N, let

$$Q_n = \{(x, y) : \quad a_n \leq x \leq b_n \quad \text{and} \quad c_n \leq y \leq d_n\}$$

where a_n, b_n, c_n, and d_n are real numbers such that Q_n is a square region containing Q_{n+1}. Then there is a point (p, q) in R^2 such that $(p, q) \in Q_n$ for each n in N.

Proof. By 1.4.13, let p and q be such that for each $n \in N$, we have $a_n \leq p \leq b_n$ and $c_n \leq q \leq d_n$. Clearly $(p, q) \in Q_n$ for each n in N.

Theorem 1.4.15 (Heine-Borel Theorem for R^2). If E is a closed and bounded subset of R^2, then E is compact.

Proof. To reach a contradiction, suppose there exists an open cover $\mathcal{C}$ for E which has no finite subcover for E. Since E is bounded, let Q_1 be a square region with center at the origin and base parallel to the x-axis such that $E \subset Q_1$. The x and y axes divide Q_1 into four square regions. At least one of these, say Q_2, contains a part of E such that this part $E \cap Q_2$ cannot be covered by a finite subfamily of $\mathcal{C}$. (Otherwise, E could be so covered.) Similarly, when Q_2 is divided into four congruent square regions, at least one of them, say Q_3, is such that $E \cap Q_3$ cannot be covered by a finite subfamily of $\mathcal{C}$. Continuing this process, we obtain for each n in N, a square region Q_n such that

$$Q_1 \supset Q_2 \supset Q_3 \supset \cdots \supset Q_n \supset Q_{n+1} \supset \cdots, \tag{1.30}$$

$$E \cap Q_n \text{ cannot be covered by a finite subfamily of } \mathcal{C}, \quad \text{and} \tag{1.31}$$

$$\text{the length of a side of } Q_n \text{ is } 2^{-n+1} \text{ times the length of a side of } Q_1. \tag{1.32}$$

By (1.30) and 1.4.14, let $z = (p, q) \in R^2$ such that $z \in Q_n$ for each n in N. Let $\epsilon > 0$. Now, $Q_{n_0} \subset N_\epsilon(z)$ if n_0 is sufficiently large by (1.32). Then $E \cap Q_{n_0}$ is infinite by (1.31). Thus $N_\epsilon(z)$ contains infinitely many points of E. Hence $z \in E^*$ by 1.4.1(a). So $z \in E$, since E is closed.

Since $\mathcal{C}$ covers E, let $G \in \mathcal{C}$ such that $z \in G$. Since G is open, let $\epsilon_1 > 0$ such that $N_{\epsilon_1}(z) \subset G$ [by 1.4.1(d)]. By (1.32), it follows that $Q_{n_1} \subset N_{\epsilon_1}(z) \subset G$ for some n_1. So $E \cap Q_{n_1} \subset Q_{n_1} \subset G$, which contradicts (1.31).

Exercises 1.4.16

In Ex. $1-8$, let E and H be subsets of R^2 and prove the result.

1. If E is compact, then E is closed and bounded.

2. The set E is closed and bounded iff each infinite subset of E has a limit point in E.

3. (**Bolzano-Weierstrass Theorem**). If E is bounded and infinite, then E has a limit point in R^2. **Hint**. Let H be closed and bounded in R^2 such that $E \subset H$. Apply Ex. 2 to H.

4. The set E is open iff $R^2 - E$ is closed.

5. If E is compact and H is a closed subset of E, then H is compact. **Hint**. Let $\mathcal{C}$ be an open cover for H and adjoin the open set $R^2 - H$ to $\mathcal{C}$ to obtain an open cover $\mathcal{D}$ for E.

6. If $H \subset E$, then $H^* \subset E^*$.

7. $(E \cup H)^* = E^* \cup H^*$

8. $(E \cup H)^- = E^- \cup H^-$ **Hint**. By 1.4.1(b), we have $(E \cup H)^- = (E \cup H) \cup (E \cup H)^* = (E \cup H) \cup (E^* \cup H^*)$ by Ex. 7.

Proof of Ex. 1. Now $\{N_k(0) : \quad k \in N\}$ is an open cover for E. Since E is compact, let $\{N_{k_1}(0), N_{k_2}(0), \cdots, N_{k_m}(0)\}$ be a finite subcover for E and let $r = \max\{k_1, k_2, \cdots, k_m\}$. Then $E \subset N_r(0)$ and hence E is bounded.

(To prove that E is closed, we shall show that each point p, not in E, is not a limit point of E, that is, $p \notin E$ implies $p \notin E^*$. This means that each limit point of E must be in E, that is, $E^* \subset E$.) Suppose $p \notin E$. For each point z in E, let $N(z)$ and $N_{r_z}(p)$ be neighborhoods of z and p, respectively, such that $N(z) \cap N_{r_z}(p) = \emptyset$. Now $\{N(z) : \quad z \in E\}$ is an open cover for E. Let

$$z_1, z_2, \cdots, z_m \in E \text{ (if } E \neq \emptyset) \text{ such that}$$

$$E \subset \bigcup\{N(z_k) : \quad k = 1, 2, \cdots, m\} = U.$$

Let $r = \min\{r_{z_1}, r_{z_2}, \cdots, r_{z_m}\}$. Then $N_r(p) \cap E \subset N_r(p) \cap U = \emptyset$. Thus $p \notin E^*$. Hence $E^* \subset E$, and E is closed.

Proof of Ex. 2. Suppose E is closed and bounded and F is an infinite subset of E. Suppose

$$F^* \cap E = \emptyset.$$

Then for each p in E, (since $p \notin F^*$), let $N(p)$ be a neighborhood of p which contains at most one point of F; namely, p if $p \in F$. Clearly, $\{N(p) : \quad p \in E\}$ is an open cover for E which has no finite subcover for E (since it has no finite subcover for $F \subset E$). This contradicts the compactness of the closed and bounded set given by 1.4.15.

To prove the converse, we want to show that if each infinite subset of E has a limit point in E, then E is bounded and closed. To do this we first prove that if E is not bounded, then there is an infinite subset of E which has no limit point in E.

$$\text{So suppose } E \text{ is not bounded.} \qquad (1.33)$$

Then for each n in N, let $z_n \in E$ such that $|z_n| > n$. The set $\{z_n : \ n \in N\}$ is an infinite subset of E which has no limit point in R^2 and hence none in E.

$$\text{Now suppose } E \text{ is not closed.} \qquad (1.34)$$

Let $p \in (E^* - E)$. Then for each n in N, let $q_n \in E \cap N\left(p, \frac{1}{n}\right)$. Again $\{q_n : \ n \in N\}$ is an infinite subset of E which has no limit point except p and hence no limit point in E.

Thus if each infinite subset of E has a limit point in E, we must reject (1.33) and (1.34).

Proof of Ex. 4.

   "If Part." Let $H = R^2 - E$ and suppose H is closed. Assume $p \in E = R^2 - H$. Then $p \notin H^*$. Thus by 1.4.1(a), let $N(p)$ be a neighborhood of p such that $H \cap N(p) = \emptyset$, that is, $N(p) \subset (R^2 - H) = E$. Hence E is open.

   "Only If Part." Suppose $p \in E$. Since E is open, let $r > 0$ such that $N_r(p) \subset E$. Then $N_r(p)$ contains no points of $H = R^2 - E$. Thus $p \notin H^*$. We have shown that any point not in H is not a limit point of H. Hence $H^* \subset H$ and H is closed.

Proof of Ex. 7. Since $E \subset (E \cup H)$, we have $E^* \subset (E \cup H)^*$ by Ex. 6. Similarly, $H^* \subset (E \cup H)^*$. Thus (1) $(E^* \cup H^*) \subset (E \cup H)^*$. To prove (2) $(E \cup H)^* \subset (E^* \cup H^*)$, we need to prove the following statement.

   (i) If $p \in (E \cup H)^*$, then $p \in (E^* \cup H^*)$.

We prove the following equivalent statement.

   (ii) If $p \notin (E^* \cup H^*)$, then $p \notin (E \cup H)^*$.

Now suppose $p \notin (E^* \cup H^*)$. Then $p \notin E^*$ and $p \notin H^*$. Since $p \notin E^*$, let $\epsilon_1 > 0$ such that $E \cap N'_{\epsilon_1}(p) = \emptyset$. Since $p \notin H^*$, let $\epsilon_2 > 0$ such that $H \cap N'_{\epsilon_2}(p) = \emptyset$. Let $\epsilon = \min\{\epsilon_1, \epsilon_2\}$. Then $(E \cup H) \cap N'_\epsilon(p) = \emptyset$. So $p \notin (E \cup H)^*$ and (ii) is proved. Thus (i) is proved, and hence (2) follows. Now $(E \cup H)^* = E^* \cup H^*$ by (1) and (2).

Definition 1.4.17 Let $\mathcal{C}$ be a nonempty collection of sets. The set $\{x : \quad x \in A$ for each A in $\mathcal{C}\}$ is called the **intersection of $\mathcal{C}$** and is denoted by $\bigcap \mathcal{C}$. If $\mathcal{C} = \{A_1, A_2, A_3, \cdots\}$, then $\bigcap \mathcal{C}$ is also denoted by $A_1 \cap A_2 \cap A_3 \cap \cdots$, or by $\bigcap \{A_n : \quad n \in N\}$.

Theorem 1.4.18 Let $A_1, A_2, A_3, \cdots$ be a sequence of nonempty closed sets such that A_1 is bounded and such that

$$R^2 \supset A_1 \supset A_2 \supset A_3 \supset \cdots. \tag{1.35}$$

Then $\bigcap \{A_n : \quad n \in N\} \neq \emptyset$.

Proof. If some A_k is finite, the conclusion is clear. Thus, we assume each A_k is infinite and let $B = \{z_1, z_2, z_3, \cdots\}$ be a set of pairwise distinct points such that

$$z_k \in A_k \quad \text{for each } k \text{ in } N. \tag{1.36}$$

By Ex. 2 in 1.4.16, let $z \in A_1$ such that

$$z \in B^*. \tag{1.37}$$

Now let $j \in N$ and let $\epsilon > 0$. By (1.37) and Ex. 4 in 1.4.10, we see that $N_\epsilon(z)$ contains infinitely many points of B. Thus $N_\epsilon(z)$ contains many points of A_j by (1.35) and (1.36). Hence $z \in A_j^*$ so that $z \in A_j$ (since A_j is closed). Thus $z \in A_j$ for each j in N. Hence $z \in \bigcap \{A_n : \quad n \in N\}$ and so this intersection is not empty.

Theorem 1.4.19 (DeMorgan's Theorem). Let $\mathcal{C}$ be a nonempty collection of subsets of some set X.

(a) $X - \bigcup \mathcal{C} = \bigcap \{X - A : \quad A \in \mathcal{C}\}$.

(b) $X - \bigcap \mathcal{C} = \bigcup \{X - A : \quad A \in \mathcal{C}\}$.

Proof of (a). A point p of X belongs to the set on the left in Part (a) iff

$$p \notin A \quad \text{for any } A \text{ in } \mathcal{C}, \tag{1.38}$$

that is, iff

$$p \in (X - A) \quad \text{for each } A \text{ in } \mathcal{C}. \tag{1.39}$$

But p belongs to the set on the right in Part (a) iff (1.39) holds. Thus Part (a) is proved.

Proof of (b). A point p of X belongs to the set on the left in Part (b) iff

$$p \text{ does not belong to all members of } \mathcal{C}, \tag{1.40}$$

that is, iff

$$\text{there is some member } A \text{ of } \mathcal{C} \text{ such that } p \in (X - A). \tag{1.41}$$

But p belongs to the set on the right in Part (b) iff (1.41) holds. Thus Part (b) is proved.

Exercises 1.4.20

1. Prove the intersection of any nonempty finite collection of open sets in R^2 is open.

2. Let C be a nonempty collection of closed subsets of R^2. Prove each of the following results.

   (a) The set $\bigcap C$ is closed. **Hint.** Use Ex. 4 in 1.4.16, 1.4.19(b), and 1.4.9.

   (b) If C is finite, then $\bigcup C$ is closed.

3. Give an example of a collection C of closed sets such that $\bigcup C$ is not closed.

4. Give an example of a collection C of open sets such that $\bigcap C$ is not open.

5. Let a and b be real numbers such that $a < b$. Prove that the closed interval $[a, b]$ is a compact subset of R^2. Use 1.4.15.

Proof of Ex. 1. Let $G = A_1 \cap A_2 \cap \cdots \cap A_n$ where each A_k is open. Suppose $z \in G$. Then $z \in A_k$ for each $k = 1, 2, \cdots, n$, and since A_k is open, let $r_k > 0$ such that

$$N_{r_k}(z) \subset A_k \quad \text{for } k = 1, 2, \cdots, n.$$

Let

$$r = \min\{r_1, r_2, \cdots, r_n\}.$$

Then $N_r(z) \subset G$. Thus G is open.

Chapter 2

Complex Functions

2.1 Functions

Definition 2.1.1 Let X and Y be sets. A **function on X to Y** (or a **mapping of X into Y**) is a correspondence f which associates with each x in X exactly one element $f(x)$ in Y.

The set X in 2.1.1 is called the **domain of f**, and $\{y : \ y = f(x)$ for some x in $X\}$ is the **range of f**.

If $E \subset X$, then the set $\{f(x) : \ x \in E\}$ is called the **image** of E under f and is denoted by $f(E)$. If $f(X) = Y$, then f is a mapping of X **onto Y**. Let f be a function whose domain is X. Then f is said to be **one-to-one** iff for each pair of **distinct** members x_1 and x_2 of X, we have $f(x_1) \neq f(x_2)$. An element which varies over a set, is called a **variable**.

A function whose domain and range are subsets of R^2 is called a **complex function of a complex variable**. We shall usually shorten this expression and refer to a complex function of a complex variable simply as a **complex function**. Let f be a mapping of X into R^2 where $X \subset R$. Then f is called a **complex function of a real variable**. If $X \subset R^2$ and $f(X) \subset R$, then f is a **real function of a complex variable**. If $X \subset R$ and $f(X) \subset R$, then f is a **real function of a real variable**.

Remarks 2.1.2 (a) We observe that a particular function f is determined when its domain X is specified and a rule is given which associates with each x in X, the image $f(x)$.

(b) Let f be a complex function. Then for each $z = (x, y)$ in the domain of f, the complex number $f(z)$ has real and imaginary parts which we denote by

$$u(x, y) = \mathcal{R}\left(f(z)\right) \quad \text{and} \quad v(x, y) = \mathcal{I}\left(f(z)\right).$$

Thus

$$f(z) = u(x, y) + iv(x, y).$$

27

For example, if $f(z) = x - iy$, then $u(x, y) = x$ and $v(x, y) = -y$. If $f(z) = x^2 + y^2$, then $u(x, y) = x^2 + y^2$ and $v(x, y) = 0$. Thus each complex function f of a complex variable z determines two real functions u and v of two real variables x and y.

(c) To represent a complex function f geometrically, it is convenient to consider the domain X of f in one plane called the z-plane, and the range $W = f(X)$ in another plane called the w-plane. Then the function is thought of as mapping the points of the domain in the z-plane onto points of the range in the w-plane. However, more information about the function is obtained by investigating the images of certain curves or regions under the mapping.

Examples 2.1.3 (a) Let f be given by $f(z) = z^2$. Then $w = z^2$ is given by

$$f(z) = f(x + iy) = (x + iy)^2 = x^2 - y^2 + i(2xy).$$

Thus

$$u(x, y) = x^2 - y^2 \quad \text{and} \quad v(x, y) = 2xy.$$

(b) Let

$$w = f(z) = u + iv = \sqrt{4x^2 + 9y^2} - 3iy. \tag{2.1}$$

Find $f(R^2)$.

Solution of (b). First let us find $f(E)$ where E is the ellipse given by

$$E = \{(x, y) : \quad 4x^2 + 9y^2 = c^2\} \text{ where } c \text{ is a non-negative real constant.} \tag{2.2}$$

From (2.2), we have $4x^2 = c^2 - (-3y)^2$. Thus $-c \le -3y = v \le c$ by (2.1). Hence

$$f(E) = \{(u, v) : \quad u = c \quad \text{and} \quad -c \le v \le c\}.$$

This means that $f(E)$ is a vertical line segment with endpoints on the rays $v = \pm u$ and $u \ge 0$. As c varies over the non-negative reals, E generates the z-plane R^2 while $f(E)$ generates the set

$$\{(u, v) : \quad |v| \le u \text{ and } u \ge 0\} \text{ in the } w\text{-plane}.$$

See Figure 2.1.

Definition 2.1.4 A **polynomial** in z is a function defined by

$$f(z) = a_0 + a_1 z + \cdots + a_n z^n$$

for each z in R^2 where the coefficients are complex constants. A **rational function** in z is the quotient of two polynomials $P(z)$ and $Q(z)$. The domain of a polynomial $P(z)$ is R^2. But the domain of a rational function $\frac{P(z)}{Q(z)}$ is the set

$$R^2 - \{z : \quad Q(z) = 0\}.$$

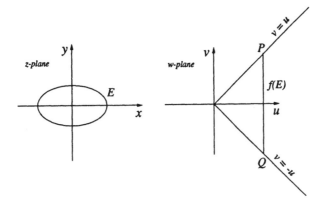

Figure 2.1: $f(E) =$ segment PQ.

2.2   Limits and Continuity

Definition 2.2.1 Let f be a complex function with domain D. Let $z_0 \in D^*$ and let $w_0 \in R^2$. Then w_0 is called the **limit of $f(z)$ as z approaches z_0** and we write

$$\lim_{z \to z_0} f(z) = w_0$$

if and only if the following condition holds.

> For each $\epsilon > 0$, there is a number $\delta > 0$ such that
> if $z \in N'_\delta(z_0) \cap D$, then $f(z) \in N_\epsilon(w_0)$.   (2.3)

Remark 2.2.2 Observe that the limit point z_0 need not be in the domain D of f. The purpose of writing $N'_\delta(z_0) \cap D$ is to make sure that z is in the domain of f. Geometrically, $f(z) \to w_0$ as $z \to z_0$ requires that for each open disk $N_\epsilon(w_0)$ centered at w_0, there is an open disk $N_\delta(z_0)$ centered at z_0 such that if z is in $N'_\delta(z_0) \cap D$, then $f(z)$ is in $N_\epsilon(w_0)$. See Figure 2.2.

Theorem 2.2.3 Let $z_0 \in D^*$ where D is the domain of a complex function f. If $\lim_{z \to z_0} f(z)$ exists, then this limit is unique.

Proof. Assume $\lim_{z \to z_0} f(z) = w_0$ and $\lim_{z \to z_0} f(z) = w_1$. Let $\epsilon > 0$. By (2.3), we let $\delta_1 > 0$ such that

$$|f(z) - w_0| < \frac{\epsilon}{2} \qquad \text{if } z \in D \cap N'(z_0, \delta_1).$$

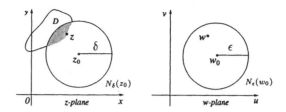

Figure 2.2: If $z \in N'_\delta(z_0) \cap D$, then $f(z) = w \in N_\epsilon(w_0)$.

Also, by (2.3), we let $\delta_2 > 0$ such that

$$|f(z) - w_1| < \frac{\epsilon}{2} \qquad \text{if } z \in D \cap N'(z_0, \delta_2).$$

Now we let $\delta = \min\{\delta_1, \delta_2\}$. If $z \in D \cap N'(z_0, \delta)$, then

$$\begin{aligned}
|w_1 - w_0| &\leq |w_1 - f(z)| + |f(z) - w_0| \quad \text{by } 1.3.1(b) \\
&< \frac{\epsilon}{2} + \frac{\epsilon}{2} = \epsilon.
\end{aligned}$$

Hence $w_1 = w_0$ since ϵ is an arbitrary positive number.

Theorem 2.2.4 Let $z_0 = (x_0, y_0) \in D^*$ and let $w = f(z) = u + iv$ be a complex function with domain D. Then

$$\lim_{z \to z_0} f(z) = u_0 + iv_0 = w_0 \tag{2.4}$$

if and only if

$$\lim_{z \to z_0} u(x, y) = u_0 \quad \text{and} \quad \lim_{z \to z_0} v(x, y) = v_0. \tag{2.5}$$

Proof of "Only If Part." Suppose (2.4) holds. Let $\epsilon > 0$. Then

there is some $\delta > 0$ such that if $z \in N'(z_0, \delta) \cap D$, then

$$\epsilon > |f(z) - w_0| = |(u - u_0) + i(v - v_0)| \geq \begin{cases} |u - u_0| \\ |v - v_0|, \end{cases}$$

and thus (2.5) follows.

Proof of "If Part." Suppose (2.5) holds. Let $\epsilon > 0$. Then there are numbers $\delta_1 > 0$ and $\delta_2 > 0$ such that

$$\begin{cases} \text{if } z \in N'(z_0, \delta_1) \cap D, \text{ then } |u - u_0| < \dfrac{\epsilon}{2} \text{ and} \\[2mm] \text{if } z \in N'(z_0, \delta_2) \cap D, \text{ then } |v - v_0| < \dfrac{\epsilon}{2}. \end{cases} \tag{2.6}$$

By (2.6) with $\delta = \min\{\delta_1, \delta_2\}$, if $z \in N'(z_0, \delta) \cap D$, then

$$|f(z) - w_0| = |(u - u_0) + i(v - v_0)| \leq |(u - u_0)| + |(v - v_0)|$$
$$< \frac{\epsilon}{2} + \frac{\epsilon}{2} = \epsilon.$$

Thus (2.4) holds.

Remarks 2.2.5 Let F be a complex function with domain D.

(a) If $\lim_{z \to z_0} F(z) = W_0$, then for some $B > 0$ there is a number $\delta > 0$ such that

$$|F(z) < B \quad \text{if } z \in D \cap N'(z_0, \delta).$$

To see this, we let $\epsilon = 1$. Then by Definition 2.2.1, we let $\delta > 0$ such that

$$|F(z)| = |[F(z) - W_0] + W_0|$$
$$\leq |F(z) - W_0| + |W_0|$$
$$< 1 + |W_0| \quad \text{if } z \in D \cap N'(z_0, \delta).$$

Now we take $B = 1 + |W_0|$.

(b) If $\lim_{z \to z_0} F(z) = W_0$ and $W_0 \neq 0$, then there is some $\delta_1 > 0$ such that $F(z)$ **is never equal to 0 for any** z **in** $N'(z_0, \delta_1)$. To verify this result, we let $\epsilon = \frac{1}{2}|W_0|$. Then $\epsilon > 0$. Clearly, $0 \notin N_\epsilon(W_0)$. See Figure 2.3. By Definition 2.2.1, we let $\delta_1 > 0$ such that

$$F(z) \in N_\epsilon(W_0) \quad \text{if } z \in D \cap N'(z_0, \delta_1).$$

Hence if $z \in D \cap N'(z_0, \delta_1)$, then $F(z) \neq 0$ since $0 \notin N_\epsilon(W_0)$. Thus F **is never zero in** $N'(z_0, \delta_1)$. Actually, we can conclude further that $|F(z)| > \frac{1}{2}|W_0|$ if $z \in D \cap N'(z_0, \delta_1)$. For if $z \in D \cap N'(z_0, \delta_1)$, then

$$|W_0| - |F(z)| \leq ||W_0| - |F(z)||$$
$$\leq |W_0 - F(z)|$$
$$< \frac{1}{2}|W_0|.$$

Now transposing terms, we have

$$|F(z)| > \frac{1}{2}|W_0| \quad \text{if } z \in D \cap N'(z_0, \delta_1).$$

(c) If $\lim_{z \to z_0} F(z) = W_0$ and $W_0 \neq 0$, then $z_0 \in E^*$ where $E = \{z : z$ is in the domain of F and $F(z) \neq 0\}$. To verify this, we first use Part (b) and let $\delta > 0$ such that F is never zero in $N'(z_0, \delta)$. In order for $\lim_{z \to z_0} F(z)$ to be defined, we know by 2.2.1 that $z_0 \in D^*$. Now let $\epsilon > 0$ and let $r = \min\{\epsilon, \delta\}$. Then $D \cap N'(z_0, r) \neq \emptyset$ since $z_0 \in D^*$. But F is never zero in $N'(z_0, r)$ since $N'(z_0, r) \subset N'(z_0, \delta)$. Thus $N'(z_0, \epsilon)$ must contain points of E. Hence $z_0 \in E^*$.

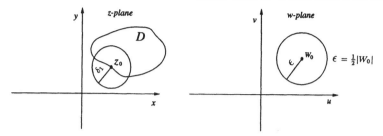

Figure 2.3: If $z \in D \cap N'(z_0, \delta_1)$, then $F(z) \in N_\epsilon(W_0)$.

Theorem 2.2.6 Let $z_0 \in D^*$ where D is the intersection of the domains of the complex functions f and F. If $\lim\limits_{z \to z_0} f(z) = w_0$ and the $\lim\limits_{z \to z_0} F(z) = W_0$, then

(a) $\lim\limits_{z \to z_0} [f(z) \pm F(z)] = w_0 \pm W_0,$

(b) $\lim\limits_{z \to z_0} f(z)\,F(z) = w_0\,W_0,$ and

(c) $\lim\limits_{z \to z_0} \dfrac{f(z)}{F(z)} = \dfrac{w_0}{W_0}$ if $W_0 \neq 0.$

Proof of (b). Let $\epsilon > 0$. By 2.2.5(a), we let $B > 0$ and $\delta_1 > 0$ such that

$$|F(z)| < B \quad \text{if } z \in D \cap N'(z_0, \delta_1). \tag{2.7}$$

By 2.2.1, we let $\delta_2 > 0$ and $\delta_3 > 0$ such that the following two inequalities hold.

$$|f(z) - w_0| \;<\; \frac{\epsilon}{2B} \qquad \text{if } z \in D \cap N'(z_0, \delta_2)$$
$$|F(z) - W_0| \;<\; \frac{\epsilon}{2\,(1 + |w_0|)} \quad \text{if } z \in D \cap N'(z_0, \delta_3) \tag{2.8}$$

[We use $1 + |w_0|$ instead of $|w_0|$ to avoid the possibility of a zero denominator.] Now we let $\delta = \min\{\delta_1, \delta_2, \delta_3\}$. If $z \in D \cap N'(z_0, \delta)$, then the inequalities in (2.7) and (2.8) hold, and we have

$$
\begin{aligned}
|f(z)F(z) - w_0 W_0| &= \;|[f(z)F(z) - w_0 F(z)] + [w_0 F(z) - w_0 W_0]| \\
&\le \;|F(z)||f(z) - w_0| + |w_0||F(z) - W_0| \\
&< \;B \cdot \frac{\epsilon}{2B} + \frac{|w_0|}{1 + |w_0|}\frac{\epsilon}{2} \\
&< \;\frac{\epsilon}{2} + \frac{\epsilon}{2} = \epsilon.
\end{aligned}
$$

Hence Part (b) is proved.

Proof of (c). Now we are given that $W_0 \neq 0$. We shall first prove

$$\lim_{z \to z_0} \frac{1}{F(z)} = \frac{1}{W_0}. \tag{2.9}$$

The domain of the function $\frac{1}{F(z)}$ is the set E of all points z in the domain of F such that $F(z) \neq 0$. By 2.2.5(c), we know that $z_0 \in E^*$. Let $\epsilon > 0$. By the last statement in 2.2.5(b), we let $\delta_1 > 0$ such that

$$|F(z)| > \frac{1}{2}|w_0| \quad \text{if } z \in E \cap N'(z_0, \delta_1).$$

By 2.2.1, we let $\delta_2 > 0$ such that

$$|F(z) - W_0| < \frac{1}{2}|W_0|^2\epsilon \quad \text{if } z \in E \cap N'(z_0, \delta_2).$$

Now we let $\delta = \min\{\delta_1, \delta_2\}$. If $z \in E \cap N'(z_0, \delta)$, then

$$\begin{aligned}
\left| \frac{1}{F(z)} - \frac{1}{W_0} \right| &= \frac{|W_0 - F(z)|}{|F(z)||W_0|} \\
&< \frac{\frac{1}{2}|W_0|^2\epsilon}{|F(z)||W_0|} \\
&= \left(\frac{\frac{1}{2}|W_0|}{|F(z)|} \right)\epsilon \\
&< \epsilon.
\end{aligned}$$

Thus (2.9) is proved. Now by Part (b) and (2.9), we have

$$\lim_{z \to z_0} \frac{f(z)}{F(z)} = \lim_{z \to z_0} f(z)\left(\frac{1}{F(z)} \right) = w_0\left(\frac{1}{W_0} \right) = \frac{w_0}{W_0}.$$

[We should notice that the domain of the function $\frac{f}{F}$ is the set H of all points z in D such that $F(z) \neq 0$. As in 2.2.5(c), we know that $z_0 \in H^*$.]

Exercises 2.2.7

When we indicate that a function f is determined by an expression for $f(z)$, unless otherwise stated, it is understood that the domain of f is the set of all points z for which $f(z)$ is a complex number.

1. State the domain and range of f where $f(z)$ is determined by the given expression.
   (a) $\bar{z}$ (b) $az^2 + bz + c$ where a, b, and c are complex constants

   (c) $\dfrac{z + 3}{z^2 + 1}$ (d) $\dfrac{5}{z^3 - 8}$
   (e) $|z|$ (f) $\mathcal{R}(z)$ (g) $\mathcal{I}(z)$

2. Evaluate each of the following limits.
   (a) $\lim_{z \to 1+i} \bar{z}$ (b) $\lim_{z \to i} \mathcal{R}(z)$
   (c) $\lim_{z \to -i} \mathcal{I}(z)$ (d) $\lim_{z \to 1} \mathcal{I}(z)$

3. Prove 2.2.6(a).

4. Prove each of the following equalities where a, b, c, and d are complex constants in Parts (a) and (d).
   (a) $\lim_{z \to z_0} c = c$ (b) $\lim_{z \to z_0} z = z_0$
   (c) $\lim_{z \to z_0} z^2 = z_0^2$ (d) $\lim_{z \to z_0} \dfrac{az+b}{cz+d} = \dfrac{az_0+b}{cz_0+d}$ if $cz+d \neq 0$
   (e) $\lim_{z \to -i} \dfrac{z^3 - i}{z + i} = -3$

5. Prove $\lim_{z \to z_0} f(z) = 0$ iff $\lim_{z \to z_0} |f(z)| = 0$. **Hint.** $|f(z) - 0| = ||f(z)| - 0|$.

Definition 2.2.8 Let f be a complex function with domain D and let $z_0 \in D$. The function f is **continuous at** z_0 [denoted by $f \in C(z_0)$] iff for each $\epsilon > 0$, there is a number $\delta > 0$ such that

$$\text{if } z \in D \cap N_\delta(z_0), \text{ then } f(z) \in N_\epsilon(f(z_0)),$$

that is, $f[D \cap N_\delta(z_0)] \subset N_\epsilon(f(z_0))$.

Remarks 2.2.9 (a) Let $z_0 \in D \cap D^*$ where D is the domain of a complex function f. Then by 2.2.1 and 2.2.8, we observe that f is continuous at z_0 iff

$$\lim_{z \to z_0} f(z) = f(z_0). \tag{2.10}$$

[This means that f is defined at z_0, that $\lim_{z \to z_0} f(z)$ exists and that the value of this limit is $f(z_0)$].

(b) Also we notice that if $z_0 \in (D - D^*)$, that is, if z_0 is an isolated point of D, then 2.2.8 implies that f is automatically continuous at z_0.

Theorem 2.2.10 Let f and g be functions on D to R^2 where $z_0 \in D \subset R^2$. If f and g are continuous at z_0, then $f+g, f-g$ and fg are continuous at z_0; and if $g(z_0) \neq 0$, then $\dfrac{f}{g}$ is continuous at z_0.

Proof. First, we suppose that $z_0 \in D^*$. Then

$$
\begin{aligned}
\lim_{z \to z_0} (f+g)(z) &= \lim_{z \to z_0} f(z) + \lim_{z \to z_0} g(z) && \text{by 2.2.9(a) and 2.2.6(a)} \\
&= f(z_0) + g(z_0) && \text{by 2.2.9(a)} \\
&= (f+g)(z_0).
\end{aligned}
$$

Thus $f+g$ is continuous at z_0 by 2.2.9(a). [If $z_0 \in (D - D^*)$, then $f+g$ is automatically continuous at z_0 by 2.2.9(b).]

The continuity of the other functions follows similarly.

Theorem 2.2.11 Let $z_0 = (x_0, y_0) \in D \subset R^2$ and let $f(z) = u(x, y) + iv(x, y)$ be a function on D to R^2. Then f is continuous at z_0 iff u is continuous at z_0 and v is continuous at z_0.

Proof. Since the proof is trivial when $z_0 \in (D - D^*)$, we assume that $z_0 \in D^*$. Let f be continuous at z_0. Now

$$
\begin{aligned}
u(x_0, y_0) + iv(x_0, y_0) \; = \; f(z_0) \; &= \; \lim_{z \to z_0} f(z) \quad \text{by 2.2.9(a)} \\
&= \; \lim_{(x,y) \to (x_0,y_0)} [u(x, y) + iv(x, y)] \\
&= \; \lim_{(x,y) \to (x_0,y_0)} u(x, y) + i \lim_{(x,y) \to (x_0,y_0)} v(x, y) \quad \text{by 2.2.4}
\end{aligned}
$$

Thus by Equation (1.1)

$$
\lim_{(x,y) \to (x_0,y_0)} u(x, y) = u(x_0, y_0) \quad \text{and} \quad \lim_{(x,y) \to (x_0,y_0)} v(x, y) = v(x_0, y_0), \qquad (2.11)
$$

and hence u and v are continuous at (x_0, y_0).

Conversely, let u and v be continuous at (x_0, y_0). That is, let (2.11) hold. Then

$$
\begin{aligned}
\lim_{z \to z_0} f(z) \; &= \; \lim_{(x,y) \to (x_0,y_0)} [u(x, y) + iv(x, y)] \\
&= \; \lim_{(x,y) \to (x_0,y_0)} u(x, y) + i \lim_{(x,y) \to (x_0,y_0)} v(x, y) \quad\quad \text{by 2.2.4} \\
&= \; u(x_0, y_0) + iv(x_0, y_0) \quad\quad\quad\quad\quad\quad \text{by hypotheses (2.11)} \\
&= \; f(z_0).
\end{aligned}
$$

Thus f is continuous at z_0.

Theorem 2.2.12 If $a_0, a_1, \cdots, a_n$ are complex constants where n is a non-negative integer, then the polynomial

$$
P(z) = a_0 + a_1 z + a_2 z^2 + \cdots + a_n z^n
$$

is continuous at each $z \in R^2$.

Proof. A constant function and the function $f(z) = z$ are continuous at each z by 2.2.8. Thus $f(z) f(z) = z^2$ is continuous at each z by 2.2.10. Also, if $a_k z^k$ is a term in $P(z)$ after the first one, then this term is continuous at each z by a finite number of applications of 2.2.10. Finally, since $P(z)$ is the sum of a finite number of functions continuous at each z, it follows that $P(z)$ is continuous at each z by a finite number of applications of 2.2.10.

Theorem 2.2.13 If $P(z)$ and $Q(z)$ are polynomials, then the rational function $\frac{P}{Q}$ is continuous at each z in R^2 at which $Q(z) \neq 0$.

Definition 2.2.14 Let g and f be functions with domains X and Y, respectively. The **composite function** $f \circ g$ is the function determined by $(f \circ g)(p) = f(g(p))$ for each p in X such that $g(p) \in Y$.

It is clear that the domain of $f \circ g$ in 2.2.14 is the set H where H is the set of all points p in X such that $g(p) \in Y$.

Theorem 2.2.15 Let g and f be complex functions with domains D and E, respectively, and let $z_0 \in D$. If g is continuous at z_0 and f is continuous at $g(z_0)$, then $f \circ g$ is continuous at z_0.

Proof. Let $\epsilon > 0$. Now f is continuous at $g(z_0)$. Thus we let $\delta_1 > 0$ such that

$$f(w) \in N_\epsilon[f(g(z_0))] \quad \text{if } w \in E \cap N_{\delta_1}(g(z_0)). \tag{2.12}$$

Also, g is continuous at z_0. Thus using δ_1 for ϵ in 2.2.8, we let $\delta > 0$ such that

$$g(z) \in N_{\delta_1}(g(z_0)) \quad \text{if } z \in D \cap N_\delta(z_0). \tag{2.13}$$

Now it follows from (2.12) and (2.13) that

$$f(g(z)) \in N_\epsilon[f(g(z_0))] \quad \text{if } z \in H \cap N_\delta(z_0)$$

where H is the domain of $f \circ g$. Thus $f \circ g$ is continuous at z_0 by Definition 2.2.8.

Exercises 2.2.16

1. Prove that $|z|$ is continuous at each z in R^2. **Hint.** Apply the definition of continuity and 1.3.1(c).

2. Prove that $|z|^2$ is continuous at each z in R^2. **Hint.** $|z|^2 = |z||z|$

3. Classify each of the following as a polynomial, a rational function or a composite function.

   (a) $(3 + 2i)z^2 + 5z + 7i$ (b) $\dfrac{iz^2 + 7}{3iz + 2}$ (c) $\left| \dfrac{z^2 - 2z + 4}{z^4 - 4} \right|$

4. For each function in Ex. 3, indicate all values of z at which the function is continuous. Explain your answer.

5. Find $\displaystyle\lim_{z \to z_0} \frac{a_0 + a_1 z + \cdots + a_m z^m}{b_0 + b_1 z + \cdots + b_n z^n}$ if the denominator is not zero at $z = z_0$.

6. Evaluate $\displaystyle\lim_{z \to i} \frac{(z^2 + 2z + 2)(z^2 - 2z + 2)}{z^4 + 4}$.

7. Prove $\lim_{z \to z_0} z^{-n} = \dfrac{1}{z_0^n}$ if $z_0 \neq 0$ and $n \in N$.

8. Let f be a complex function which is continuous at the point z_0 and suppose $f(z_0) \neq 0$. Prove the following results.

   (a) There is some $\delta > 0$ such that f is never zero in $N_\delta(z_0)$. See Figure 2.3.

   (b) There is some $\delta > 0$ such that $|f(z)| > \frac{1}{2}|f(z_0)|$ if $z \in D \cap N_\delta(z_0)$ where D is the domain of f.

   Hint. For Part (a), see the first part of 2.2.5(b). For Part (b), see the final part of 2.2.5(b).

2.3   The Derivative

Definition 2.3.1 Let f be a complex function with domain D and let $z \in D^\circ$ (where D° is the interior of D). If

$$\lim_{t \to z} \frac{f(t) - f(z)}{t - z} \tag{2.14}$$

exists, then this limit is called the **derivative of f at z** and is denoted by $f'(z)$ or by $\dfrac{df(z)}{dz}$.

When $f'(z)$ exists, we say that f **is differentiable at z** or that f **has a derivative at z**. If $f'(z)$ exists at each point z in a set S, then we say f **is differentiable in S**.

Remarks 2.3.2 We notice that $f'(z)$ is defined only if z is an interior point of the domain of f. Also, $f'(z)$ is a complex number if this derivative exists. Let

$$E = \{z : \ f'(z) \text{ exists}\}.$$

The function which associates with each z in E, the value $f'(z)$, is called the **derived function** or **derivative of f** and is denoted by f'. The function f'' is defined by replacing f by f' in (2.14), and f''' is defined by replacing f by f'' in (2.14). **Higher derivatives** are defined in like manner. We also observe that the limit in (2.14) may be expressed as

$$\lim_{h \to 0} \frac{f(z+h) - f(z)}{h}.$$

Theorem 2.3.3 Let f be a complex function. If $f'(z)$ exists, then f is continuous at z.

Proof. If t is a point in the domain of f and $t \neq z$, then

$$f(t) - f(z) = \frac{f(t) - f(z)}{t - z}(t - z).$$

Hence

$$\lim_{t \to z}[f(t) - f(z)] = \lim_{t \to z}\left(\frac{f(t) - f(z)}{t - z}(t - z)\right) = f'(z) \cdot 0 = 0.$$

Thus we have

$$\lim_{t \to z} f(t) = \lim_{t \to z}[\{f(t) - f(z)\} + f(z)] = 0 + f(z) = f(z).$$

Therefore, f is continuous at z by 2.2.9 (a).

Remark 2.3.4 The converse to 2.3.3 is not true. For consider the function

$$f(z) = |z|^2, \tag{2.15}$$

which is continuous at each point z in R^2. Assume $f'(z)$ exists where $z = (x, y)$. Let $t = (u, v) \neq z$. Then

$$
\begin{aligned}
f'(z) &= \lim_{t \to z}\frac{|t|^2 - |z|^2}{t - z} = \lim_{t \to z}\frac{t\bar{t} - z\bar{z}}{t - z} \\
&= \lim_{t \to z}\frac{t\bar{t} - z\bar{t} + z\bar{t} - z\bar{z}}{t - z} \\
&= \lim_{t \to z}\bar{t} + \lim_{t \to z}\frac{z(\bar{t} - \bar{z})}{t - z} \\
&= \lim_{v \to y}(x - iv) + z\lim_{v \to y}\frac{i(y - v)}{i(v - y)} \quad \text{putting } u = x \\
&= x - iy + z(-1) = \bar{z} - z = -2iy \tag{2.16}
\end{aligned}
$$

Here we have used the hypothesis that the above limit exists as t approaches z to conclude that the special limit exists as (x, v) approaches (x, y). We sometimes say that if the limit of $g(t)$ exists **as t approaches z in general**, then the special limit of $g(t)$ exists **as t approaches z in some subset** of the domain of g where z is a limit point of this subset. In our present example, the subset which we used in the above display was the vertical line $u = x$ through the fixed point (x, y). Next, we shall consider the special limit as t approaches z along the horizontal line $v = y$ through the point (x, y). Thus we have

$$
\begin{aligned}
f'(z) &= \lim_{t \to z}\bar{t} + \lim_{t \to z}\frac{z(\bar{t} - \bar{z})}{t - z} \\
&= \lim_{u \to x}(u - iy) + z\lim_{u \to x}\frac{u - x}{u - x} \quad \text{putting } v = y \\
&= \bar{z} + z = 2x \tag{2.17}
\end{aligned}
$$

Now by (2.16) and (2.17), we have

$$-2iy = 2x. \tag{2.18}$$

This means that $y = 0$ and $x = 0$. Hence if $f'(z)$ exists, then $z = 0$. But as observed above, f is continuous at each point in R^2.

Theorem 2.3.5 Let f and g be complex functions. If $f'(z)$ and $g'(z)$ exist at some point z, then the following formulas hold.

$$\begin{aligned}
(f + g)'(z) &= f'(z) + g'(z) \\
(f - g)'(z) &= f'(z) - g'(z) \\
(fg)'(z) &= f(z)g'(z) + g(z)f'(z) \\
\left(\frac{f}{g}\right)'(z) &= \frac{g(z)f'(z) - f(z)g'(z)}{g^2(z)} \quad \text{if } g(z) \neq 0
\end{aligned} \tag{2.19}$$

Proof for $(fg)'(z)$. Let $P = fg$. Then

$$\begin{aligned}
\frac{P(z + h) - P(z)}{h} &= \frac{f(z + h)g(z + h) - f(z)g(z)}{h} \\
&= \frac{f(z + h)g(z + h) - f(z + h)g(z) + f(z + h)g(z) - f(z)g(z)}{h} \\
&= f(z + h)\frac{g(z + h) - g(z)}{h} + g(z)\frac{f(z + h) - f(z)}{h}. \tag{2.20}
\end{aligned}$$

By hypothesis, 2.3.3, and 2.2.9(a),

$$\lim_{h \to 0} f(z + h) = f(z), \quad \lim_{h \to 0} \frac{g(z + h) - g(z)}{h} = g'(z),$$

$$\lim_{h \to 0} g(z) = g(z) \quad \text{and} \quad \lim_{h \to 0} \frac{f(z + h) - f(z)}{h} = f'(z).$$

Thus from (2.20), by 2.2.6(a) and (b),

$$(fg)'(z) = \lim_{h \to 0} \frac{P(z + h) - P(z)}{h} = f(z)\,g'(z) + g(z)f'(z).$$

Theorem 2.3.6 (Chain Rule). Let f and g be complex functions such that $g'(z)$ and $f'(g(z))$ exist for some point z. Then $(f \circ g)'(z)$ exists, and $(f \circ g)'(z) = f'(g(z))g'(z)$.

Proof. Let $w = f(g(t))$ for each t in the domain of $f \circ g$. Then from the definition of the derivative, there exist functions $\theta_1(t)$ and $\theta_2(t)$ such that

$$\Delta g = [g'(z) + \theta_1(t)]\Delta t \quad \text{where } \Delta g = g(t) - g(z), \ \Delta t = t - z, \ \text{and} \tag{2.21}$$

$$\Delta w = [f'(g(z)) + \theta_2(t)]\Delta g \quad \text{where } \Delta w = f(g(t)) - f(g(z)), \tag{2.22}$$

and

$$\lim_{t \to z} \theta_1(t) = \lim_{t \to z} \theta_2(t) = 0. \tag{2.23}$$

Substituting Δg from (2.21) into (2.22) and dividing by Δt, we have

$$\frac{\Delta w}{\Delta t} = [f'(g(z)) + \theta_2(t)][g'(z) + \theta_1(t)]. \tag{2.24}$$

Hence by (2.23) and (2.24),

$$(f \circ g)'(z) = \lim_{\Delta t \to 0} \frac{\Delta w}{\Delta t} = f'(g(z))g'(z).$$

[We should notice that z is in the interior of the domain of $f \circ g$ since $g'(z)$ and $f'(g(z))$ exist and since g is continuous at z. See 2.3.1, 2.3.3, and 2.2.8.]

Theorem 2.3.7 Let f be a complex function and let $u = \mathcal{R}(f)$ and $v = \mathcal{I}(f)$. If f' exists at $z = x + iy$, then the partial derivatives

$$u_x, \ u_y, \ v_x \text{ and } v_y \text{ exist at } (x, y)$$

and satisfy the Cauchy-Riemann equations

$$u_x = v_y \quad \text{and} \quad u_y = -v_x \quad \text{at } (x, y). \tag{2.25}$$

Also,

$$f'(z) = u_x + iv_x = v_y - iu_y \quad \text{at } (x, y). \tag{2.26}$$

Proof. Let $h = h_1 + ih_2$. Then

$$\begin{aligned} z + h &= x + h_1 + i(y + h_2), \text{ and} \\ f(z + h) &= u(x + h_1, y + h_2) + iv(x + h_1, y + h_2). \end{aligned} \tag{2.27}$$

By hypothesis, $f'(z)$ exists. Hence

$$f'(z) = a + ib \quad \text{where } a \text{ and } b \text{ are real.} \tag{2.28}$$

Thus by (2.27) and (2.28),

$$\begin{aligned} a + ib = f'(z) &= \lim_{h \to 0} \frac{f(z + h) - f(z)}{h} \\ &= \lim_{h \to 0} \left[\frac{u(x + h_1, y + h_2) - u(x, y)}{h_1 + ih_2} + i\frac{v(x + h_1, y + h_2) - v(x, y)}{h_1 + ih_2} \right]. \end{aligned} \tag{2.29}$$

In (2.29), we first let $z + h \to z$ along the horizontal line through z (that is, we take $h_2 = 0$). Thus by 2.2.4

$$f'(z) = a + ib = u_x + iv_x. \tag{2.30}$$

Next in (2.29), we let $z + h \to z$ along the vertical line through z (that is, we take $h_1 = 0$). Then

$$f'(z) = a + ib = \frac{1}{i} u_y + v_y = v_y - iu_y. \tag{2.31}$$

Thus (2.30) and (2.31) imply (2.25).

Remarks 2.3.8 It is important to notice that 2.3.7 states that (2.25) is **necessary** for $f'(z)$ to exist. Thus to prove that $f'(z)$ does not exist, we may prove that (2.25) does not hold. However, to prove that $f'(z)$ does exist it is not sufficient to show that (2.25) holds. The sufficient conditions are given in 2.3.11.

Example 2.3.9 Show that $f'(0)$ does not exist and yet the Cauchy-Riemann equations are satisfied at $z = 0$ if

$$f(z) = \begin{cases} \dfrac{ax^3 - by^3}{ax^2 + by^2} + i\, \dfrac{ax^3 + by^3}{ax^2 + by^2} & \text{if } z \neq 0 \\[2mm] 0 & \text{if } z = 0 \quad \text{where } ab > 0. \end{cases}$$

Solution. From the definition of the partial derivatives of real-valued functions,

$$u_x(0,0) = \lim_{h \to 0} \frac{u(h,0) - u(0,0)}{h} = 1 \quad \text{and} \quad u_y(0,0) = \lim_{h \to 0} \frac{u(0,h) - u(0,0)}{h} = -1.$$

Similarly, $v_x(0,0) = v_y(0,0) = 1$. Thus $u_x = v_y$ and $u_y = -v_x$ at $(0,0)$.

Next, we show that $f'(0)$ does not exist. We first let z approach zero along the line $y = x$. Thus $z = x + iy = x(1+i)$, and

$$\lim_{x \to 0} \frac{f(z) - f(0)}{z - 0} = \lim_{x \to 0} \frac{\frac{a-b}{a+b} + i}{1 + i} = \frac{a}{a+b} + i \frac{b}{a+b}.$$

Now we let $z \to 0$ along the line $y = 0$. Thus $z = x$, and

$$\lim_{x \to 0} \frac{f(z) - f(0)}{z - 0} = \lim_{x \to 0} \frac{x + ix}{x} = 1 + i.$$

Hence $f'(0)$ does not exist.

Theorem 2.3.10 Let u be a function on D to R where $D \subset R^2$ and let $z_0 = (x_0, y_0) \in D$. Suppose there is a neighborhood $N_r(z_0) \subset D$ such that u_x and u_y exist at each point of $N_r(z_0)$. If u_x and u_y are continuous at z_0, then

$$\Delta u = u(x_0 + h, y_0 + k) - u(x_0, y_0) \tag{2.32}$$

is given by

$$\Delta u = u_x(x_0, y_0)h + u_y(x_0, y_0)k + \alpha h + \beta k \tag{2.33}$$

where $\alpha = \alpha(h, k) \to 0$ and $\beta = \beta(h, k) \to 0$ as $(h, k) \to (0, 0)$.

Proof. By (2.32), $\Delta u = (u(x_0+h, y_0+k) - u(x_0, y_0+k)) + (u(x_0, y_0+k) - u(x_0, y_0))$; and by the Mean Value Theorem of the calculus,

$$\Delta u = u_x(x_1, y_0 + k)h + u_y(x_0, y_1)k, \qquad (2.34)$$

where x_1 is between x_0 and $x_0 + h$, and y_1 is between y_0 and $y_0 + k$. Let $\alpha = \alpha(h, k)$, $\beta = \beta(h, k)$ be defined by

$$\begin{aligned} u_x(x_1, y_0 + k) &= u_x(x_0, y_0) + \alpha \\ u_y(x_0, y_1) &= u_y(x_0, y_0) + \beta. \end{aligned} \qquad (2.35)$$

Since u_x and u_y are continuous at z_0, it follows by (2.35) that $\alpha \to 0$ and $\beta \to 0$ as $(h, k) \to (0, 0)$. Thus (2.33) follows by (2.34) and (2.35).

Theorem 2.3.11 Let f be a complex function with $u = \mathcal{R}(f)$ and $v = \mathcal{I}(f)$. Let u_x, u_y, v_x, and v_y exist at each point **in some neighborhood of** z_0, where $z_0 = x_0 + iy_0$. If u_x, u_y, v_x, and v_y are continuous at z_0 and satisfy the Cauchy-Riemann equations

$$u_x = v_y \quad \text{and} \quad u_y = -v_x \quad \text{at } z_0, \qquad (2.36)$$

then $f'(z_0)$ exists.

Proof. By (2.33),

$$\begin{aligned} f(z_0 + \Delta z) - f(z_0) &= [u(x_0 + h, y_0 + k) - u(x_0, y_0)] \\ &+ i[v(x_0 + h, y_0 + k) - v(x_0, y_0)] \\ &= u_x h + u_y k + i(v_x h + v_y k) + \epsilon \end{aligned} \qquad (2.37)$$

where

$$\epsilon = (\alpha h + \beta k) + i(\gamma h + \delta k), \qquad (2.38)$$

and

$$\alpha, \beta, \gamma, \delta \to 0 \text{ as } h, k \to 0. \qquad (2.39)$$

By (2.36) and (2.37),

$$f(z_0 + \Delta z) - f(z_0) = (h + ik)u_x + i(h + ik)v_x + \epsilon. \qquad (2.40)$$

From (2.38) and (2.40), with $\Delta z = h + ik$,

$$\frac{f(z_0 + \Delta z) - f(z_0)}{\Delta z} = u_x + iv_x + (\alpha + i\gamma)\frac{h}{\Delta z} + (\beta + i\delta)\frac{k}{\Delta z}. \qquad (2.41)$$

But

$$\left| \frac{h}{\Delta z} \right| \leq 1 \quad \text{and} \quad \left| \frac{k}{\Delta z} \right| \leq 1. \qquad (2.42)$$

Thus by (2.39), (2.41), and (2.42)

$$f'(z_0) = \lim_{\Delta z \to 0} \frac{f(z_0 + \Delta z) - f(z_0)}{\Delta z} = u_x + iv_x.$$

Hence $f'(z_0)$ exists.

Definition 2.3.12 Let f be a complex function on D to R^2 where $D \subset R^2$. Then f is **analytic at** z_0 iff

$$f'(z) \text{ exists for each } z \text{ in some neighborhood of } z_0.$$

Also, f is **analytic in** S (denoted by $f \in A(S)$) iff f is analytic at each point of S. If $S = \{z_0\}$, then we shall write $f \in A(z_0)$. The function f is an **entire function** iff $f \in A(R^2)$.

Theorem 2.3.13 If $|n| \in N$, $f'(z)$ exists and C is a complex constant, then

(a) $(z^n)' = nz^{n-1}$ on $\begin{cases} R^2 & \text{if } n > 1 \\ R^2 - \{0\} & \text{if } n < 0, \end{cases}$

(b) $(f^n)'(z) = nf^{n-1}(z)f'(z)$ if $f(z) \neq 0$ when $n < 0$,

(c) $\dfrac{dc}{dz} = 0$ and $\dfrac{dz}{dz} = 1$ for each $z \in R^2$ and

(d) $(cf)'(z) = cf'(z)$.

In Part (b), if $n = -1$, then we must interpret f^n as $\frac{1}{f}$ — not the inverse function.

Proof of (a) when $n > 1$. If $h \neq 0$, then

$$\frac{(z+h)^n - z^n}{h} = nz^{n-1} + \left(\frac{n(n-1)}{2!} z^{n-2}h + \cdots + h^{n-1} \right).$$

Thus by (2.14),

$$(z^n)' = \lim_{h \to 0} \frac{(z+h)^n - z^n}{h} = nz^{n-1} \quad \text{for each } z \text{ in } R^2.$$

Thus z^n is an entire function if $n > 1$. By (c), the function $g(z) = z$ is also an entire function.

Theorem 2.3.14 A polynomial is an entire function.

Proof. The function z^n is an entire function for each integer $n \geq 0$ by 2.3.13(a) and (c). Thus by 2.3.5, it follows that a polynomial is an entire function.

Theorem 2.3.15 Let f be analytic in S where S is a domain (i.e., an open connected set).

(a) If $f'(z) \equiv 0$ on S then f is constant on S.

(b) If one of the four real-valued functions

$$\mathcal{R}(f), \quad \mathcal{I}(f), \quad |f|, \quad \text{or} \quad \text{Arg}\, f$$

is constant on S, then f is constant on S.

Proof of (a). If $f = u + iv$, then $f'(z) = 0 + 0i = u_x + iv_x = v_y - iu_y$ by (2.26). Thus $u_x \equiv u_y \equiv 0$ on S. Since $u_x \equiv u_y \equiv 0$ on S, it follows that u is constant on any line segment in S which is parallel to either coordinate axis. Thus u is constant on S. Similarly v is constant on S. Hence f is constant on S.

Proof of (b). Let $\mathcal{R}(f(z)) = u = C$ (constant) on S. Then $u_x \equiv u_y \equiv 0$ on S. Hence $v_x \equiv -u_y \equiv 0$ on S by (2.25). Thus by (2.26), we have $f'(z) = u_x + iv_x \equiv 0$ on S. Hence f is constant on S by Part (a). The proof is similiar if $\mathcal{I}(f(z))$ is constant on S.

Next, let $|f| = \gamma$ (constant) on S. (In case $\gamma = 0$, it follows that $f \equiv 0$ on S.) Now suppose $\gamma \neq 0$. Then

$$\overline{f} = \frac{\gamma^2}{f} \in A(S).$$

Hence f is constant on S by Ex. 17 in 2.3.16.

Finally, suppose $\text{Arg}\, f(z) = \gamma$ (constant) on S. Then

$$f(z) = |f(z)| \operatorname{cis} \gamma.$$

Thus $f(z) \operatorname{cis}(-\gamma) = |f(z)|$. Hence $|f| \in A(S)$. Thus $|f(z)|$ is constant on S by Ex. 16 in 2.3.16. Hence f is constant on S by the third part of Part (b).

Exercises 2.3.16

1. Apply 2.3.5 and 2.3.13 to find the derivative of each of the following functions where a, b, c, and d are complex constants.

   (a) $az^3 + bz^2 + cz + d$ (b) $(-bz^n)^p$ where n and p are positive integers

   (c) $\dfrac{az + b}{cz + d}$ (d) $(az + b)(cz + d)$

2. Prove 2.3.13(a) for $n < 0$. (Apply 2.3.1 since the quotient formula has not been proved; see Ex. 4.)

3. Prove 2.3.13(b)–(d). In Part (b) use 2.3.13(a) and 2.3.6.

4. Let $f'(z)$ and $g'(z)$ exist where $g(z) \neq 0$. Prove

$$\left(\frac{f}{g}\right)'(z) = \frac{g(z)f'(z) - f(z)g'(z)}{g^2(z)}.$$

Hint. $\frac{f}{g} = f \cdot \frac{1}{g}$. Apply 2.3.13(b) and obtain $\left(\frac{1}{g}\right)(z) = -\frac{g'(z)}{g^2(z)}$.

In Ex. 5–12, if $f(z)$ is the given function,
(a) state where $f'(z)$ exists and
(b) find $f'(z)$ at these points.

5. $y^2 - 3ix^2$

  Ans. (a) By 2.3.11, $f'(z)$ exists for each z on the line $y = 3x$; by 2.3.7, $f'(z)$ does not exist for z not on this line.

  Ans. (b) $f'(z) = -6ix$ for each point on this line.

6. $ax + by + i(ay - bx)$ where a and b are real   (a) R^2

7. $e^x(\cos y + i \sin y)$ (a) R^2

8. $e^x(\cos y - i \sin y)$ (a) nowhere

9. $2xy + 2iy$ (a) $z = i$

10. $\sin x \cosh y + i \cos x \sinh y$

                 $f'(z) = \cos x \cosh y - i \sin x \sinh y$ for each z in R^2

11. $(y - 2)^3 + ix^3$ $f'(2i) = 0$

12. $z\mathcal{R}(z)$ $f'(0) = 0$

13. State where each function in Ex. 5–12 is analytic.   In Ex. 6, 7 and 10, $f \in A(R^2)$. In Ex. 5, 8, 9, 11 and 12, f is nowhere analytic.

14. Which functions in Ex. 5–12 are entire functions?

15. Use 2.3.7 to show that f is not differentiable at any z in R^2 if $f(z)$ is
  (a) $\bar{z}$,   (b) $\mathcal{R}(z)$,   (c) $\mathcal{I}(z)$,   (d) $|z|$.
  Hint. To show that $u_x(0,0)$ does not exist in Part (d), we should notice that $u(x,0) = |x|$. Then we should write the difference quotient $\frac{u(t,0)-u(0,0)}{t-0}$ for $t > 0$ and for $t < 0$ and notice that the right- and left-hand limits are different.

16. Prove each of the following results.

  (a) If a real-valued function f of a complex variable is analytic on a domain S, then f is constant on S. **Hint.** $f = u + iv$ where $v \equiv 0$ on S. Use 2.3.15(b).

  (b) If a pure imaginary function f is analytic on a domain S, then f is constant on S.

17. Prove if f and $\overline{f}$ are analytic on a domain S, then f is constant on S. **Hint.** Let $f = u + iv$. Then $f + \overline{f} = 2u \in A(S)$. Hence u is constant on S by Ex. 16. Hence f is constant on S by the first part of 2.3.15(b).

18. Let $f, g \in A(D)$ where D is a domain. If $z_0 \in D$ such that $f(z_0) = g(z_0) = 0$ and $g'(z_0) \neq 0$, prove $\lim_{z \to z_0} \left(\frac{f(z)}{g(z)} \right) = \frac{f'(z_0)}{g'(z_0)}$ (L'Hôpital's rule). **Hint.** $\lim_{z \to z_0} \frac{f(z)}{g(z)} = \lim_{z \to z_0} \frac{\frac{f(z)-f(z_0)}{z-z_0}}{\frac{g(z)-g(z_0)}{z-z_0}}$. Now use 2.2.6(c).

19. Prove if P and Q are polynomials, then the rational function $\frac{P(z)}{Q(z)}$ is analytic in $R^2 - \{z : Q(z) = 0\}$.

20. Prove if $f'(z) = g'(z)$ on a domain D, then $f - g$ is constant on D. **Hint.** See 2.3.15(a).

2.4   Compactness and Continuity

Definition 2.4.1 Let f be a complex function with domain D and let $S \subset D$.

(a) The function f is **continuous on S** iff the following condition holds.

> For each z_0 in S and for each $\epsilon > 0$, there is a number $\delta > 0$ such that if $z \in N_\delta(z_0) \cap S$, then $f(z) \in N_\epsilon(f(z_0))$.

We write $f \in C(S)$ to indicate that f is continuous on S. We say "f is continuous," "f is a continuous function," or "f is a continuous mapping," to mean that f is continuous on its domain D.

(b) The function f is **uniformly continuous on S** iff the following condition holds.

> For each $\epsilon > 0$, there is a number $\delta > 0$ such that if z_1 and z_2 are in S and $|z_1 - z_2| < \delta$, then $|f(z_1) - f(z_2)| < \epsilon$.

We write $f \in UC(S)$ to indicate that f is uniformly continuous on S.

Remarks 2.4.2 **(a)** We should observe that $f \in UC(S)$, then $f \in C(S)$. However, 2.4.3 shows that the converse is not true.

(b) By comparing 2.2.8 and 2.4.1(a), we see that $f \in C(D)$ iff f is continuous at each point of D (where D is the domain of f). Furthermore, if f is continuous at each point of S where $S \subset D$, then $f \in C(S)$. However, the converse is not true if $S \neq D$. (See 2.4.8.)

Example 2.4.3 By 2.2.13, the function $\frac{1}{z}$ is continuous on D where

$$D = \{z : \; 0 < |z| \leq 1\}.$$

To see that $\frac{1}{z}$ is not uniformly continuous on D, we let $\epsilon = \frac{1}{10}$ and we let $0 < \delta < 1$. Next, we take $z_1 = \delta$ and $z_2 = \frac{9}{10}\delta$. Then

$$|z_1 - z_2| = \frac{1}{10}\delta < \delta.$$

But

$$\left|\frac{1}{z_1} - \frac{1}{z_2}\right| = \frac{1}{9\delta} > \frac{1}{10} = \epsilon.$$

Theorem 2.4.4 Let f be a complex function which is continuous on a nonempty compact set S.

(a) Then f is uniformly continuous on S.

(b) If $f(S) \subset R$, then f assumes a **maximum** on S and a **minimum** on S. That is, if f is real-valued and continuous on the nonempty compact set S, then there are points z_1 and z_2 in S such that $f(z_1) \leq f(z) \leq f(z_2)$ for each z in S.

If in Chapter 1, we combine 1.4.15 with Ex. 1 in 1.4.16, we conclude that a subset S of R^2 is compact iff S is closed and bounded. Thus 2.4.4(b) may be stated as follows

> If a real-valued function f of two real variables is
> continuous on a nonempty closed and bounded set, S,
> then f assumes its **maximum** and **minimum** on S. $\qquad$ (2.43)

Now (2.43) may be recalled from calculus or advanced calculus.

Also, a proof of 2.4.4(a) may be made to depend upon the analogous theorem for real functions of two real variables as follows.

If $f(z) = u(x, y) + iv(x, y)$ is continuous on S, then $u \in C(S)$ and $v \in C(S)$. $\quad$ (2.44)

Now S is closed and bounded. Thus from calculus or advanced calculus, (2.44) implies

$$u \in UC(S) \quad \text{and} \quad v \in C(S). \qquad (2.45)$$

But since

$$|f(z_2) - f(z_1)| \leq |u(x_2, y_2) - u(x_1, y_1)| + |v(x_2, y_2) - v(x_1, y_1)|,$$

it follows from (2.45) that f is uniformly continuous on S.

Thus for a short course, one may accept 2.4.4 as proved and omit the rest of this section. However, the rest of this section (except 2.4.8) is devoted to a careful, detailed proof of 2.4.4. For a thorough course in complex analysis, it is desirable to consider this detailed proof.

Proof of 2.4.4(a). Let $\epsilon > 0$. For each z in S, let $\delta_z > 0$ such that

$$t \in N(z, \delta_z) \cap S \text{ implies } f(t) \in N\left(f(z), \frac{\epsilon}{2}\right). \tag{2.46}$$

Since S is compact, let

$$\left\{N\left(z_1, \frac{1}{2}\delta_{z_1}\right), N\left(z_2, \frac{1}{2}\delta_{z_2}\right), \cdots, N\left(z_n, \frac{1}{2}\delta_{z_n}\right)\right\} \tag{2.47}$$

be a finite subfamily of

$$\left\{N\left(z, \frac{1}{2}\delta_z\right) : \ z \in S\right\}$$

such that the family in (2.47) covers S. Let

$$\delta = \min\left\{\frac{1}{2}\delta_{z_1}, \frac{1}{2}\delta_{z_2}, \cdots, \frac{1}{2}\delta_{z_n}\right\}. \tag{2.48}$$

Then $\delta > 0$. Now let z and w be points in S such that

$$|z - w| < \delta. \tag{2.49}$$

Since the sets in (2.47) cover S,

$$z \in N\left(z_k, \frac{1}{2}\delta_{z_k}\right) \text{ for some } k \text{ such that } 1 \le k \le n. \tag{2.50}$$

Now

$$\begin{aligned}
|w - z_k| &\le |w - z| + |z - z_k| \\
&< \delta + \frac{1}{2}\delta_{z_k} && \text{by (2.49) and (2.50)} \\
&\le \frac{1}{2}\delta_{z_k} + \frac{1}{2}\delta_{z_k} = \delta_{z_k} && \text{by (2.48).}
\end{aligned}$$

Thus

$$w \in N(z_k, \delta_{z_k}). \tag{2.51}$$

Now

$$|f(z) - f(w)| \le |f(z) - f(z_k)| + |f(z_k) - f(w)| < \frac{\epsilon}{2} + \frac{\epsilon}{2} = \epsilon \tag{2.52}$$

by (2.50), (2.51), and (2.46). Thus $f \in UC(S)$ by (2.49) and (2.52).

Theorem 2.4.5 Let f be a complex function which is continuous on a nonempty compact set S. Then $f(S)$ is compact.

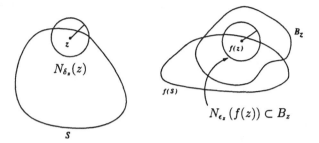

Figure 2.4: $f(N_{\delta_z}(z) \cap S) \subset N_{\epsilon_z}(f(z)) \subset B_z$.

Proof. Let $\mathcal{C}$ be an open cover for $f(S)$. For each z in S, let $B_z \in \mathcal{C}$ such that $f(z) \in B_z$. Since B_z is open, let $\epsilon_z > 0$ such that $N_{\epsilon_z}(f(z)) \subset B_z$. But f is continuous on S. So let $\delta_z > 0$ such that

$$f(N_{\delta_z}(z) \cap S) \subset N_{\epsilon_z}(f(z)) \subset B_z. \tag{2.53}$$

See Figure 2.4. Now

$$\mathcal{D} = \{N(z, \delta_z) : \ z \in S\} \text{ is an open cover for } S.$$

Since S is compact, let

$$\{N(z_1, \delta_{z_1}), N(z_2, \delta_{z_2}), \cdots, N(z_k, \delta_{z_k})\}$$

be a finite subcover of $\mathcal{D}$ for S. Thus

$$S \subset [N(z_1, \delta_{z_1}) \cap S] \cup [N(z_2, \delta_{z_2}) \cap S] \cup \cdots \cup [N(z_k, \delta_{z_k}) \cap S].$$

Then by Ex. 1 in 2.4.9,

$$
\begin{aligned}
f(S) \ &\subset \ f([N(z_1, \delta_{z_1}) \cap S] \ \cup \ [N(z_2, \delta_{z_2}) \cap S] \ \cup \ \cdots \ \cup \ [N(z_k, \delta_{z_k}) \cap S]) \\
&= \ f[N(z_1, \delta_{z_1}) \cap S] \ \cup \ f[N(z_2, \delta_{z_2}) \cap S] \ \cup \ \cdots \ \cup \ f[N(z_k, \delta_{z_k}) \cap S] \\
&\subset \ B_{z_1} \ \cup \ B_{z_2} \ \cup \ \cdots \ \cup \ B_{z_k} \quad \text{by (2.53)}.
\end{aligned}
$$

Thus $\{B_{z_1}, B_{z_2}, \cdots, B_{z_k}\}$ is a finite subcover of $\mathcal{C}$ for $f(S)$. Hence $f(S)$ is compact.

Remark 2.4.6 Theorem 2.4.5 may be stated as follows.

A continuous image (in R^2) of a nonempty compact subset of R^2 is compact.

Theorem 2.4.7 If E is a nonempty compact subset of R, then $(\text{lub}\, E) \in E$ and $(\text{glb}\, E) \in E$.

Proof. Since E is compact, it is bounded. Also $E \neq \emptyset$. Then $\text{lub}\,E$ and $\text{glb}\,E$ exist by 1.4.12. Now

$$(\text{lub}\,E) \in E^* \quad \text{or} \quad (\text{lub}\,E) \notin E^*.$$

If $(\text{lub}\,E) \in E^*$, then $(\text{lub}\,E) \in E$ since E is closed. [See 1.4.1(c).] But if $(\text{lub}\,E)$ $\notin E^*$, it follows from the definition of $\text{lub}\,E$ that $(\text{lub}\,E) \in E$. Similiarly $(\text{glb}\,E) \in E$.

Proof of 2.4.4(b). By 2.4.6, the set $f(S)$ is compact. Thus, we let z_1 and z_2 be points in S such that $f(z_1) = \text{glb}\,f(S)$ and $f(z_2) = \text{lub}\,f(S)$. Hence

$$f(z_1) \leq f(z) \leq f(z_2) \quad \text{for each } z \text{ in } S.$$

Example 2.4.8 Let $S = \{(x,y) : \quad x \in Q \text{ and } y \in Q\}$ where Q is the set of all rational numbers. Let f be the complex function given by

$$f(z) = \begin{cases} 1 & \text{if } z \in S \\ 0 & \text{if } z \in (R^2 - S). \end{cases}$$

Clearly f is not continuous at any point of R^2. However by 2.4.1(a), we see that $f \in C(S)$ since f is constant on S.

Exercises 2.4.9

1. Let f be a function on X to Y where X and Y are sets. Let $B_k \subset X$ for $k = 1$ to n. Prove $f(B_1 \cup B_2 \cup \cdots \cup B_n) = f(B_1) \cup f(B_2) \cup \cdots \cup f(B_n)$. **Hint.** Let p be an element of the set on the left and show that p is in the set on the right. Then prove the converse.

2. Let f be a function on D to R^2 where $D \subset R^2$ and let $f \in C(D)$.

   (a) Prove $|f| \in C(D)$.

   (b) Prove if D is compact, then there is some z_0 in D such that $|f(z)| \leq |f(z_0)|$ for each z in D. **Hint.** Use Part (a) and 2.4.4(b).

   (c) Prove if D is compact and f is never zero on D, then there is some point z_1 in D such that

$$0 < |f(z_1)| \leq |f(z)| \quad \text{for each } z \text{ in } D.$$

Chapter 3

Complex Integrals

3.1 Introduction

Definition 3.1.1 Let f be a bounded real-valued function on a closed interval $[a, b]$. Let $x \in [a, b]$ and let

$$\mathcal{P} = \{x_0, x_1, \cdots, x_n\}$$

be a partition of $[a, x]$ such that $a = x_0 \leq x_1 \leq \cdots \leq x_{n-1} \leq x_n = x$. Then

$$\sum_{k=1}^{n} f(t_k) \Delta x_k \quad \text{where } t_k \in [x_{k-1}, x_k] \text{ and } \Delta x_k = x_k - x_{k-1},$$

is called a **Riemann sum** for $\mathcal{P}$. If

$$\lim_{|\mathcal{P}| \to 0} \sum_{k=1}^{n} f(t_k) \Delta x_k$$

exists where

$$|\mathcal{P}| = \max\{\Delta x_k : \quad k = 1, 2, \cdots, n\},$$

then f is **Riemann integrable on** $[a, x]$. This limit is called the Riemann (definite) integral of f over $[a, x]$ and is denoted by

$$\int_a^x f(t) \, dt. \tag{3.1}$$

We complete the definition of the definite integral by

$$\int_x^a f(t) \, dt = - \int_a^x f(t) \, dt. \tag{3.2}$$

Observe that

$$\int_a^x f(t) \, dt = 0 \quad \text{if } x = a.$$

51

The precise meaning of (3.1) follows

for each $\epsilon > 0$, there is some $\delta > 0$ such that

$$\left| \int_a^x f(t)\, dt - \sum_{k=1}^n f(t_k)\Delta x_k \right| < \epsilon$$

for each $\mathcal{P}$ such that $|\mathcal{P}| < \delta$ and for each choice of points t_k in $[x_{k-1}, x_k]$. We shall need the following theorems, 3.1.2, 3.1.3, and 3.1.7, from the calculus.

Theorem 3.1.2 Let $c \in [a, b]$ and let f_c be defined by

$$f_c(x) = \left\{ \begin{array}{ll} 0 & \text{if } x \neq c \\ m_c & \text{if } x = c \text{ where } m_c \in R. \end{array} \right.$$

Then

$$\int_a^b f_c(x)\, dx = 0.$$

Theorem 3.1.3 Let f be a real-valued function which is Riemann integrable on $[a, b]$. For each $j = 1, 2, \cdots, n$, let $x_j \in [a, b]$ and let f_{x_j} be defined as in 3.1.2. Let $g = f + f_{x_1} + \cdots + f_{x_n}$. Then g is Riemann integrable on $[a, b]$ and

$$\int_a^b g(x)\, dx = \int_a^b f(x)\, dx.$$

Remark 3.1.4 Theorem 3.1.3 states that changing the values of a function f at a finite number of points in $[a, b]$ neither affects the integrability of f nor the value of

$$\int_a^b f(x)\, dx.$$

Thus it is natural to extend the definition of $\int_a^b f(x)\, dx$ to include a function which is defined on $[a, b]$ except at a finite number of points.

Definition 3.1.5 Let f be a real-valued function which is defined on $[a, b]$ except at a finite number of points. Let g be defined by

$$g(x) = \left\{ \begin{array}{ll} f(x) & \text{if } f(x) \text{ is defined} \\ 0 & \text{if } f(x) \text{ is not defined.} \end{array} \right.$$

Then $\int_a^b f(x)\, dx$ is defined by $\int_a^b f(x)\, dx = \int_a^b g(x)\, dx$ if g is Riemann integrable on $[a, b]$. (In view of 3.1.3, it is clear that g could have had an arbitrary value at each point where f is not defined.)

Definition 3.1.6 A complex function f of a real variable is **piecewise continuous** on $[a, b]$ iff $[a, b]$ is the union of a finite number of subintervals I_n such that f is continuous on each I_n° and f has finite limits from the interior of each I_n at both endpoints. (We should note that f may fail to be defined at a finite number of points of $[a, b]$.)

Sometimes we may indicate that f is piecewise continuous on $[a, b]$ by writing $f \in PC[a, b]$. We shall state without proof the following theorems from the calculus.

Theorem 3.1.7 (a) If $f \in PC[a, b]$, then $\int_a^x f(t)\, dt$ exists for each x in $[a, b]$.

(b) Fundamental Theorem of Calculus. If $f \in C([a, b])$ and if g is given by

$$g(x) = \int_a^x f(t)\, dt, \text{ then } g'(x) = f(x) \quad \text{for each } x \text{ in } [a, b].$$

(c) Let f and h be Riemann integrable on $[a, b]$ and let $f(x) \le h(x)$ on $[a, b]$. Then

$$\int_a^b f(x)\, dx \le \int_a^b h(x)\, dx.$$

Also, $|f|$ is Riemann integrable on $[a, b]$, and $\left| \int_a^b f(x)\, dx \right| \le \int_a^b |f(x)|\, dx$.

(d) If f is bounded, if f is continuous except at a finite number of points in $[a, b]$, and if

$$g(x) = \int_a^x f(t)\, dt,$$

then $g \in C([a, b])$, and $g'(x) = f(x)$ for each x in $[a, b]$ at which f is continuous.

3.2 The Definite Integral of a Complex Function

Definition 3.2.1 Let f be a complex-valued function such that $u = \mathcal{R}(f)$ and $v = \mathcal{I}(f)$ are piecewise continuous on $[a, b]$. The **definite integral of f on $[a, b]$** is defined by

$$\int_a^b f(t)\, dt = \int_a^b u(t)\, dt + i \int_a^b v(t)\, dt. \tag{3.3}$$

Remark 3.2.2 We observe that the integrands u and v in (3.3) are piecewise continuous on $[a, b]$ and thus may be undefined at a finite number of points in $[a, b]$. Nevertheless, by 3.1.7(a), the integrals in the right member of (3.3) exist and are real numbers. We also notice that

$$\mathcal{R}\left[\int_a^b f(t)\, dt \right] = \int_a^b u(t)\, dt = \int_a^b \mathcal{R}[f(t)]\, dt. \tag{3.4}$$

Many of the properties of real integrals are also valid for the definite integral of a complex function f of a real variable t. For example, we have the following theorem.

Theorem 3.2.3 Let $\gamma = \alpha + i\beta$ where α and β are real and let $f = u + iv$ where u and v are piecewise continuous on $[a, b]$. Then

$$\int_a^b \gamma f(t)\, dt = \gamma \int_a^b f(t)\, dt. \tag{3.5}$$

Proof. From (3.3),

$$
\begin{aligned}
\int_a^b \gamma f(t)\, dt &= \int_a^b (\alpha + i\beta)(u + iv)\, dt = \int_a^b (\alpha u - \beta v)\, dt + i \int_a^b (\alpha v + \beta u)\, dt \\
&= \alpha \int_a^b u\, dt - \beta \int_a^b v\, dt + i\alpha \int_a^b v\, dt + i\beta \int_a^b u\, dt \\
&= (\alpha + i\beta)\left(\int_a^b u\, dt + i \int_a^b v\, dt \right) = \gamma \int_a^b f(t)\, dt.
\end{aligned}
$$

Theorem 3.2.4 Let $f = u + iv$ where u and v are piecewise continuous on $[a, b]$. Then

$$\left| \int_a^b f(t)\, dt \right| \leq \int_a^b |f(t)|\, dt. \tag{3.6}$$

Proof. Now (3.3) implies that $\int_a^b f(t)\, dt$ is a complex number. Thus if $\int_a^b f(t)\, dt \neq 0$, let

$$\int_a^b f(t)\, dt = r \operatorname{cis} \theta \tag{3.7}$$

where

$$r = \left| \int_a^b f(t)\, dt \right| \quad \text{and} \quad \theta = \arg\left(\int_a^b f(t)\, dt \right). \tag{3.8}$$

Then by (3.5),

$$
\begin{aligned}
\int_a^b [\operatorname{cis}(-\theta)] f(t)\, dt &= \operatorname{cis}(-\theta) \int_a^b f(t) dt \\
&= r \qquad \text{by (3.7).} \tag{3.9}
\end{aligned}
$$

But $r \in R$ and $r > 0$. Thus

$$
\begin{aligned}
r &= \mathcal{R}\left(\int_a^b [\operatorname{cis}(-\theta)] f(t)\, dt \right) \qquad \text{by (3.9)} \\
&= \int_a^b \mathcal{R}\,[\operatorname{cis}(-\theta) f(t)]\, dt \qquad \text{by (3.4)} \\
&\leq \int_a^b |\mathcal{R}\,[\operatorname{cis}(-\theta) f(t)]|\, dt \qquad \text{by 3.1.7(c)} \\
&\leq \int_a^b |\operatorname{cis}(-\theta) f(t)|\, dt = \int_a^b |f(t)|\, dt. \tag{3.10}
\end{aligned}
$$

Thus (3.6) follows by (3.8) and (3.10). If $\int_a^b f(t)dt = 0$, then

$$\left| \int_a^b f(t)\, dt \right| = 0,$$

so that (3.6) is obviously true by the definition of a Riemann integral.

Exercises 3.2.5

1. Let $f_k(t) = u_k(t) + iv_k(t)$ on $[a, b]$ for $k = 1,\ 2$. Prove

$$\int_a^b [f_1(t) + f_2(t)]\, dt = \int_a^b f_1(t)\, dt + \int_a^b f_2(t)\, dt$$

   when the last two integrals exist.

2. Let $f(t) = u(t) + iv(t)$ on $[a, b]$ and let $a < c < b$. Prove

$$\int_a^b f(t)\, dt = \int_a^c f(t)\, dt + \int_c^b f(t)\, dt$$

   when the last two integrals exist.

3.3 Curves in R^2

Definition 3.3.1 (a) Let $C \subset R^2$. Then C is a **curve** iff there is a continuous function f whose domain is a closed interval $[a, b]$ in R and whose range is C. Thus a curve in R^2 is a continuous image in R^2 of a closed interval in R.

A **parametrization** of a curve C is a continuous function

$$\begin{aligned} f(t) &= (x(t), y(t)) \\ &= x(t) + iy(t) \quad \text{for each } t \text{ in } [a, b] \end{aligned} \tag{3.11}$$

such that $f([a, b]) = C$.

(b) A curve C is an **arc** iff there is a parametrization of C which is a one-to-one function.

(c) A curve C in R^2 is **smooth** iff C has a parametrization f given by (3.11) such that $x'(t)$ and $y'(t)$ are continuous on $[a, b]$ and such that $f'(t)$ is never zero in the open interval (a, b). [We use $f'(t)$ to denote $x'(t) + iy'(t)$.] Whenever we choose a parametrization $f(t) = (x(t), y(t))$ on $[a, b]$ for a smooth curve, it will be understood that x' and y' are continuous on $[a, b]$ and that $f'(t)$ is never zero on (a, b).

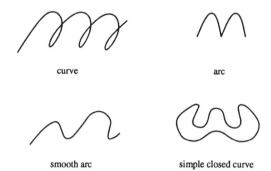

Figure 3.1: Curves

(d) A curve C is a **simple closed curve** (or a **Jordan curve**) iff there exists a parametrization (3.11) of C such that f is one-to-one on $[a,b]$ and $f(b) = f(a)$. See Figure 3.1.

Example. (Simple Closed Curve). If

$$C = \{(a\cos t, b\sin t): \ t \in [0, 2\pi]\},$$

then C is an ellipse if $ab \neq 0$.

Remark 3.3.2 A curve may have different parametrizations. For example, the arc of the parabola $\{(x, y): \ y = x^2 \text{ and } |x| \le 8\}$ has the parametrizations

$$z(t) = t + it^2 \quad \text{on } [-8, 8]$$

and also

$$z(t) = t^3 + it^6 \quad \text{on } [-2, 2].$$

The word "smooth" implies that the motion of a point which traces the curve has no abrupt change in direction. Definition 3.3.1(c) excludes this for a smooth curve. The need for the condition $f'(t) \neq 0$ is shown by the following example. Let C be given by

$$C: \quad z(t) = t^3 + i|t|^3 = x(t) + iy(t) \quad \text{on } [-1, 1]. \tag{3.12}$$

This curve C is not smooth. We see that

$$x'(t) = 3t^2 \text{ on } [-1, 1] \quad \text{and} \quad y'(t) = \begin{cases} 3t^2 & \text{on } [0, 1] \\ -3t^2 & \text{on } [-1, 0]. \end{cases}$$

For this parametrization, $z'(0) = 0$. However, this does not prove that C is not smooth since (3.12) is only one parametrization of C. The fact that C has a sharp

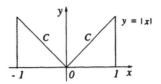

Figure 3.2: C is not smooth.

turning point at the origin shows that C has no parametrization $z(t) = x(t) + iy(t)$ such that $x'(t)$ and $y'(t)$ are continuous at the origin and such that $z' \neq 0$ at the origin. See Figure 3.2. Another parametrization of C is $z = t + i|t|$ for each t in $[-1, 1]$. For this parameter

$$y'(t) = \begin{cases} 1 & \text{if } t \in (0, 1] \\ -1 & \text{if } t \in [-1, 0). \end{cases}$$

For this parametrization, $y'(t)$ is not continuous at $t = 0$ since $y'(0)$ does not exist.

We shall state the following theorem without proof.

Theorem 3.3.3 (The Jordan Curve Theorem). Let C be a Jordan curve in R^2. Then

$$R^2 - C = A \cup B \tag{3.13}$$

where A and B are domains, exactly one of which is **bounded**. Furthermore, C is the boundary of A and also the boundary of B.

The bounded domain in (3.13) is called the **interior of** C and is denoted by $I(C)$. The unbounded domain in (3.13) is called the **exterior of** C and is denoted by $E(C)$. [This interior $I(C)$ is not to be confused with the interior C° of the set defined in 1.4.6(c).]

The Jordan Curve Theorem is intuitively evident for such Jordan curves as circles, ellipses, and simple closed polygons. Jordan was the first to observe that the statement requires a proof, but his proof was incomplete. It was first proved correctly by Veblen in 1905.

Remark 3.3.4 The following question arises relative to the definition of a curve given in 3.3.1(a). Does there exist a continuous function which maps the interval $[a, b]$ onto a given subset of R^2? This is answered by the following theorem.

Theorem 3.3.5 (Hahn-Mazurkiewicz for R^2). A nonempty subset S of R^2 is a curve iff S is compact, connected, and locally connected.

[A set S is said to be **locally connected** iff for each point p in S and for each neighborhood $N_r(p)$, there is an open set G containing p such that $G \cap S$ is connected and $G \cap S \subset N_r(p)$.]

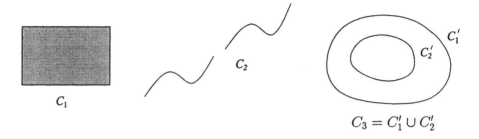

$$C_3 = C_1' \cup C_2'$$

Figure 3.3: C_1 is a curve by 3.3.5. C_2 is not connected. C_3 is not connected.

Remark 3.3.6 By 3.3.5, any rectangle together with its interior is a curve (since it is compact, connected, and locally connected). But the sets C_2 and C_3 in Figure 3.3 are not curves (since they are not connected).

Definition 3.3.1(a) is much too general for our intuitive concept of thinness or of the property of being one dimension. In 1890, Peano surprised his colleagues by showing the existence of a continuous mapping of a closed unit interval onto a closed unit square. Such a mapping is called a space-filling curve. However, L. E. J. Brouwer proved that there exists no one-to-one continuous mapping of a closed unit interval onto a square.

Jordan removed this difficulty of generality by excluding curves with "multiple points," that is, by requiring that their parametrizations not only be continuous but that they also be one-to-one.

3.4   Contour Integrals

Definition 3.4.1 (a) A **parametrized curve** is a pair $(C, z(t))$ where C is a curve and $z(t)$ is a parametrization of C. [We shall denote this parametrized curve by C without explicit mention of $z(t)$.]

(b) A **contour** is a parametrized curve $(C, z(t))$ where the domain $[a, b]$ of $z(t) = x(t) + iy(t)$ can be divided into a finite number of subintervals such that $x'(t)$ and $y'(t)$ are continuous on each of these closed subintervals and $z'(t)$ is never zero on the interiors of these subintervals. [This means that the curve C is the union of a finite number of adjoining smooth curves. Also, at the ends of the subintervals, $x'(t)$ and $y'(t)$ refer to the appropriate left- and right-hand derivatives.] A curve C is **piecewise smooth** iff $(C, z(t))$ is a contour for some $z(t)$.

(c) A contour $(C, z(t))$ is a **simple closed contour** iff $z(t)$ is one-to-one on $[a, b)$
    and $z(b) = z(a)$ (where $[a, b]$ is the domain of $z(t)$).

(d) A simple closed contour $(C, z(t))$ is **counterclockwise oriented (cco, or pos-**
    itively oriented) iff as t varies over its domain from a to b, the point $z(t)$
    traverses C so that $I(C)$ is on the left. [The contour C is **clockwise oriented**
    (or **co**) iff $z(t)$ traverses C in the opposite manner as t varies from a to b.]

We recall that in Definition 2.3.1, we defined the derivative $f'(z)$ only for points
z in D°. Thus when the domain of $f(t) = u(t) + iv(t)$ is an interval $[a, b]$ in R, the
derivative $f'(t)$ is not defined in the sense of Definition 2.3.1. For no point in $[a, b]$ is
an interior point of the domain of f when $[a, b]$ is considered as a subset of R^2. Hence
we are led to the following definition.

Definition 3.4.2 Let

$$z(t) = x(t) + iy(t)$$

be defined on the interval $[a, b]$ in R. Then

$$z'(t) = x'(t) + iy'(t) \qquad \text{for each } t \text{ in } (a, b)$$

if $x'(t)$ and $y'(t)$ exist. [The right-hand derivative $z'(a)$ and the left-hand derivative
$z'(b)$ as well as the right- and left-hand derivatives at t are defined in the usual way.]

Remark 3.4.3 The definite integral of a complex function $f = u + iv$ of the real
variable t has been defined in terms of real integrals with respect to the real variable
t on $[a, b]$. In this case u and v were piecewise continuous on $[a, b]$. This concept of
a definite integral of f is generalized to a contour (or complex line) integral of f by
replacing $[a, b]$ by a contour C contained in R^2.

Definition 3.4.4 (a) Let $(C, z(t))$ be a contour where the domain of $z(t)$ is $[a, b]$.
    Let f be a function on C to R^2 such that f is piecewise continuous on C, i.e.,
    C can be divided into a finite number of arcs on each of which f is continuous.
    Then the **contour (or complex line) integral** of f along C is defined by

$$\int_C f(z)\, dz = \int_a^b f(z(t))\, z'(t)\, dt, \tag{3.14}$$

where this definite integral is defined in 3.2.1.

> **Note.** Here $z'(t)$ may be undefined at a finite number of points of
> $[a, b]$. However, $f(z(t))z'(t)$ is piecewise continuous on $[a, b]$. Thus by
> 3.1.7 and 3.2.1, the integral on the right in (3.14) does exist.

(b) For $k = 1, 2$, let

$$z_k(t) = (x_k(t), y_k(t)) \qquad \text{for each } t \text{ in } [a_k, b_k]$$

be a parametrization of a curve C. Then $z_1(t)$ and $z_2(t)$ are **equivalent parametrizations** of C iff the following condition is satisfied.

There is a strictly increasing function h from $[a_2, b_2]$ onto $[a_1, b_1]$ such that h' is piecewise continuous on $[a_2, b_2]$ and such that $z_2(t) = z_1(h(t))$ for each t in $[a_2, b_2]$.

Theorem 3.4.5 The contour integral $\int_C f(z)\, dz$ is independent of the parametrization of the curve C as long as equivalent parametrizations are used.

Proof. Let $z_1(t)$ and $z_2(t)$ be equivalent parametrizations of C with domains $[a_1, b_1]$ and $[a_2, b_2]$, respectively. Then by 3.4.4(b), we let

$$h \text{ be a strictly increasing function from } [a_2, b_2] \text{ onto } [a_1, b_1] \tag{3.15}$$

such that

$$z_2(\tau) = z_1(h(\tau)) \qquad \text{for each } \tau \text{ in } [a_2, b_2]. \tag{3.16}$$

The function in (3.15) may be expressed by

$$t = h(\tau) \qquad \text{for each } \tau \text{ in } [a_2, b_2]. \tag{3.17}$$

Then by (3.16),

$$z_2'(\tau) = z_1'(h(\tau))h'(\tau) \qquad \text{for each } \tau \text{ in } [a_2, b_2]. \tag{3.18}$$

Now

$$\int_{a_1}^{b_1} f(z_1(t))\, z_1'(t)\, dt \;=\; \int_{a_2}^{b_2} f(z_1(h(\tau)))\, z_1'(h(\tau))h'(\tau)\, d\tau \qquad \text{by (3.17)}$$

$$=\; \int_{a_2}^{b_2} f(z_2(\tau))\, z_2'(\tau)\, d\tau \qquad \text{by (3.16) and (3.18).}$$

Thus $\int_C f(z)\,dz$ has the same value whether $z_1(t)$ or $z_2(t)$ is used as the parametrization of C.

Definition 3.4.6 (a) Let $f(t)$ be a parameterization of the contour C_1 with domain $[a, b]$, and let $g(t)$ be a parameterization of the contour C_2 with domain $[b, c]$ where $f(b) = g(b)$. Then the **contour** C represented by

$$h(t) = \begin{cases} f(t) & \text{for each } t \text{ in } [a, b] \\ g(t) & \text{for each } t \text{ in } [b, c], \end{cases}$$

is called the **sum of C_1 and C_2** and we write $C = C_1 + C_2$. Also, $C_1 + C_2 + \cdots + C_n$ is defined similarly for any finite number of adjoining contours.

(b) If $C_1, C_2, \cdots, C_n$ are disjoint contours, then we shall use $\sum_{k=1}^{n} C_k$ as a purely **formal** notation and we shall use

$$\int_{\sum_{k=1}^{n} C_k} f(z)\, dz$$

to denote $\displaystyle\sum_{k=1}^{n} \int_{C_k} f(z)\, dz.$

Remark 3.4.7 Let $f(z) = u(x,y) + iv(x,y)$ and let C be a contour with parametrization

$$C: \quad z(t) = x(t) + iy(t) \qquad \text{for each } t \text{ in } [a,b].$$

Then

$$\begin{aligned}
\int_C f(z)\, dz &= \int_a^b f(z(t))\, z'(t)\, dt \qquad \text{by 3.4.4(a)} \\
&= \int_a^b (u + iv)(x' + iy')\, dt \qquad \text{by 3.4.2} \\
&= \int_a^b (ux' - vy')\, dt + i \int_a^b (vx' + uy')\, dt \qquad \text{by 3.2.1}
\end{aligned}$$

where

$$u = u(x(t), y(t)), \ \ v = v(x(t), y(t)), \ \ x' = x'(t) \text{ and } y' = y'(t).$$

Definition 3.4.8 Let the contour C have the parametrization

$$C: \quad z(t) = x(t) + iy(t) \quad \text{for each } t \text{ in } [a,b].$$

Then the **oppositely oriented** contour $-C$ is the contour given by

$$-C: \quad z_1(t) = z(-t) \quad \text{for each } t \text{ in } [-b, -a].$$

We see that the initial point of $-C$ is $z[-(-b)] = z(b)$, which is the terminal point of C. Also, the terminal point of $-C$ is $z[-(-a)] = z(a)$, which is the initial point of C.

Remark 3.4.9 The circle

$$C_0 = \{z: \ |z - z_0| = r\}$$

is not **oriented**, whereas the circles

$$C: \quad z = z_0 + r\operatorname{cis} t \qquad \text{for each } t \text{ in } [0, 2\pi]$$

and

$$-C: \quad z = z_0 + r\operatorname{cis}(-t) \quad \text{for each } t \text{ in } [-2\pi, 0]$$

are oppositely **oriented**.

Theorem 3.4.10 Let C be a contour with length L and parametrization

$$z(t) = x(t) + iy(t) \quad \text{on } [a, b].$$

Then each of the following holds.

(a) $\displaystyle \int_{-C} f(z)\, dz = -\int_C f(z)\, dz$

(b) $\displaystyle \int_C Kf(z)\, dz = K \int_C f(z)\, dz \quad$ where K is a complex constant in R^2

(c) $\displaystyle \int_C [f(z) + g(z)]\, dz = \int_C f(z)\, dz + \int_C g(z)\, dz$

(d) $\displaystyle \int_C f(z)\, dz = \int_{C_1} f(z)\, dz + \int_{C_2} f(z)\, dz \quad$ where $C = C_1 + C_2$

(e) $\displaystyle 0 \neq L = \int_a^b |z'(t)|\, dt = \int_a^b \sqrt{[x'(t)]^2 + [y'(t)]^2}\, dt$

(f) $\displaystyle \left| \int_C f(z)\, dz \right| \leq \int_a^b |f(z(t))||z'(t)|\, dt$

(g) $\displaystyle \left| \int_C f(z)\, dz \right| \leq ML \quad$ if $|f(z)| \leq M$ for each z on C

Proof of (a). By 3.4.8 and 3.4.4(a),

$$
\begin{aligned}
\int_{-C} f(z)\, dz &= \int_{-b}^{-a} f[z(-t)]\, [z(-t)]'\, dt \\
&= \int_{-b}^{-a} f[z(-t)]\, [-z'(-t)]\, dt \\
&= \int_b^a f[z(\tau)]\, z'(\tau)\, d\tau \quad \text{by substituting } \tau = -t \\
&= -\int_a^b f[z(t)]\, z'(t)\, dt \quad \text{by 3.2.1 and (3.2)} \\
&= -\int_C f(z)\, dz \quad \text{by 3.4.4(a).}
\end{aligned}
$$

Proof of (b).

$$
\begin{aligned}
\int_C Kf(z)\, dz &= \int_a^b Kf[z(t)]\, z'(t)\, dt \quad \text{by 3.4.4(a)} \\
&= K \int_a^b f[z(t)]\, z'(t)\, dt \quad \text{by 3.2.3} \\
&= K \int_C f(z)\, dz \quad \text{by 3.4.4(a).}
\end{aligned}
$$

Proof of (c).

$$\int_C [f(z) + g(z)]\, dz \;=\; \int_a^b [f(z(t)) + g(z(t))]\, z'(t)\, dt \quad \text{by 3.4.4(a)}$$

$$=\; \int_a^b f(z(t))\, z'(t)\, dt + \int_a^b g(z(t))\, z'(t)\, dt \quad \text{by Ex. 1 in 3.2.5}$$

$$=\; \int_C f(z)\, dz + \int_C g(z)\, dz \quad \text{by 3.4.4(a)}$$

Proof of (d). This follows from 3.4.6 and Ex. 2 in 3.2.5.

Proof of (e). This is proved in the calculus.

Proof of (f).

$$\left| \int_C f(z)\, dz \right| \;=\; \left| \int_a^b f(z(t))\, z'(t)\, dt \right|$$

$$\leq\; \int_a^b |f(z(t))|\, |z'(t)|\, dt \quad \text{by 3.2.4}$$

Proof of (g).

$$\left| \int_C f(z)\, dz \right| \;\leq\; \int_a^b |f(z(t))|\, |z'(t)|\, dt \quad \text{by (f)}$$

$$\leq\; \int_a^b M\, |z'(t)|\, dt = ML \quad \text{by (e)}$$

Example 3.4.11 Evaluate $\int_C f(z)\, dz$ if f and C are as given.

(a) The contour C is the part of the parabola $y = x^2$ from $(0,0)$ to $(1,1)$, and $f(z) = 3z$.

(b) The contour C is the cco boundary of the triangle with vertices at $(0,0), (1,0)$, and $(1,2)$, and $f(z) = z^2$. See Figure 3.4.

(c) The contour C is the cco unit circle $|z| = 1$, and $f(z) = x + 2yi$.

Solution of (a). Let C be given by $x = t$, $y = t^2$ for each t in $[0,1]$. Then $x' = 1$, $y' = 2t$, $u = 3x = 3t$, and $v = 3y = 3t^2$. Thus by 3.4.7,

$$\int_C 3z\, dz = \int_0^1 \left(3t - 6t^3 \right) dt + i \int_0^1 \left(3t^2 + 6t^2 \right) dt = 3i.$$

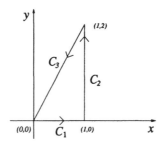

Figure 3.4: C = cco boundary of triangle.

Remark. In Part (a), we should notice that we may parameterize C by writing

$$x = x \quad \text{and} \quad y = x^2 \quad \text{for each } x \text{ in } [0,1].$$

This means that x may be used as the parameter.

Solution of (b). By 3.4.6(a) and 3.4.8, we see that $C = C_1 + C_2 + (-C_3)$ where C_1, C_2, and C_3 are as follows

$$
\begin{aligned}
C_1 : & \quad x = x, \; y = 0 \quad \text{for each } x \text{ in } [0,1] \\
C_2 : & \quad x = 1, \; y = y \quad \text{for each } y \text{ in } [0,2] \\
C_3 : & \quad x = x, \; y = 2x \quad \text{for each } x \text{ in } [0,1].
\end{aligned}
$$

Also $f(z) = z^2 = x^2 - y^2 + 2xyi = u + iv$. By 3.4.4(a) with x as the parameter for C_1,

$$\int_{C_1} z^2 \, dz = \int_0^1 (u + iv)(x' + iy') \, dx = \int_0^1 x^2 \, dx = \frac{1}{3}.$$

By 3.4.4(a) with y as the parameter for C_2,

$$
\begin{aligned}
\int_{C_2} z^2 \, dz &= \int_0^2 (u + iv)(x' + iy) \, dy \\
&= \int_0^2 \left((1 - y^2) + 2yi \right) i \, dy \\
&= i \int_0^2 (1 - y^2 + 2yi) \, dy \qquad \text{by 3.4.10(b)} \\
&= i \left[y - \frac{y^3}{3} + y^2 i \right]_0^2 \\
&= i \left(-\frac{2}{3} + 4i \right) = -4 - \frac{2}{3} i.
\end{aligned}
$$

By 3.4.7 with x as the parameter for C_3,

$$\int_{C_3} z^2 \, dz = \int_a^b (ux' - vy') \, dx + i \int_a^b (vx' + uy') \, dx$$

$$= \int_0^1 \left((x^2 - 4x^2)1 - 4x^2 \cdot 2 \right) dx + i \int_0^1 \left(4x^2 + (x^2 - 4x^2)2 \right) dx$$

$$= \left[-\frac{11x^3}{3} \right]_0^1 + i \left[\frac{4x^3}{3} - 2x^3 \right]_0^1$$

$$= -\frac{11}{3} - \frac{2}{3}i.$$

Now by 3.4.10(a) and (d),

$$\int_C z^2 \, dz = \int_{C_1} z^2 \, dz + \int_{C_2} z^2 \, dz - \int_{C_3} z^2 \, dz$$

$$= \frac{1}{3} - 4 - \frac{2}{3}i - \left(-\frac{11}{3} - \frac{2}{3}i \right) = 0.$$

Note that we may use 3.4.7 or 3.4.4(a), depending upon which is more convenient.

Solution of (c). $f(z) = x + 2yi = u + iv$. Now C is given by

$$z = x(\theta) + iy(\theta) = \cos\theta + i\sin\theta \quad \text{for } \theta \text{ in } [0, 2\pi].$$

Also,

$$x' = -\sin\theta \text{ and } y' = \cos\theta.$$

By 3.4.7,

$$\int_C f(z)\, dz = \int_a^b (ux' - vy')\, d\theta + i \int_a^b (vx' + uy')\, d\theta$$

$$= \int_0^{2\pi} (-\sin\theta\cos\theta - 2\sin\theta\cos\theta)\, d\theta$$

$$\qquad + i \int_0^{2\pi} (-2\sin^2\theta + \cos^2\theta)\, d\theta$$

$$= 3 \int_0^{2\pi} \cos\theta\,(-\sin\theta)\, d\theta + i \int_0^{2\pi} \left(-(1 - \cos 2\theta) + \frac{1 + \cos 2\theta}{2} \right) d\theta$$

$$= \frac{3}{2} \left[\cos^2\theta \right]_0^{2\pi} + i \left[-\frac{\theta}{2} + \frac{3}{4}\sin 2\theta \right]_0^{2\pi} = -\pi i.$$

Exercises 3.4.12

Evaluate $\int_C f(z)dz$ for each function f and contour C given in Ex. 1–9.

1. $f(z) = x - y + 3y^2 i$; C is the line segment from the origin to the point $1 + 2i$.

$$-\frac{17}{2} + 3i$$

2. $f(z) = x - y + 3y^2 i$; C consists of the line segment from $z = 0$ to $z = 1$ followed by the line segment from $z = 1$ to $z = 1 + 2i$.
$$\frac{1}{2} - 8 = -\frac{15}{2}$$

3. $f(z) = \overline{z}$; C is given by $z = 2 \operatorname{cis} \theta$ for θ in $[0, \pi]$.
$$4\pi i$$

4. $f(z) = \overline{z}$; $C = -C_1$ where C_1 is given by $z = 2 \operatorname{cis} \theta$ for θ in $[-\pi, 0]$.
$$-4\pi i$$

5. $f(z) = \overline{z}$; C is the circle $z = 2 \operatorname{cis} \theta$ for θ in $[-\pi, \pi]$.
$$-1(-4\pi i) + 4\pi i = 8\pi i$$

6. $f(z) = 3z + 2$; C consists of the line segment from $(1, 1)$ to $(3, 1)$.
$$16 + 6i$$

7. $f(z) = e^x(\cos y + i \sin y)$; C is the line segment from the point $(0, 0)$ to the point $(2, 0)$.
$$e^2 - 1$$

8. $f(z) = e^x(\cos y + i \sin y)$; C is the line segment from the point $(0, 0)$ to the point $(0, \pi)$.
$$-2$$

9. $f(z) = \dfrac{z + 3}{z}$; C is the cco circle $|z| = 3$. See 3.4.11(c).
$$6\pi i$$

10. Without evaluating the integral, show that $\left| \int_C \frac{dz}{z^2 + b^2} \right| \leq \frac{2\pi a}{a^2 - |b|^2}$ if C is the cco circle $|z| = a$ where $|b| < a$. **Hint.** $|z^2 + b^2| = |z^2 - (-b^2)| \geq ||z|^2 - |b|^2|$. Use 3.4.10(g).

11. Prove if C is any smooth contour from z_1 to z_2, then $\int_C dz = z_2 - z_1$.

    Solution. Let C be given by $z(t) = x(t) + iy(t)$ for t in $[a, b]$ such that $z_1 = z(a)$ and $z_2 = z(b)$. Then
    $$\int_C dz = \int_a^b (x' + iy') \, dt = [x(t) + iy(t)]_a^b = [z(t)]_a^b = z(b) - z(a) = z_2 - z_1.$$

3.5 The Cauchy-Goursat Theorem

Definition 3.5.1 (a) A subset S of R^2 is **arcwise connected** iff for each pair of distinct points z_1 and z_2 in S, there is an arc in S with endpoints at z_1 and z_2.

(b) Let D be a domain in R^2. The domain D is **simply connected** iff D has the following property.

If C is a simple closed curve contained in D, then $I(C)$ is contained in D.

(c) A **multiply connected** domain is a domain which is not simply connected.

Remarks 3.5.2 It can be shown that every arcwise connected set in R^2 is connected. See 1.4.5(b). But the converse is not true. For example, the set

$$\left\{(x, y): \ x \in [-1, 1], \ y = 0 \text{ if } x \in [-1, 0] \text{ and } y = \sin\frac{1}{x} \text{ if } x \in (0, 1]\right\}$$

is connected but not arcwise connected. However, it can be shown that every domain (open connected set) in R^2 is arcwise connected. Intuitively speaking, a **simply connected** domain has no holes in it.

Example 3.5.3 The interiors of triangles, circles, squares, and sectors of circles are simply connected domains. Also the quadrant $\{(x, y): \ x > 0 \text{ and } y > 0\}$ and R^2 are simply connected domains. An annulus (a region bounded between two concentric circles) and a deleted neighborhood are examples of multiply connected domains.

We shall state Green's Theorem without proof.

Theorem 3.5.4 (Green's Theorem). Let C be a cco simple closed contour in R^2 and let $D = C \cup I(C)$. If $P(x, y)$ and $Q(x, y)$ are continuous real-valued functions on D such that P_y and Q_x are continuous on D, then

$$\int_C (P\,dx + Q\,dy) = \iint_D (Q_x - P_y)\,dx\,dy. \tag{3.19}$$

Theorem 3.5.5 (Cauchy's Integral Theorem). Let C be a cco simple closed contour in R^2 and let $D = C \cup I(C)$. If $f = u + iv$ is analytic in D and f' is continuous in D, then

$$\int_C f(z)\,dz = 0. \tag{3.20}$$

Proof. By 3.4.7,

$$\begin{aligned}
\int_C f(z)\,dz &= \int_a^b (ux' - vy')\,dt + i\int_a^b (vx' + uy')\,dt \\
&= \int_C (u\,dx - v\,dy) + i\int_C (v\,dx + u\,dy),
\end{aligned} \tag{3.21}$$

when expressed in real line integrals. By Theorem 2.3.7,

$$f'(z) = u_x + iv_x = v_y - iu_y. \tag{3.22}$$

Thus by Theorem 2.2.11, it follows that u_x, v_x, $-u_y$ and v_y are continuous on D, since f' is continuous on D by hypothesis. Hence we may apply Green's Theorem, first with $P = u$ and $Q = -v$. In this case,

$$P_y = u_y \quad \text{and} \quad Q_x = -v_x. \tag{3.23}$$

Next, we apply Green's Theorem with $P = v$ and $Q = u$. In this case,

$$P_y = v_y \quad \text{and} \quad Q_x = u_x. \tag{3.24}$$

Hence by (3.21)–(3.24),

$$\int_C f(z)\,dz = -\iint_D (v_x + u_y)\,dx\,dy + i\iint_D (u_x - v_y)\,dx\,dy = 0.$$

It is clear that the functions $f(z) = 1$, $f(z) = z$, and $f(z) = z^2$ satisfy the hypotheses of 3.5.5 in R^2. Thus 3.5.5 gives the following results.

Theorem 3.5.6 If C is any simple closed contour in R^2, then

$$\int_C 1\,dz = 0, \quad \int_C z\,dz = 0, \quad \text{and} \quad \int_C z^2\,dz = 0.$$

Remarks 3.5.7 The statement, obtained by deleting from Cauchy's Integral Theorem the hypothesis that f' is continuous on D, is known as the Cauchy-Goursat Theorem. Goursat was the first to prove this stronger theorem. The Cauchy-Goursat Theorem is extremely important in the theory of analytic functions. From this theorem, it is shown that if f is analytic on a domain D, then f' is analytic and hence continuous on D. This leads to the conclusion that if f is analytic on D, then all of the higher derivatives of f are analytic on D.

Lemma 3.5.8 Let C be a simple closed contour, let $D = C \cup I(C)$, and let $\epsilon > 0$. If $f \in A(D)$, then there exists a subdivision of D into a finite number n of squares (square regions) R_j and partial squares P_j such that there exists a point z_j in R_j (or in P_j) for which

$$\left| \frac{f(z) - f(z_j)}{z - z_j} - f'(z_j) \right| < \epsilon \text{ for each } z \text{ in } R_j \text{ (or } P_j) \text{ such that } z \neq z_j. \tag{3.25}$$

Proof. Let D be divided into a finite number n of congruent squares R_j and partial squares P_j by lines drawn parallel to the coordinate axes. Assume for this subdivision of D, there is at least one R_j (or P_j) for which (3.25) does not hold for any point z_j in R_j (or z_j in P_j). If this is the case, we subdivide R_j into four congruent squares (or we subdivide P_j into appropriate squares and partial squares).

If after this subdivision of R_j (or P_j), there is a square or a partial square in which there exists no point z_j such that (3.25) holds, then we continue the subdivision of this square or partial square. Continuing this process of division of D, two cases arise. Either after a finite number of steps, we arrive at a subdivision for which (3.25) holds on each R_j (or P_j), and this lemma is proved; or however far we go in this process, there always remains a square or partial square for which (3.25) does not hold.

In the latter case we assume that Q_1 is a square (or partial square) in which there exists no point z_j such that (3.25) holds. Subdividing Q_1, we choose a subsquare

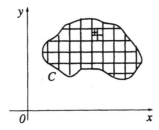

Figure 3.5: A subdivision of $C \cup I(C)$

(or a partial subsquare) Q_2 for which, (3.25) does not hold. Continuing in this manner, we obtain a sequence $Q_1 \supset Q_2 \supset Q_3 \supset \cdots$ of squares (or partial squares) whose dimensions approach zero as n becomes infinite and for each of which (3.25) does not hold. Thus by Theorem 1.4.18, there exists a point z_0 such that $z_0 \in (D \cap Q_1 \cap Q_2 \cap Q_3 \cap \cdots)$. Now since f is analytic at z_0, there is a number $\delta > 0$ such that for each z in $N'_\delta(z_0)$,

$$\left| \frac{f(z) - f(z_0)}{z - z_0} - f'(z_0) \right| < \epsilon. \tag{3.26}$$

Let $j \in N$ such that $s_j < \frac{\delta}{\sqrt{2}}$ where s_j is the length of a side of Q_j (or the length of a side of the completed square containing Q_j). Then

$$Q_j \subset N_\delta(z_0) \tag{3.27}$$

(since $z_0 \in Q_j$ and the diagonal $\sqrt{2}\, s_j$ of Q_j is less than δ). Thus if we take $z_j = z_0$ it follows by (3.26) and (3.27) that (3.25) holds for Q_j, contrary to our assumption. This proves the lemma. See Figure 3.5.

Remarks 3.5.9 If in 3.5.8, the contour C is cco, then

$$\int_C f(z)\, dz = \sum_{j=1}^{n} \int_{C_j} f(z)\, dz \tag{3.28}$$

where C_j is the cco boundary of R_j (or P_j). For if ABEG and GEHF are two adjacent squares, then the side EG is traversed from E to G in the first square and from G to E in the second. Thus the two integrals along EG cancel. Hence the integrals along sides in $I(C)$ cancel, leaving in the right member of (3.28) only $\int_C f(z)\, dz$. See Figure 3.6.

Theorem 3.5.10 (Cauchy-Goursat Theorem). Let C be a cco simple closed contour in R^2 and let $D = C \cup I(C)$. If $f \in A(D)$, then

$$\int_C f(z)\, dz = 0. \tag{3.29}$$

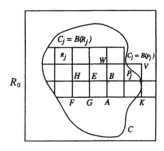

Figure 3.6: $\int_C f(z)\,dz = \sum_{j=1}^n \int_{C_j} f(z)\,dz$.

Proof. Let R_0 be a square region with side of length S such that $C \subset R_0$. Let $\epsilon > 0$ and let

$$T = \sqrt{2}\,S(4S + L) \tag{3.30}$$

where L is the length of C. By 3.5.8, let

$$R_1, \cdots, R_m, P_{m+1}, \cdots, P_n$$

be a subdivision of $C \cup I(C)$ into squares R_j and partial squares P_j such that for some z_j in R_j (or P_j),

$$|\phi_j(z)| < \frac{\epsilon}{T} \qquad \text{for each } z \text{ in } R_j \text{ (or } P_j) \tag{3.31}$$

where

$$\phi_j(z) = \begin{cases} \dfrac{f(z) - f(z_j)}{z - z_j} - f'(z_j) & \text{if } z \neq z_j \\ 0 & \text{if } z = z_j. \end{cases} \tag{3.32}$$

From (3.32) for each z in D,

$$f(z) = f(z_j) - z_j f'(z_j) + z f'(z_j) + (z - z_j)\phi_j(z). \tag{3.33}$$

Let C_j be the cco boundary of R_j (or P_j). By (3.33), 3.4.10(c), and 3.5.6,

$$\left| \int_{C_j} f(z)\,dz \right| = \left| \int_{C_j} (z - z_j)\phi_j(z)\,dz \right|$$

$$\leq \sqrt{2}\,s_j \left(\frac{\epsilon}{T} \right) L_j \quad \text{by 3.4.10(g) and (3.31)} \tag{3.34}$$

where s_j is the length of a side of R_j (or P_j) and L_j is the length of C_j (since $|z - z_j| \leq \sqrt{2}\,s_j$ for each z on C_j). Now

$$\begin{aligned} L_j &= 4s_j & \text{if } j \leq m \text{ and} \\ L_j &\leq 4s_j + k_j & \text{if } m < j \leq n, \end{aligned} \tag{3.35}$$

where k_j is the length of $C \cap P_j$. (See Figure 3.6.) By (3.28),

$$\left| \int_C f(z)\, dz \right| = \left| \sum_{j=1}^{n} \int_{C_j} f(z)\, dz \right| \le \sum_{j=1}^{n} \left| \int_{C_j} f(z)\, dz \right|$$

$$\le \sum_{j=1}^{m} \sqrt{2}\, s_j \left(\frac{\epsilon}{T} \right) 4s_j + \sum_{j=m+1}^{n} \sqrt{2}\, s_j \left(\frac{\epsilon}{T} \right)(4s_j + k_j)$$

$$\text{by (3.34) and (3.35)}$$

$$\le \frac{\epsilon \sqrt{2}}{T} \left[\sum_{j=1}^{m} 4s_j^2 + \sum_{j=m+1}^{n} 4s_j^2 + \sum_{j=m+1}^{n} s_j k_j \right]$$

$$\le \frac{\epsilon \sqrt{2}}{T} [4S^2 + SL] = \epsilon \quad \text{by (3.30).} \tag{3.36}$$

Since ϵ is arbitrary, (3.29) follows. [In (3.36) the number s_j^2 is the area of R_j^2 and thus $\sum_{j=1}^{n} s_j^2 \le S^2$, the area of R_0, where it is understood that S is sufficiently large so that completed squares (such as AKVW in Figure 3.6) are contained in R_0. Also,

$$\sum_{j=m+1}^{n} s_j k_j \le \sum_{j=m+1}^{n} S k_j = S \sum_{j=m+1}^{n} k_j \le SL.]$$

Theorem 3.5.11 Let C, $C_1, \cdots, C_n$ be cco simple closed contours in R^2 such that

$$(a)\ C_j \subset I(C) \quad \text{for each } j, \quad \text{and} \quad (b)\ C_j \subset E(C_k) \quad \text{if } j \ne k.$$

If $f \in A(D)$ where

$$D = C \cup I(C) - \bigcup \{ I(C_j): \ j = 1, 2, \cdots, n \},$$

then

$$\int_C f(z)\, dz = \sum_{j=1}^{n} \int_{C_j} f(z)\, dz. \tag{3.37}$$

Proof. (See Figure 3.7 for the case in which $n = 2$.) By 3.5.10,

$$0 = \int_{K'} f(z)\, dz$$

$$= \int_{C'} + \int_{-L_1} + \int_{-C_1'} + \int_{-L_2} + \int_{-C_2'} + \int_{-L_3} \quad \text{by 3.4.10(d)} \tag{3.38}$$

where the integrand in each term is $f(z)$. Also,

$$0 = \int_{K''} f(z)\, dz \quad \text{by 3.5.10}$$

$$= \int_{C''} + \int_{L_3} + \int_{-C_2''} + \int_{L_2} + \int_{-C_1''} + \int_{L_1} \quad \text{by 3.4.10(d).} \tag{3.39}$$

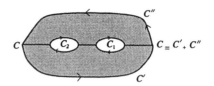

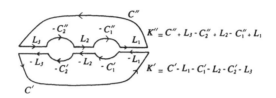

Figure 3.7: $D = \overline{I(C)} - [I(C_1) \cup I(C_2)]$.

By (3.38) and (3.39),

$$0 = \left(\int_{C'} + \int_{C''}\right) + \left(\int_{-C_1'} + \int_{-C_1''}\right) + \left(\int_{-C_2'} + \int_{-C_2''}\right)$$
$$= \int_C + \int_{-C_1} + \int_{-C_2}$$

where $C = C' + C''$ and $-C_j = -C_j' + -C_j''$ for $j = 1, 2$. Now (3.37) follows from the last displayed equation if $n = 2$. If $n > 2$, the conclusion follows in the same way.

Remark 3.5.12 From (3.37),

$$\int_C f(z)\,dz - \sum_{j=1}^n \int_{C_j} f(z)\,dz = 0$$

or

$$\int_C f(z)\,dz + \sum_{j=1}^n \int_{-C_j} f(z)\,dz = 0 \quad \text{by 3.4.10(a).} \tag{3.40}$$

If we now let B denote the formal sum

$$B = C + \sum_{j=1}^n (-C_j)$$

given in 3.4.6(b), then (3.40) may be written as

$$\int_B f(z)\,dz = 0.$$

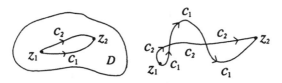

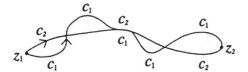

Figure 3.8: $\int_{C_1} f(z)\,dz = \int_{C_2} f(z)\,dz$ is simply connected.

3.6 Applications of the Cauchy-Goursat Theorem

Theorem 3.6.1 Let D be a simply connected domain in R^2, let $f \in A(D)$ and let C_1 and C_2 be any contours in D with initial points at z_1 and terminal points at z_2. Then

$$\int_{C_1} f(z)\,dz = \int_{C_2} f(z)\,dz. \tag{3.41}$$

Proof. Case 1. Suppose $C_1 \cap C_2 = \{z_1, z_2\}$. Then by 3.5.10,

$$\int_{C_1} f(z)\,dz + \int_{-C_2} f(z)\,dz = 0,$$

which implies

$$\int_{C_1} f(z)\,dz = -\int_{-C_2} f(z)\,dz = \int_{C_2} f(z)\,dz \quad \text{by 3.4.10(a)}.$$

Case 2. If C_1 intersects C_2 in more than two points, we apply Case 1 to the appropriate parts of C_1 and C_2 to obtain the desired conclusion. [For if parts of C_1 and C_2 coincide as in Figure 3.8(c), then the integrands along these parts have the same value.] See Figure 3.8.

Remark 3.6.2 In 3.6.1,

$$\int_C f(z)\,dz$$

is independent of the contour $C\,(\subset D)$ from z_1 to z_2. Thus we may denote this **contour integral** by

$$\int_{z_1}^{z_2} f(z)\,dz.$$

Theorem 3.6.3 Let $f \in A(D)$ and let $z_0 \in D$, where D is a simply connected domain in R^2.

(a) Let

$$F(z) = \int_{z_0}^{z} f(z^*) \, dz^* \quad \text{for each } z \text{ in } D. \tag{3.42}$$

Then $F \in A(D)$, and

$$F'(z) = f(z) \quad \text{for each } z \text{ in } D. \tag{3.43}$$

(b) If $G'(z) = f(z)$ for each z in D, then

$$\int_{z_0}^{z} f(z^*) \, dz^* = G(z) - G(z_0) \quad \text{for each } z \text{ in } D.$$

Proof of (a). From (3.42),

$$\begin{aligned} F(z + k) - F(z) &= \int_{z_0}^{z+k} f(z^*) \, dz^* - \int_{z_0}^{z} f(z^*) \, dz^* \\ &= \int_{z}^{z+k} f(z^*) \, dz^*, \end{aligned} \tag{3.44}$$

where $|k|$ is sufficiently small that the line segment S_k from z to $z + k$ is contained in D. Thus by 3.6.1, the contour of integration may be taken to be S_k. From (3.44),

$$\frac{F(z + k) - F(z)}{k} = \frac{1}{k} \int_{z}^{z+k} f(z^*) \, dz^*.$$

Thus

$$\frac{F(z + k) - F(z)}{k} - f(z) = \frac{1}{k} \int_{z}^{z+k} [f(z^*) - f(z)] \, dz^* \tag{3.45}$$

by 3.4.10(b) and (c) and by Ex. 11 in 3.4.12. Now let $\epsilon > 0$. By Theorem 2.3.3, we know that $f \in C(D)$ since $f \in A(D)$. Hence we let $\delta > 0$ such that

$$|f(z^*) - f(z)| < \epsilon \quad \text{if } z^* \in D \cap N_\delta(z). \tag{3.46}$$

We now let the contour S_k be such that $|k| < \delta$. Then it follows by 3.4.10(g), (3.45), and (3.46) that

$$\left| \frac{F(z + k) - F(z)}{k} - f(z) \right| \leq \frac{1}{|k|} \epsilon |k| = \epsilon.$$

This proves (3.43). See Figure 3.9.

Proof of (b). Part (b) follows from Part (a) and Ex. 20 in 2.3.16.

The Cauchy Integral Formula 3.6.4 Let C be a cco simple closed contour in R^2 and let $D = C \cup I(C)$. If $f \in A(D)$ and $z_0 \in I(C)$, then

$$f(z_0) = \frac{1}{2\pi i} \int_C \frac{f(z)}{z - z_0} \, dz. \tag{3.47}$$

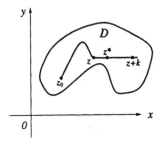

Figure 3.9: $S_k \subset D$ where S_k is the straight line segment from z to $z + k$.

Proof. Let γ be a cco circle contained in the interior of C which is given by

$$\gamma: \quad z - z_0 = r \operatorname{cis} \theta \quad \text{for each } \theta \text{ in } [0, 2\pi]. \tag{3.48}$$

By 3.5.11 with $n = 1$,

$$\int_C \frac{f(z)\,dz}{z - z_0} = \int_\gamma \frac{f(z)\,dz}{z - z_0} = f(z_0) \int_\gamma \frac{dz}{z - z_0} + \int_\gamma \frac{f(z) - f(z_0)}{z - z_0}\,dz. \tag{3.49}$$

From (3.48), we have $dz = r i \operatorname{cis} \theta \, d\theta$. Thus

$$\int_\gamma \frac{dz}{z - z_0} = \int_0^{2\pi} i \, d\theta = 2\pi i. \tag{3.50}$$

Now let $\epsilon > 0$. Since $f \in C(D)$, we let $\delta > 0$ such that

$$\text{if } |z - z_0| < \delta, \text{ then } |f(z) - f(z_0)| < \epsilon. \tag{3.51}$$

In (3.48), we take $r < \delta$. Then by Theorem 3.4.10(g) and (3.51), we have

$$\left| \int_\gamma \frac{f(z) - f(z_0)}{z - z_0}\,dz \right| \le \frac{\epsilon}{r}(2\pi r) = 2\pi\epsilon. \tag{3.52}$$

By (3.49) and (3.50),

$$\int_C \frac{f(z)\,dz}{z - z_0} - (2\pi i)f(z_0) = \int_\gamma \frac{f(z) - f(z_0)}{z - z_0}\,dz,$$

which by (3.52) implies

$$\left| \frac{1}{2\pi i} \int_C \frac{f(z)\,dz}{z - z_0} - f(z_0) \right| = \frac{1}{2\pi} \left| \int_\gamma \frac{f(z) - f(z_0)}{z - z_0}\,dz \right| \le \epsilon.$$

Thus (3.47) holds.

Cauchy's Integral Formula (3.47) gives the value of $f(z)$ for each point z of $I(C)$ in terms of an integral along C.

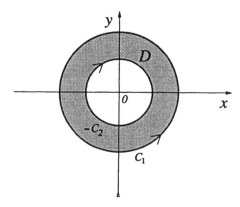

Figure 3.10: $D = \overline{I(C_1)} - I(C_2)$.

Exercises 3.6.5

Evaluate the following contour integrals.

1. $\displaystyle\int_B \frac{dz}{z^2(z^2 + 16)}$ where $B = C_1 - C_2$ and C_1 and C_2 are, respectively, the cco
   circles given by $z = 3\operatorname{cis}\theta$ and $z = 2\operatorname{cis}\theta$ for θ in $[0, 2\pi]$.

   Solution. By 2.3.5 and 2.3.14, the integrand $I(z) = \frac{1}{z^2(z^2+16)}$ is analytic in D
   where $D = \{z:\ 2 \le |z| \le 3\}$. See Figure 3.10. By 3.4.6(b),

   $$
   \begin{aligned}
   \int_B I(z)\,dz &= \int_{C_1} I(z)\,dz + \int_{-C_2} I(z)\,dz \\
   &= \int_{C_1} I(z)\,dz - \int_{C_2} I(z)\,dz \qquad \text{by 3.4.10(a)} \\
   &= 0 \quad \text{by 3.5.12.}
   \end{aligned}
   $$

2. $\displaystyle\int_C \frac{z\,dz}{(9 - z^2)(z + i)}$ where C is the cco circle $z = 2\operatorname{cis}\theta$ for each θ in $[0, 2\pi]$.

   Solution. $I = \int_C \frac{f(z)}{z+i}\,dz$ where $f(z) = \frac{z}{9-z^2}$ is analytic in D and $D = \{z:$
   $|z| \le 2\}$. Thus by 3.6.4, we have $I = 2\pi i\, f(-i) = 2\pi i\left(\frac{-i}{9+1}\right) = \frac{\pi}{5}$.

3. $\displaystyle\int_C \frac{2z^2 - z - 2}{z - 2}\,dz$ where C is the cco circle $z = 3\operatorname{cis}\theta$ for each θ in $[0, 2\pi]$.

Solution. Note that $f(z) = (2z^2 - z - 2)$ is analytic in D where $D = \{z : |z| \leq 3\}$. Thus by 3.6.4,

$$\int_C \frac{2z^2 - z - 2}{z - 2} \, dz = 2\pi i \, f(2) = 2\pi i (4) = 8\pi i.$$

4. (a) $\int_C \frac{dz}{z - \frac{\pi}{2}i}$ (b) $\int_C \frac{dz}{z(z^2 + 8)}$ (c) $\int_C \frac{z \, dz}{2z + 1}$

where C is the cco boundary of the square S whose sides lie along the lines $x = \pm 2$, $y = \pm 2$.

Solution (a). We apply 3.6.4 with $f(z) \equiv 1$, which is analytic in $\overline{I(C)}$. Thus we obtain

$$\int_C \frac{dz}{z - \frac{\pi}{2}i} = 2\pi i(1) = 2\pi i.$$

Solution (b). We apply 3.6.4 with $f(z) = \frac{1}{z^2+8}$, which is analytic in $\overline{I(C)}$. Thus we obtain

$$\int_C \frac{f(z)}{z - 0} \, dz = 2\pi i \left(\frac{1}{8}\right) = \frac{\pi i}{4}.$$

Solution (c). By 3.6.4,

$$\int_C \frac{z \, dz}{2z + 1} = \frac{1}{2} \int_C \frac{z \, dz}{z + \frac{1}{2}} = \frac{1}{2} 2\pi i \, [z]_{z=-\frac{1}{2}} = -\frac{\pi i}{2}.$$

5. $\int_C (z - a)^n \, dz$ where $a \in R^2$, $n \in N$, and C is a cco simple closed contour.

Solution. Since $(z - a)^n$ is a polynomial in z and hence an entire function, it follows by 3.5.10 (or by 3.5.5) that

$$\int_C (z - a)^n \, dz = 0.$$

Also by 3.6.3,

$$\int_{z_0}^{z_1} (z - a)^n \, dz = \frac{(z - a)^{n+1}}{n + 1} \Bigg|_{z_0}^{z_1} = 0 \quad \text{if } z_1 = z_0 \, .$$

6. $\int_C \frac{dz}{(z - a)^n}$ where C is a cco simple closed contour in R^2 if

(a) $a \in E(C)$ and $n \in N$ (b) $a \in I(C)$ and $n = 1$.

Solution (a). By 3.5.10 (or by 3.5.5), we have $\int_C \frac{dz}{(z-a)^n} = 0$. (Also, this follows by 3.6.3 if $n \neq 1$.)

Solution (b). By 3.6.4 with $f(z) \equiv 1$, we have $\int_C \frac{dz}{z-a} = 2\pi i$.

7. $\int_C \frac{dz}{z^2 - 1}$ where C is the cco circle $z = 2\operatorname{cis}\theta$ for each θ in $[0, 2\pi]$.

Solution.

$$
\begin{aligned}
\int_C \frac{dz}{z^2 - 1} &= \int_C \left(\frac{\frac{1}{2}}{z - 1} - \frac{\frac{1}{2}}{z + 1} \right) dz \quad \text{(partial fractions)} \\
&= \frac{1}{2} \int_C \frac{dz}{z - 1} - \frac{1}{2} \int_C \frac{dz}{z + 1} \\
&= \frac{1}{2} 2\pi i \left[1\right]_{z=1} - \frac{1}{2} 2\pi i \left[1\right]_{z=-1} \quad \text{by 3.6.4} \\
&= 0.
\end{aligned}
$$

We have used $[1]_{z=1}$ to stand for $[f(z)]_{z=1}$ where $f(z)$ is the constant function $f(z) \equiv 1$.

Alternate Solution. Let C_1 be the cco circle $z = 1 + \frac{1}{4}\operatorname{cis}\theta$ on $[0, 2\pi]$ and let C_2 be the cco circle $z = -1 + \frac{1}{4}\operatorname{cis}\theta$ on $[0, 2\pi]$. By 3.5.11, we have

$$
\begin{aligned}
\int_C \frac{dz}{z^2 - 1} &= \int_{C_1} \frac{(z+1)^{-1}\, dz}{z - 1} + \int_{C_2} \frac{(z-1)^{-1}\, dz}{z + 1} \\
&= 2\pi i \left[\frac{1}{z+1} \right]_{z=1} + 2\pi i \left[\frac{1}{z-1} \right]_{z=-1} \quad \text{by 3.6.4} \\
&= 0.
\end{aligned}
$$

Thus in this case,

$$
\int_C \frac{dz}{z^2 - 1} = 0,
$$

even though the function $\frac{1}{z^2-1}$ is not analytic in $\overline{I(C)}$. See Figure 3.11.

Remark 3.6.6 By Ex. 5 and 6 in 3.6.5, if C is any cco simple closed contour, then

$$
\int_C \frac{dz}{(z - a)^n} = \begin{cases} 0 & \text{if } -n \in N \\ 0 & \text{if } a \in E(C) \text{ and } n \in N \\ 2\pi i & \text{if } a \in I(C) \text{ and } n = 1. \end{cases}
$$

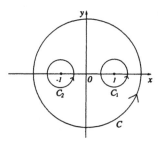

Figure 3.11: $(z+1)^{-1}$ is analytic in $\overline{I(C_1)}$; $(z-1)^{-1}$ is analytic in $\overline{I(C_2)}$.

Exercises 3.6.7

In Ex. 1–9, let C be the cco circle $|z| = 3$ and determine the value of each integral.

1. $\displaystyle \int_C \frac{z^3\,dz}{z^2 + 16}$ 0   by 3.5.10

2. $\displaystyle \int_C \frac{(z^2+4)\,dz}{z^2 - 5z + 4}$ $-\dfrac{10\pi i}{3}$ by 3.6.4

3. $\displaystyle \int_C \frac{z^3\,dz}{z^2 - 4}$ **Hint.** $\frac{1}{z^2-4} = \frac{\frac14}{z-2} - \frac{\frac14}{z+2}$ $8\pi i$

4. $\displaystyle \int_C \frac{dz}{2z + 3i}$ πi

5. $\displaystyle \int_C (z - 1 - 2i)^9\,dz$ 0

6. $\displaystyle \int_C \frac{dz}{(z - 3 + 4i)^9}$ 0

7. $\displaystyle \int_C \frac{dz}{z}$ $2\pi i$

8. $\displaystyle \int_C \frac{e^x \operatorname{cis} y\,dz}{z}$ Use 2.3.11 to show that $e^x \operatorname{cis} y$ is an entire function and use 3.6.4.
                                                                                                                            $2\pi i$

9. $\displaystyle \int_C \frac{dz}{z^2 + 1}$ Evaluate this integral in two ways as in Ex. 7 of 3.6.5.

   Note. $\dfrac{1}{(z - i)(z + i)} = \dfrac{-\frac{i}{2}}{z - i} + \dfrac{\frac{i}{2}}{z + i}$ 0.

In Ex. 10–14, let C be any cco simple closed contour and determine the value of the given integral.

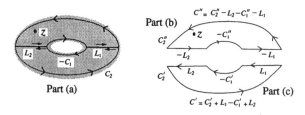

Figure 3.12: $f \in A[\overline{I(C_2)} - I(C_1)]$.

10. $\int_C (\cos x \cosh y - i \sin x \sinh y)\, dz$ **Hint.** Use 2.3.11 to show that the integrand is an entire function, and then use 3.5.10.

11. $\int_C \dfrac{e^x \operatorname{cis} y\, dz}{z - \pi i}$ where $\pi i \in I(C)$ $-2\pi i$

12. $\int_C \dfrac{e^x \operatorname{cis} y\, dz}{z - \pi i}$ where $\pi i \in E(C)$ 0

13. $\int_C \dfrac{dz}{z^2 + \pi^2}$ where $\pi i \in I(C)$ and $-\pi i \in E(C)$ 1

14. $\int_C \dfrac{dz}{z^2 + \pi^2}$ where $-\pi i$ and πi are in $I(C)$

 Hint. $\dfrac{1}{(z-\pi i)(z+\pi i)} = \dfrac{\frac{1}{2\pi i}}{z - \pi i} - \dfrac{\frac{1}{2\pi i}}{z + \pi i}$ 0

15. Evaluate $\displaystyle\int_1^i z^{18}\, dz$. $-\dfrac{1+i}{19}$

16. Evaluate $\displaystyle\int_0^{\frac{\pi}{2}i} e^x \operatorname{cis} y\, dz$ **Hint.** By 2.3.11, the function $(e^x \operatorname{cis} y)$ is analytic in R^2. By (2.26), we have $\frac{d}{dz}(e^x \cos y + i e^x \sin y) = e^x \operatorname{cis} y$. $i - 1$

17. Let C be a cco simple closed contour in R^2. Prove $\displaystyle\int_C x\, dy = A$ where A is the area of $I(C)$. **Hint.** Let $P = 0$ and $Q = x$ in Green's Theorem. See 3.5.4.

Theorem 3.6.8 (The Generalized Cauchy Integral Formula). Let C_1 and C_2 be cco simple closed contours in R^2 such that $C_1 \subset I(C_2)$. If $f \in A[\overline{I(C_2)} - I(C_1)]$, then for each z in $I(C_2) \cap E(C_1)$,

$$f(z) = \frac{1}{2\pi i}\left(\int_{C_2} \frac{f(t)\, dt}{t - z} - \int_{C_1} \frac{f(t)\, dt}{t - z} \right).$$

Proof. Let $z \in I(C_2) \cap E(C_1)$. Let L_1 and L_2 be line segments as in Figure 3.12 Part (a) such that $z \notin (L_1 \cup L_2)$. See Figure 3.12 Part (c). Now we have

$$0 = \int_{C'} \frac{f(t)\,dt}{t-z} \quad \text{by 3.5.10}$$

$$= \int_{C_2'} + \int_{L_1} - \int_{C_1'} + \int_{L_2}.$$

Next we apply 3.6.4 where the simple closed contour C'' is shown in Figure 3.12 Part (b), and we conclude that

$$f(z) = \frac{1}{2\pi i} \int_{C''} \frac{f(t)\,dt}{t-z}$$

$$= \frac{1}{2\pi i} \left(\int_{C_2''} - \int_{L_2} - \int_{C_1''} - \int_{L_1} \right).$$

Combining the last two sentences, we obtain

$$f(z) = 0 + f(z) = \frac{1}{2\pi i} \left(\int_{-C_1'-C_1''} \frac{f(t)\,dt}{t-z} + \int_{C_2'+C_2''} \frac{f(t)\,dt}{t-z} \right)$$

$$= \frac{1}{2\pi i} \left(\int_{C_2} \frac{f(t)\,dt}{t-z} - \int_{C_1} \frac{f(t)\,dt}{t-z} \right).$$

Theorem 3.6.9 Let C be a cco simple closed contour in R^2 and let $f \in A(D)$ where $D = \overline{I(C)}$. Then $f^{(n)} \in A(I(C))$ for each n in $N \cup \{0\}$, and

$$f^{(n)}(z) = \frac{n!}{2\pi i} \int_C \frac{f(t)\,dt}{(t-z)^{n+1}} \quad \text{for each } z \text{ in } I(C), \tag{3.53}$$

where $f^{(0)}$ stands for f.

Proof (by induction). Let $z \in I(C)$. Since $I(C)$ is open, let $r > 0$ such that the cco circle

$$\gamma = \{t : \; |t-z| = r\} \subset I(C). \tag{3.54}$$

See Figure 3.13.

By 3.5.11, we may replace C in (3.53) by γ. Now $f \in A(C \cup I(C))$ implies $f \in C(\gamma)$. Also, γ is compact (and hence is closed and bounded). Thus by Ex. 2(b) in 2.4.9, let $M \in R$ such that $|f(t)| \leq M$ for each t on γ.

When $n = 0$, equation (3.53) is the Cauchy Integral Formula (since $f^{(0)} = f$ and $0! = 1$ by definition). Thus (3.53) is true for $n = 0$. Assume (3.53) holds for $n = k$. Then for each h in R^2 such that $h \neq 0$ and such that $(z+h) \in I(\gamma)$,

$$\frac{\Delta f^{(k)}}{h} = \frac{1}{h} \left(f^{(k)}(z+h) - f^{(k)}(z) \right)$$

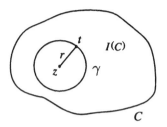

Figure 3.13: $\gamma = \{t : \ |t - z| = r\} \subset I(C)$.

$$
= \frac{k!}{2\pi i h} \int_\gamma f(t) \left(\frac{1}{(t - z - h)^{k+1}} - \frac{1}{(t - z)^{k+1}} \right) dt
$$

$$
= \frac{k!}{2\pi i h} \int_\gamma f(t) \frac{q^{k+1} - (q - h)^{k+1}}{q^{k+1}(q - h)^{k+1}}\, dt \quad \text{where } q = t - z
$$

$$
= \frac{k!}{2\pi i h} \int_\gamma f(t) h \frac{(k+1)q^k + A_1 q^{k-1}h + \cdots + A_k h^k}{q^{k+1}(q - h)^{k+1}}\, dt,
$$

where the numbers A_j are the coefficients in the binomial expansion of $(q - h)^{k+1}$. Hence

$$
\left| \frac{\Delta f^{(k)}}{h} - \frac{(k+1)!}{2\pi i} \int_\gamma \frac{f(t)\, dt}{(t - z)^{k+2}} \right|
$$

$$
= \frac{k!}{2\pi} \left| \int_\gamma f(t) \left(\frac{(k+1)\, q^k + A_1 q^{k-1} h + \cdots + A_k h^k}{q^{k+1}(q - h)^{k+1}} - \frac{k+1}{q^{k+2}} \right) dt \right|
$$

$$
= \frac{k!}{2\pi} \left| \int_\gamma f(t) \frac{(k+1)\, q^{k+1} + A_1 q^k h + \cdots + A_k q h^k - (k+1)(q - h)^{k+1}}{q^{k+2}(q - h)^{k+1}}\, dt \right|
$$

$$
= \frac{k!}{2\pi} \left| \int_\gamma f(t) h \frac{B_0 q^k + B_1 q^{k-1} h + \cdots + B_k h^k}{q^{k+2}(q - h)^{k+1}}\, dt \right| \quad \text{for some numbers } B_j
$$

$$
\leq \frac{k!}{2\pi} M |h| \frac{|B_0| r^k + |B_1| r^{k-1}(\frac{r}{2}) + \cdots + |B_k|(\frac{r}{2})^k}{r^{k+2}(r - \frac{r}{2})^{k+1}} (2\pi r) \quad \left\{ \begin{array}{l} \text{by 3.4.10(g)} \\ \text{if } |h| < \frac{r}{2} \end{array} \right.
$$

$$
= M_0 |h| \quad \text{where } M_0 \text{ is independent of } h \text{ when } |h| < \frac{r}{2}. \tag{3.55}
$$

[Note $|q| = |t - z| = r$ for each t on γ by (3.54). Also, $|q - h| \geq |q| - |h| \geq r - \frac{r}{2}$ if $|h| < \frac{r}{2}$.]

Thus by (3.55),

$$
\left| \frac{\Delta f^{(k)}}{h} - \frac{(k+1)!}{2\pi i} \int_\gamma \frac{f(t)\, dt}{(t - z)^{k+2}} \right| \to 0 \text{ as } h \to 0.
$$

Therefore,

$$
f^{(k+1)}(z) = \frac{(k+1)!}{2\pi i} \int_\gamma \frac{f(t)\, dt}{(t - z)^{k+2}},
$$

and the theorem is proved.

Remarks 3.6.10 (a) As in the proof of 3.6.8, we may extend 3.6.9 to a multiply connected domain as follows. Let C, C_j, and D be given as in 3.5.11. If $f \in A(D)$, then for each z in D° and for each integer $k \geq 0$, we have

$$f^{(k)}(z) = \frac{k!}{2\pi i}\left(\int_C \frac{f(t)\,dt}{(t-z)^{k+1}} - \sum_{j=1}^{n}\int_{C_j}\frac{f(t)dt}{(t-z)^{k+1}}\right) = \frac{k!}{2\pi i}\int_B \frac{f(t)\,dt}{(t-z)^{k+1}}$$

where B is given in 3.5.12.

(b) If $f \in A(D)$ where D is a domain in R^2, then $f^{(n)} \in A(D)$ for each positive integer n.

Corollary 3.6.11 If $a \in I(C)$ and C is a cco simple closed contour, then

$$\int_C \frac{dz}{(z-a)^n} = 0 \quad \text{if } n \in N \text{ such that } n \geq 2.$$

Proof. In 3.6.9, let $f(t) \equiv 1$. Then $f \in A(R^2)$. Thus by (3.53),

$$\int_C \frac{1\,dz}{(z-a)^n} = \frac{2\pi i}{(n-1)!}\,f^{(n-1)}(a) = 0, \text{ since } f^{(n-1)}(a) = 0 \quad \text{for each } n \geq 2.$$

Theorem 3.6.12 (Morera's Theorem). Let $f \in C(D)$ where D is a domain in R^2. If

$$\int_C f(z)\,dz = 0 \tag{3.56}$$

for each simple closed contour C contained in D, then $f \in A(D)$.

Proof. Let $z_0 \in D$ and let $r > 0$ such that $N_r(z_0) \subset D$. Let

$$G(z) = \int_{z_0}^z f(w)\,dw \quad \text{for each } z \text{ in } N_r(z_0).$$

Now $G(z)$ is well defined since (3.56) implies

$$\int_{z_0}^z f(w)\,dw$$

is independent of the path in $N_r(z_0)$. In 3.6.3, we used the hypothesis that $f \in A(D)$ to conclude (i) $f \in C(D)$ and (ii) $\int_C f(z)\,dz = 0$ for each simple closed contour C in D. By the hypotheses of our present theorem, (i) and (ii) hold in $N_r(z_0)$. Thus, as in the proof of 3.6.3, we have $G'(z) = f(z)$ for each z in $N_r(z_0)$. Thus G' is analytic in $N_r(z_0)$ by 3.6.10(b). That is, f is analytic in $N_r(z_0)$ since $f = G'$ on this neighborhood. Since z_0 was an arbitrary point in D, it follows that f is analytic in D.

We may restate the Cauchy-Goursat Theorem as follows.

Let D be a simply connected domain in R^2. If $f \in A(D)$, then $\int_C f(z) \, dz = 0$ for each (cco simple closed contour) $C \subset D$.

Thus the Cauchy-Goursat Theorem gives a necessary condition that $f \in A(D)$, where D is a simply connected domain in R^2. Morera's Theorem gives a sufficient condition that a continuous function f be analytic on D. In this sense, Morera's Theorem is a converse of the Cauchy-Goursat Theorem.

Cauchy's Inequalities 3.6.13 If $f \in A(\overline{I(\gamma)})$, where γ is the circle given by

$$\gamma : \qquad z = z_0 + r \operatorname{cis} \theta \quad \text{for each } \theta \text{ in } [0, 2\pi]$$

and if $|f(z)| \leq M$ on γ, then

$$\left| f^{(n)}(z_0) \right| \leq \frac{n! M}{r^n} \quad \text{for each } n \text{ in } N. \tag{3.57}$$

Proof. By 3.6.9,

$$
\begin{aligned}
|f^{(n)}(z_0)| &= \frac{n!}{2\pi} \left| \int_\gamma \frac{f(t) \, dt}{(t - z_0)^{n+1}} \right| \\
&\leq \frac{n!}{2\pi} \cdot \frac{M}{r^{n+1}} (2\pi r) \quad \text{by 3.4.10(g)} \\
&\leq \frac{n! M}{r^n}.
\end{aligned}
$$

[In 3.6.13, we know that γ is compact, since γ is closed and bounded. Hence by Ex. 2(b) in 2.4.9, there does exist some number M such that $|f(z)| \leq M$ on γ.]

Theorem 3.6.14 (Liouville's Theorem). If $f \in A(R^2)$ and f is bounded on R^2, then f is constant on R^2.

Proof. By hypothesis, f is bounded on R^2. Hence we let M be a number such that

$$|f(z)| < M \quad \text{for each } z \text{ in } R^2.$$

Thus by (3.57) with $n = 1$,

$$|f'(z)| \leq \frac{M}{r} \quad \text{for each } z \text{ in } R^2 \quad \text{and} \quad \text{for each } r > 0.$$

Since r may be taken arbitrarily large, it follows that $f'(z) = 0$ for each z in R^2. Hence f is constant on R^2 by 2.3.15(a).

Theorem 3.6.15 (Fundamental Theorem of Algebra). Let

$$P(z) = \sum_{k=0}^{n} a_k z^k \quad \text{for each } z \text{ in } R^2$$

be a polynomial in z of degree $n \geq 1$ with complex coefficients. Then the equation $P(z) = 0$ has a root in R^2.

Proof. Suppose for each z in R^2, we have $P(z) \neq 0$. Then

$$f = \frac{1}{P} \in A(R^2). \tag{3.58}$$

Now

$$|f(z)| = \frac{1}{|P(z)|} \quad = \quad \frac{1}{|z|^n} \frac{1}{|a_0 \frac{1}{z^n} + \cdots + a_n|}$$
$$\rightarrow \quad 0 \left(\frac{1}{|0 + \cdots + a_n|} \right) = 0 \quad \text{as } |z| \rightarrow \infty. \tag{3.59}$$

Thus there exists a number $r > 0$ such that

$$|f(z)| \leq 1 \quad \text{if } |z| > r. \tag{3.60}$$

Also,

$$f \in C(\{z : \; |z| \leq r\}) \quad \text{by (3.58).}$$

Hence there is a number M such that

$$|f(z)| \leq M \quad \text{if } |z| \leq r \quad \text{by Ex. 2(b) in 2.4.9.} \tag{3.61}$$

By (3.60) and (3.61),

$$|f(z)| \leq M + 1 \quad \text{for each } z \text{ in } R^2.$$

Thus by Liouville's Theorem, f is constant on R^2. Hence

$$f(z) \equiv f(0) = \frac{1}{P(0)} \neq 0,$$

since the reciprocal of a complex number is never zero. Thus $f(z) = c$ on R^2 where c is a nonzero constant. This contradicts (3.59). Because of this contradiction, we must reject our supposition that $P(z)$ is never 0. Thus $P(z_0) = 0$ for at least one number z_0 in R^2.

Lemma 3.6.16 Let $0 \leq g(x) \leq M$ for each x in $[a, b]$. If $g \in C([a, b])$ and if there is some x_0 in $[a, b]$ such that $g(x_0) < M$, then

$$\int_a^b g(x) \, dx < M(b - a).$$

Proof. Without loss of generality, we may assume that $x_0 \in (a, b)$. By continuity of g at x_0, there is a number $\delta > 0$ such that

$$g(x) < M \quad \text{on } (x_0 - \delta, x_0 + \delta). \tag{3.62}$$

[For let $\epsilon = \frac{1}{2}(M - g(x_0)) > 0$. Then there is some $\delta > 0$ such that on $(x_0 - \delta, x_0 + \delta)$, we have $|g(x) - g(x_0)| < \epsilon = \frac{M}{2} - \frac{g(x_0)}{2}$, which implies $g(x) < \frac{M}{2} + \frac{g(x_0)}{2} < \frac{M}{2} + \frac{M}{2} = M$.]
Then

$$
\begin{aligned}
\int_a^b g(x)\,dx &= \int_a^{x_0-\delta} g(x)\,dx + \int_{x_0-\delta}^{x_0+\delta} g(x)\,dx + \int_{x_0+\delta}^b g(x)\,dx \\
&\leq M(x_0 - \delta - a) + \int_{x_0-\delta}^{x_0+\delta} g(x)\,dx + M(b - x_0 - \delta) \quad \text{by 3.1.7(c)} \\
&< M(b - a) - 2M\delta + 2M\delta \quad \text{by (3.62)} \\
&= M(b - a).
\end{aligned}
$$

Theorem 3.6.17 Let $f \in A(D)$ where D is the closed disk

$$D = \{z : \ |z - z_0| \leq r_0\} \quad \text{and where } r_0 > 0.$$

If

$$|f(z)| \leq |f(z_0)| = M \quad \text{for each } z \text{ in } D, \tag{3.63}$$

then f is constant on D. (That is, if $|f|$ assumes its maximum M on D at the center of D, then f is constant on D.)

Proof. Suppose

$$|f(z_1)| \neq M \quad \text{for some } z_1 \text{ in } D. \tag{3.64}$$

Then by (3.63),

$$|f(z_1)| < M. \tag{3.65}$$

Let γ be the circle centered at z_0 and passing through z_1. Then γ is given by

$$\gamma: \quad z = z_0 + r \operatorname{cis} \theta \quad \text{for each } \theta \text{ in } [0, 2\pi] \text{ where } r = |z_1 - z_0| > 0.$$

Now $\gamma \subset D$. By (3.63) and the Cauchy Integral Formula,

$$
\begin{aligned}
|f(z_0)| = \frac{1}{2\pi} \left| \int_\gamma \frac{f(z)\,dz}{z - z_0} \right| &= \frac{1}{2\pi} \left| \int_0^{2\pi} \frac{f(z_0 + r \operatorname{cis} \theta)\, ir \operatorname{cis} \theta\, d\theta}{r \operatorname{cis} \theta} \right| \\
&\leq \frac{1}{2\pi} \int_0^{2\pi} |f(z_0 + r \operatorname{cis} \theta)|\, d\theta \quad \text{by 3.2.4} \\
&< \frac{1}{2\pi} M(2\pi) = M \quad \text{by 3.6.16}, \tag{3.66}
\end{aligned}
$$

since $|f(z_1)| < M$ and since $z_1 \in \gamma$, so that $z_1 = z_0 + r \operatorname{cis} \theta_0$ for some θ_0 in $[0, 2\pi]$. But (3.66) contradicts (3.63). Thus assumption (3.64) cannot hold. Hence $|f|$ is constant on D° (the interior of D). By 2.3.15(b), we see that f is constant on D° (since D° is a domain). Finally, $f \in C(D)$ implies f is constant on D.

Exercises 3.6.18

1. Show that if $f = u + iv$ is analytic on a domain D, then u_{xx}, u_{xy}, u_{yx}, and u_{yy} are continuous in D.

2. Let $f \in C(D)$ where D is a domain. Prove that if $\int_C f(w)\, dw = 0$ for each simple closed contour C such that $\overline{I(C)} \subset D$, then $f \in A(D)$. **Hint.** Let $z_0 \in D$ and let $r > 0$ such that $N_r(z_0) \subset D$. Observe if $C \subset N_r(z_0)$, then $\overline{I(C)} \subset D$ so that $\int_C f(w)\, dw = 0$. Apply 3.6.12 to $N_r(z_0)$ and conclude that $f \in A(N_r(z_0))$.

3. Prove that if C is a curve in R^2, then C is compact. Use Ex. 5 in 1.4.20 and 2.4.6.

Solution of Ex. 1. By 2.3.7,

$$f'(z) = u_x + iv_x = v_y - iu_y.$$

Thus by 3.6.10(b) and 2.3.7,

$$f''(z) = u_{xx} + iv_{xx} = u_{xx} + i(-u_y)_x.$$

Also, $f''(z) = (-u_y)_y - iv_{yy} = -u_{yy} - i(u_x)_y$. Now by 3.6.10(b), 2.3.3, and 2.2.11, it follows that each of the real and imaginary parts of $f''(z)$ shown is continuous on D. This gives the desired results.

Theorem 3.6.19 Let C be a contour in R^2 and let D be a domain in R^2. Let $F(z,t)$ be a complex-valued function such that if t is any fixed point on C, then $F \in A(D)$ and such that if z is fixed in D, then $F(z,t)$ is piecewise continuous on C. Let

$$g(z) = \int_C F(z,t)\, dt \quad \text{for each } z \text{ in } D.$$

If for each z_0 in D,

$$\frac{F(z,t) - F(z_0,t)}{z - z_0} \xrightarrow{\; U\,C\;} F_z(z_0,t), \tag{3.67}$$

then $g \in A(D)$ and

$$g'(z_0) = \int_C F_z(z_0,t)\, dt, \tag{3.68}$$

where $F_z(z_0,t) = \frac{d}{dz}F(z,t)|_{z=z_0}$ for each fixed t on C and where the notation used in (3.67) means for each $\epsilon' > 0$, there is a number $\delta > 0$ (independent of t) such that

$$\left| \frac{F(z,t) - F(z_0,t)}{z - z_0} - F_z(z_0,t) \right| < \epsilon' \text{ for each } t \text{ on } C \tag{3.69}$$

whenever $z \in D$ and $0 < |z - z_0| < \delta$.

Proof. Let $\epsilon > 0$ and let L be the length of C. Now

$$\left| \frac{g(z) - g(z_0)}{z - z_0} - \int_C F_z(z_0, t)\, dt \right| = \left| \int_C \left(\frac{F(z, t) - F(z_0, t)}{z - z_0} - F_z(z_0, t) \right) dt \right| \quad (3.70)$$

$< L\epsilon' = \frac{L\epsilon}{2L} < \epsilon$ if $z \in D$ and $0 < |z - z_0| < \delta$, where δ is chosen so that (3.69) holds with $\epsilon' = \frac{\epsilon}{2L}$. Thus (3.69) implies (3.70), and hence (3.68) follows.

Exercise 3.6.20

Let h be piecewise continuous on the cco circle $C = \{t : \ |t| = r_0\}$. Use 3.6.19 to prove $g \in A(I(C))$ where

$$g(z) = \int_C \frac{t + z}{t - z}\, h(t)\, dt \quad \text{for each } z \text{ in } I(C).$$

Hint. Let $F(z, t) = \frac{t+z}{t-z} h(t)$. Then

$$\left| \frac{F(z, t) - F(z_0, t)}{z - z_0} - \frac{2t}{(t - z_0)^2} h(t) \right| = \frac{2|t\, h(t)|}{|t - z_0|^2} \frac{|z - z_0|}{|t - z|} \leq M \frac{|z - z_0|}{r_0 - |z|},$$

where M is the maximum of $\frac{2|t\, h(t)|}{|r_0 - |z_0||^2}$ on C. Thus $g \in A(I(C))$.

3.7   Miscellaneous Results on Integrals

We shall conclude this chapter by proving that under certain conditions, if

$$g(z) = \int_C f(z, w)\, dw,$$

where for each fixed w on C, the function f is analytic in a domain D, then $g \in A(D)$, and

$$g'(z) = \int_C f_z(z, w)\, dw, \quad \text{where } f_z(z, w) = \frac{\partial}{\partial z} f(z, w) \quad \text{for each fixed } w.$$

For this result we shall need the following definitions.

Definition 3.7.1 (a) Let A and B be sets. The set

$$\{(a, b) : \ a \in A \text{ and } b \in B\}$$

is called the **cartesian product** of A and B, and is denoted by $A \times B$.

(b) For each (z, w) in $A \times B$, let $F(z, w) \in R^2$ where A is contained in R^2 and B is contained in R^2. The function F is said to be **continuous on** $A \times B$ [indicated by $F \in C(A \times B)$] iff for each (z_0, w_0) in $A \times B$ and for each $\epsilon > 0$, there is a number $\delta > 0$ such that $|F(z, w) - F(z_0, w_0)| < \epsilon$ when $(z, w) \in A \times B$, $|z - z_0| < \delta$, and $|w - w_0| < \delta$.

Theorem 3.7.2 Let f be continuous on $D \times K$ where K is a contour and D is a domain in R^2 and let

$$g(z) = \int_K f(z, w) \, dw \quad \text{for each } z \text{ in } D.$$

Then $g \in C(D)$.

Proof. Let $a \in D$ and let $\epsilon > 0$. Since f is continuous on $D \times K$, for each w' on K, let $\delta_{w'} > 0$ such that

$$|f(z, w) - f(a, w')| < \frac{\epsilon}{2L} \tag{3.71}$$

if $(z, w) \in D \times K$, $|z - a| < \delta_{w'}$, and $|w - w'| < \delta_{w'}$ where L is the length of K. Now K is compact by Ex. 3 in 3.6.18. Thus let

$$\{N(w_1, \delta_{w_1}), N(w_2, \delta_{w_2}) \cdots, N(w_n, \delta_{w_n})\}$$

be a finite subcover of $\{N(w', \delta_{w'}) : \ w' \in K\}$ for K. Let

$$\delta = \min\{\delta_{w_1}, \delta_{w_2}, \cdots, \delta_{w_n}\}.$$

Now let $w \in K$. Then

$$w \in N(w_j, \delta_{w_j}) \text{ for some } j = 1, 2, \cdots, n. \tag{3.72}$$

Now when $z \in D$ such that $|z - a| < \delta \leq \delta_{w_j}$,

$$\begin{aligned} |f(z, w) - f(a, w)| &\leq |f(z, w) - f(a, w_j)| + |f(a, w_j) - f(a, w)| \\ &< \frac{\epsilon}{2L} + \frac{\epsilon}{2L} = \frac{\epsilon}{L} \quad \text{by (3.71) and (3.72).} \end{aligned} \tag{3.73}$$

Thus we have proved that

$$\text{there is some } \delta > 0 \text{ such that if } z \in D \text{ and } |z - a| < \delta, \tag{3.74}$$

then (3.73) holds for each w on K (where δ **is independent** of w). Now for each z in D such that $|z - a| < \delta$, we have

$$\begin{aligned} |g(z) - g(a)| &= \left| \int_K [f(z, w) - f(a, w)] \, dw \right| \\ &< L \cdot \frac{\epsilon}{L} = \epsilon. \end{aligned}$$

Thus $g \in C(a)$ and hence $g \in C(D)$ since a was an arbitrary point of D.

Remark 3.7.3 By 3.7.2,

$$\lim_{z \to a} \int_K f(z, w) \, dw = \lim_{z \to a} g(z) = g(a) = \int_K f(a, w) \, dw$$

$$= \int_K \left[\lim_{z \to a} f(z, w) \right] dw.$$

Exercises 3.7.4

In Ex. 1, prove if F and G are continuous on $A \times B$, where A and B are subsets of R^2, then the given function is continuous on $A \times B$ where in Part (c), $G \neq 0$ in $A \times B$.

1. (a) $F \pm G$ (b) $F \cdot G$ (c) $\dfrac{F}{G}$

2. Prove if $G(z, w) = (w - z)^n$ where $n \in N$, then $G \in C(A \times B)$, where A and B are subsets of R^2.

3. Let A and B be subsets of R^2 such that $A \cap B = \emptyset$ and let $Q(z, w) = \dfrac{h(w)}{(w-z)^n}$ where $n \in N$ and $h \in C(B)$. Prove $Q \in C(A \times B)$.

Hints for Ex. 1(b). Let $P = (z, w)$, $P_0 = (z_0, w_0)$, $F(P) = F$, $G(P) = G$, $F(P_0) = F_0$, and $G(P_0) = G_0$. In 3.7.1(b), take $\epsilon' > 0$ such that $0 < \epsilon' < 1$. Then $|F - F_0| < \epsilon'$ and $|G - G_0| < \epsilon'$ when $|z - z_0| < \delta$ and $|w - w_0| < \delta$ for some $\delta > 0$. Observe that

$$|F| \leq |F_0| + |F - F_0| < |F_0| + \epsilon'.$$

Also note that

$$\begin{aligned}
|FG - F_0G_0| &= |[FG - G_0F] + [G_0F - F_0G_0]| \\
&\leq |F||G - G_0| + |F - F_0||G_0| \\
&< (|F_0| + \epsilon')\epsilon' + \epsilon'|G_0| \\
&< (|F_0| + |G_0| + 1)\epsilon'.
\end{aligned}$$

Take $\epsilon' = \frac{\epsilon}{|F_0| + |G_0| + 1}$ where $0 < \epsilon < 1$.

Hints for Ex. 1(c). Let G and G_0 be given as in the Hints for Ex. 1(b). Let $\epsilon' = \frac{|G_0|}{2} > 0$. Show there is some $\delta > 0$ such that $|G| \geq \frac{|G_0|}{2} > 0$ when $|z - z_0| < \delta$ and $|w - w_0| < \delta$. See the last statement in 2.2.5(b). Note that

$$\left| \frac{1}{G} - \frac{1}{G_0} \right| = |G_0 - G| \cdot \frac{1}{|G|} \cdot \frac{1}{|G_0|} < \epsilon' \cdot \frac{2}{|G_0|} \cdot \frac{1}{|G_0|} = \epsilon$$

if $\epsilon' = |G_0|^2 \frac{\epsilon}{2}$. Finally, note that $\frac{F}{G} = F \cdot \frac{1}{G}$ and apply Part (b).

Hints for Ex. 2. If $H(z, w) = w - z$, then $H \in C(A \times B)$ by Ex. 1(a) or 3.7.1(b). By Ex. 1(b), we conclude that $(w - z)^n \in C(A \times B)$.

Hints for Ex. 3. Let $H(z, w) = h(w)$. Then $h \in C(B)$ implies $H \in C(A \times B)$ by 3.7.1(b). Now use Ex. 1(c) and Ex. 2. Now $(w - z)^n \neq 0$ on $A \times B$ since $A \cap B = \emptyset$.

Theorem 3.7.5 If $F \in C(H \times K)$ where H and K are contours in R^2, then

$$\int_H \left[\int_K F(z, w) \, dw \right] dz = \int_K \left[\int_H F(z, w) \, dz \right] dw.$$

Proof. First suppose H and K are smooth. Let H have the smooth parameterization $z(s)$ for s in $[a, b]$, and let K have the smooth parameterization $w(t)$ for t in $[c, d]$. Then

$$
\begin{aligned}
\int_H \left[\int_K F(z, w) \, dw \right] dz &= \int_a^b \left[\int_c^d F(z(s), w(t)) \, w'(t) \, dt \right] z'(s) \, ds \\
&= \int_c^d \left[\int_a^b F(z(s), w(t)) \, z'(s) \, ds \right] w'(t) \, dt \\
&= \int_K \left[\int_H F(z, w) \, dz \right] dw.
\end{aligned}
$$

If H and K are merely piecewise smooth, then the intervals $[a, b]$ and $[c, d]$ can be broken up into subintervals on which $z'(s)$ and $w'(t)$ are continuous respectively.

Theorem 3.7.6 Let $h \in C(H)$ where H is a contour in R^2 and let

$$g(z) = \int_H \frac{h(w)}{w - z} \, dw \quad \text{for each } z \text{ in } (R^2 - H). \tag{3.75}$$

Then $g \in A(R^2 - H)$, and for each z in $(R^2 - H)$,

$$g^{(n)}(z) = n! \int_H \frac{h(w) \, dw}{(w - z)^{n+1}} \quad \text{for each } n \text{ in } N. \tag{3.76}$$

Proof for $n = 1$. The proof for $n > 1$ is given after the proof of 3.7.7. Note by (3.71), for a fixed z in $(R^2 - H)$,

$$g(z') - g(z) = \int_H h(w) \left[\frac{1}{w - z'} - \frac{1}{w - z} \right] dw$$

so that

$$\frac{g(z') - g(z)}{z' - z} = \int_H \frac{h(w) \, dw}{(w - z)(w - z')}.$$

Thus by 3.7.3 and 3.7.4,

$$
\begin{aligned}
g'(z) &= \lim_{z' \to z} \frac{g(z') - g(z)}{z' - z} = \int_H \lim_{z' \to z} \frac{h(w) \, dw}{(w - z)(w - z')} \\
&= \int_H \frac{h(w) \, dw}{(w - z)^2} \quad \text{for each } z \text{ in } (R^2 - H).
\end{aligned}
$$

Thus $g \in A(R^2 - H)$ and (3.76) holds with $n = 1$.

Theorem 3.7.7 Let $f(z, w)$ be continuous on $D \times K$ where K is a contour and D is a domain in R^2. If

> for each fixed w on K, the function $f(z, w)$[as a function of z]
> is analytic in D (3.77)

and if

$$g(z) = \int_K f(z, w)\, dw \quad \text{for each } z \text{ in } D, \qquad (3.78)$$

then $g \in A(D)$ and

$$g'(z) = \int_K f_z(z, w)\, dw \quad \text{for each } z \text{ in } D. \qquad (3.79)$$

Proof. Let $z_0 \in D$. Since D is open, let H be a cco circle centered at z_0 such that $\overline{I(H)} \subset D$. By (3.78) for each z in $I(H)$,

$$
\begin{aligned}
g(z) &= \int_K f(z, w)\, dw \\
&= \int_K \frac{1}{2\pi i} \left[\int_H \frac{f(t, w)}{t - z}\, dt \right] dw \quad \left\{ \begin{array}{l} \text{by (3.77) and the Cauchy} \\ \text{Integral Formula} \end{array} \right. \\
&= \frac{1}{2\pi i} \int_H \left[\int_K f(t, w)\, dw \right] \frac{dt}{t - z} \quad \text{by 3.7.5} \\
&= \frac{1}{2\pi i} \int_H \frac{g(t)\, dt}{t - z} \quad \text{by (3.78).} \qquad (3.80)
\end{aligned}
$$

Now $g \in C(D)$ by (3.78) and 3.7.2. Thus by (3.80) and 3.7.6, we have $g \in A(I(H))$ and for each z in $I(H)$,

$$
\begin{aligned}
g'(z) &= \frac{1}{2\pi i} \int_H \frac{g(t)\, dt}{(t - z)^2} \quad \text{by 3.7.6 with } n = 1 \\
&= \frac{1}{2\pi i} \int_H \left[\int_K f(t, w)\, dw \right] \frac{dt}{(t - z)^2} \quad \text{by (3.78)} \\
&= \int_K \left[\frac{1}{2\pi i} \int_H \frac{f(t, w)}{(t - z)^2}\, dt \right] dw \quad \text{by 3.7.5} \\
&= \int_K f_z(z, w)\, dw \quad \text{by (3.77) and 3.6.9 with } n = 1.
\end{aligned}
$$

But z_0 is arbitrary. Thus (3.79) holds.

We now apply 3.7.7 to prove (3.76) with $n > 1$.

Proof of (3.76) with $n > 1$. By 3.7.6 with $n = 1$, we have

$$g'(z) = \int_H \frac{h(w)}{(w - z)^2}\, dw \quad \text{for each } z \text{ in } (R^2 - H). \qquad (3.81)$$

By 3.7.7 and (3.81),

$$
\begin{aligned}
g''(z) &= \int_H \frac{\partial}{\partial z}\left(\frac{h(w)}{(w-z)^2}\right) dw \\
&= 2\int_H \frac{h(w)}{(w-z)^3} dw.
\end{aligned}
$$

By 3.7.7 and mathematical induction, we obtain (3.76) for each $n \in N$.

Theorem 3.7.8 Let $F \in C(D \times [m, M])$, where D is a domain in R^2. If $z_0 \in D$, then

$$
\lim_{z \to z_0} \int_m^M F(z, r)\, dr = \int_m^M F(z_0, r)\, dr.
$$

Proof. Let K be the segment $[m, M]$ parametrized by $w(r) = r$ for r in $[m, M]$. Then $\lim_{z \to z_0} \int_m^M F(z, r)\, dr = \lim_{z \to z_0} \int_K F(z, w)\, dw = \int_K F(z_0, w)\, dw = \int_m^M F(z_0, r)\, dr$ by 3.7.3 (and 3.7.2).

Chapter 4

Series

4.1 Sequences

In the calculus, we studied sequences of real numbers. If we referred to a given sequence, we understood that some rule was given which would associate with each positive integer n, a unique real number which we might denote by x_n or by $x(n)$. Thus in the calculus, we recognized that each sequence of real numbers was an example of a function whose domain was the set N of all positive integers and whose range was contained in R. The notation $x(n)$ would be consistent with our usual functional notation. However, due to tradition, we usually use x_n to denote the image of n (if the name of the function is x). Of course, we use a_n or b_n or S_n to denote the image of n if the name of the function is a or b or S, etc.

We sometimes denote a sequence by $a_0, a_1, a_2, \cdots$ or by $b_4, b_5, b_6, \cdots$. This suggests that we might define a sequence as a function whose domain is the set of all integers greater than some given integer. This would be a good definition. But for each example of a sequence, there is a first member, a second member, a third member, etc. Furthermore, the sequence denoted by $b_4, b_5, b_6, \cdots$ may be recognized as the composite function $(b \circ g)(n) = (b \circ g)_n = b_{n+3}$ for each n in N. Now the domain of the composite function is N. Hence it is also appropriate to define a sequence as a function whose domain is the set N of all positive integers.

Definitions 4.1.1 (a) **A sequence in R^2** is a function z whose domain is the set N of all positive integers and whose range is contained in R^2. We often denote such a sequence z by

$$\{z_n\} \quad \text{or by} \quad z_1, z_2, z_3, \cdots.$$

(b) A sequence $\{z_n\}$ **converges to b** in R^2 iff

for each $\epsilon > 0$, there is an integer M such that
if $n \geq M$, then $|z_n - b| < \epsilon$.

95

(c) A sequence **diverges** iff it does not converge to any point b in R^2.

We usually write

$$b = \lim_{n \to \infty} z_n$$

to indicate that $\{z_n\}$ converges to b. However, sometimes we may indicate this briefly by writing

$$b = \lim z_n, \quad \text{or} \quad z_n \to b \text{ as } n \to \infty, \quad \text{or} \quad z_n \to b.$$

Theorem 4.1.2 Let $\{s_n\}$ and $\{t_n\}$ be sequences in R^2 such that

$$\lim s_n = a \quad \text{and} \quad \lim t_n = b.$$

Then

$$\lim(s_n + t_n) = a + b, \quad \lim(s_n - t_n) = a - b,$$

$$\lim(s_n t_n) = ab, \quad \text{and} \quad \text{if } b \neq 0, \lim \frac{s_n}{t_n} = \frac{a}{b}.$$

Also, if $|s_n| \leq |t_n|$ for each n in N, then $|a| \leq |b|$. Finally, if $s_n = c$ for each n in N, then $s_n \to c$ as $n \to \infty$.

The proofs are similar to the proofs of the companion statements in 2.2.6.

4.2   Series of Complex Numbers

If $\{a_n\}$ is a sequence of real numbers, then in the calculus we refer to the series $\sum_{n=1}^{\infty} a_n$. Thus we use the notation $\sum_{n=1}^{\infty} a_n$ to stand for "something which we call a series." But what is the mathematical concept of a series? To answer this question, we recall that in the calculus, we say: "The series $\sum_{n=1}^{\infty} a_n$ converges to the number S iff the sequence $\{S_n\}$ of partial sums converges to S." This quotation suggests that it would be appropriate to refer to the sequence of partial sums as the series. Certainly this sequence is of utmost importance when discussing the infinite series. Now if we want the technical term "infinite series" to have a conceptual interpretation which goes beyond the notation or expression on paper, then we may define the term by saying the "infinite series $\sum_{n=1}^{\infty} a_n$" is the sequence $\{S_n\}$ of partial sums. If we do this, then we have a mathematical definition of "infinite series" in terms of "sequence," which has already been defined mathematically, as a function whose domain is the set of positive integers.

It is important to note that the algebraic operation of addition in R or in R^2, is a **binary operation**. However, due to the general associative law of addition, we may extend the algebraic concept of the sum of two numbers to the idea of the sum of any finite number of numbers. We do this without appealing to any structure in R or in R^2 other than the algebraic operation of addition. Thus the **sum** of a finite number of numbers is a pure **algebraic** concept. On the other hand, the "sum" of an infinite series is actually the **limit** of a sequence. Now the limit of a sequence is a topological concept, since it is defined in terms of the topological structure on R, or on R^2. Although we use the word "sum" in the expression "sum of a series," this sum concept is quite different from the algebraic concept of the sum of a finite number of numbers. The **sum** of a finite number of numbers is an **algebraic** concept, whereas the **sum** of an infinite series is a **topological** concept.

Finally, before we begin our systematic study of series of complex numbers, we recall from the calculus that we often use the summation notation to stand for a series. For example, we may say the series $\sum_{n=1}^{\infty} \frac{1}{n^3}$ converges. On the other hand, we often use the summation notation to denote the number which is the sum of the series. For example, we may write $e^3 = \sum_{n=0}^{\infty} \frac{3^n}{n!}$. Although we frequently use the same summation notation to stand for two different things, the context always clearly indicates whether the notation stands for the series, or for the number which is the sum of the series.

Definitions 4.2.1 Let $\{a_n\}$ be a sequence in R^2 and let $S_n = \sum_{k=1}^{n} a_k$ for each n in N.

(a) The sequence $\{S_n\}$ is called an **infinite series** of complex numbers and is denoted by

$$\sum_{n=1}^{\infty} a_n \quad \text{or} \quad \sum a_n \quad \text{or} \quad a_1 + a_2 + a_3 + \cdots.$$

(b) The number S_n is called the nth **partial sum** of the series $\sum a_n$.

(c) The series $\sum a_n$ **converges to S** iff

$$\lim S_n = S.$$

(d) If $\lim S_n = S$ where $S \in R^2$, then S is called the **sum** of the series, and we write

$$S = \sum a_n = a_1 + a_2 + a_3 + \cdots.$$

(Note that $\sum a_n$ is used to denote both the series and its sum, if the sum exists. However, the context always clearly indicates whether the notation denotes the series or the number which is the sum of the series.)

(e) If $\lim S_n$ does not exist, then we say the series **diverges**.

(f) The series $\sum a_n$ **converges absolutely** iff $\sum |a_n|$ converges.

Theorem 4.2.2 If $z_n = a_n + ib_n$, then

$$\sum z_n = \sum (a_n + ib_n) = a + ib$$

if and only if

$$\sum a_n = a \quad \text{and} \quad \sum b_n = b.$$

Proof. Let

$$\alpha_n = \sum_{k=1}^{n} a_k \quad \text{and} \quad \beta_n = \sum_{k=1}^{n} b_k. \tag{4.1}$$

"If Part." Let

$$\lim \alpha_n = a \quad \text{and} \quad \lim \beta_n = b.$$

Now

$$S_n = \sum_{k=1}^{n} z_k = \sum_{k=1}^{n} (a_k + ib_k) = \alpha_n + i\beta_n \quad \text{by (4.1)}. \tag{4.2}$$

Thus by the appropriate three parts of 4.1.2,

$$\lim S_n = a + ib.$$

"Only If Part." Suppose $\sum z_n = a + ib$. Then

$$\lim S_n = a + ib \tag{4.3}$$

where S_n is given by (4.2). Thus by (4.3), for each $\epsilon > 0$, there is an integer M such that if $n > M$, then

$$\epsilon > |S_n - (a + ib)| = \sqrt{(\alpha_n - a)^2 + (\beta_n - b)^2} \geq \left\{ \begin{array}{l} |\alpha_n - a| \\ |\beta_n - b|. \end{array} \right.$$

Hence

$$\lim \alpha_n = a \quad \text{and} \quad \lim \beta_n = b.$$

Theorem 4.2.3 (Ratio Test). Let $\sum a_n$ be a series of nonzero complex numbers and let

$$\lim \left| \frac{a_{n+1}}{a_n} \right| = \rho. \tag{4.4}$$

Then $\sum a_n$

converges absolutely if $\rho < 1$,

diverges if $\rho > 1$, and

may converge or may diverge if $\rho = 1$.

Proof. Let $\rho < 1$. Let x be such that $\rho < x < 1$ and let $\epsilon = x - \rho$. Then $\epsilon > 0$. By 4.1.1(b), let M be such that if $n > M$, then

$$\left| \frac{a_{n+1}}{a_n} \right| < \rho + \epsilon = x \quad \text{and} \quad 0 < x < 1.$$

Thus for each $n > M$,

$$|a_{n+1}| < |a_n|\, x = |a_n| \frac{x^{n+1}}{x^n}$$

which implies

$$\frac{|a_{n+1}|}{x^{n+1}} < \frac{|a_n|}{x^n} < \frac{|a_{n-1}|}{x^{n-1}} < \cdots < \frac{|a_M|}{x^M} = c$$

where c is a constant. Hence

$$|a_n| < cx^n \quad \text{for each } n > M.$$

But $\sum_{n=0}^{\infty} x^n$ is a geometric series with $0 < x < 1$. Hence $\sum_{n=0}^{\infty} cx^n$ converges. Thus by the comparison test, $\sum a_n$ converges absolutely.

Let $\rho > 1$. Let x be such that $1 < x < \rho$ and let $\epsilon = \rho - x$. Then $\epsilon > 0$. Thus there is an integer M such that whenever $n > M$, we have

$$\left|\frac{a_{n+1}}{a_n}\right| > \rho - \epsilon = x > 1.$$

Hence

$$|a_{n+1}| > |a_n| \quad \text{for each } n > M.$$

Therefore $|a_n|$ does not approach 0 as $n \to \infty$, and thus $\sum a_n$ diverges. [For $\sum a_n$ converges implies $\lim a_n = \lim (S_n - S_{n-1}) = S - S = 0$.]

Let $\rho = 1$. Now $\sum \frac{1}{n}$ diverges and $\sum \frac{1}{n^2}$ converges, but $\rho = 1$ for each of these series.

Definition 4.2.4 Let $\{x_n\}$ be a sequence in R (that is, a sequence of real numbers) and let $L \in R$.

(a) The number L is called the **upper limit** (or **limit superior**) of $\{x_n\}$ iff for each $\epsilon > 0$, the following two conditions hold.

(i) There is an integer M such that $x_n < L + \epsilon$ for each $n \geq M$.

(ii) For each k in N, there is an integer $n > k$ such that $x_n > L - \epsilon$.

(b) If for each M in R there is an integer n such that $x_n > M$, then the **upper limit** of $\{x_n\}$ is $+\infty$.

(c) If for each M in R, we have $x_n < M$ for all but a finite number of values of n, then the **upper limit** of $\{x_n\}$ is $-\infty$.

Remarks 4.2.5 (a) Let $\{x_n\}$ be **any** sequence in R. By 4.2.4, the upper limit of $\{x_n\}$ exists and is unique. (See the solution of Ex. 3 in 4.5.6.) We denote this unique upper limit by $\overline{\lim} x_n$.

(b) Condition (i) in 4.2.4 means that

$$x_n < L + \epsilon \quad \text{for all but a finite number of values of } n.$$

Condition (ii) in 4.2.4 means that

$$x_n > L - \epsilon \quad \text{for infinitely many values of } n.$$

(c) Part (b) in 4.2.4 means that $\overline{\lim}\, x_n = +\infty$ iff the sequence $\{x_n\}$ is not bounded above. Also, 4.2.4(c) means that $\overline{\lim}\, x_n = -\infty$ iff $\lim x_n = -\infty$.

(d) By 4.2.4, it is easy to verify each of the following results.

(i) $\overline{\lim}\,[n + (-1)^n\, n] = +\infty$

(ii) $\overline{\lim}\,[(-1)^n\, n\, -\, n] = 0$

(iii) $\overline{\lim}\, \sin \frac{n\pi}{2} = 1$

(iv) $\overline{\lim}\, n \sin \frac{n\pi}{2} = +\infty$

(v) $\overline{\lim}\,(-n) = -\infty$

Theorem 4.2.6 (Root Test). Let $\sum a_n$ be a series of complex numbers and let

$$\rho = \overline{\lim} \sqrt[n]{|a_n|}. \qquad (4.5)$$

Then $\sum a_n$

converges absolutely if $\rho < 1$,

diverges if $\rho > 1$, and

may converge or may diverge if $\rho = 1$.

Proof. Let $\rho < 1$. Take x such that $\rho < x < 1$ and let $\epsilon = x - \rho > 0$. By (4.5), let M be such that if $n > M$, then

$$0 \le \sqrt[n]{|a_n|} < \rho + \epsilon = x < 1.$$

Thus if $n > M$, then
$$|a_n| < x^n.$$

But the geometric series $\sum x^n$ converges since $|x| < 1$. Thus $\sum a_n$ converges absolutely by the comparison test.

Let $\rho > 1$. Let x be such that $1 < x < \rho$ and let $\epsilon = \rho - x$. By 4.2.5(b),

$$\sqrt[n]{|a_n|} > \rho - \epsilon = x > 1 \quad \text{for infinitely many values of } n.$$

Hence $|a_n| > 1$ for infinitely many values of n. Thus a_n does not approach zero as $n \to \infty$, and so $\sum a_n$ diverges.

Let $\rho = 1$. Use $\sum \frac{1}{n}$ and $\sum \frac{1}{n^2}$. (See Ex. 2 and Ex. 3 in 4.2.7.)

Exercises 4.2.7 In Ex. 1 and 2, let $\{a_n\}$ be a sequence in R and prove the stated results.

1. If $\lim a_n = a$, then for each $\epsilon > 0$, there is an integer M such that

(a) $a_n < a + \epsilon$ for each $n > M$ and

(b) $a_n > a - \epsilon$ for each $n > M$.

2. If $\lim a_n = a$, then $\overline{\lim} a_n = a$. **Hint.** Compare Part (a) and Part (b) in Ex. 1 with (i) and (ii) in 4.2.4 if $a \in R$. If $a = \pm\infty$, see 4.2.4(b) and (c).

3. Prove that $\lim \sqrt[n]{n} = 1$. **Hint.** Notice that $n^{\frac{1}{n}} = e^{\frac{1}{n}\text{Log}\,n}$ and apply L'Hôpital's rule to $\lim\limits_{x\to\infty} \frac{\text{Log}\,x}{x}$.

4. Let $\{a_n\}$ and $\{b_n\}$ be sequences in R such that $\infty > \lim a_n = a > 0$ and such that $b_n \geq 0$ for each n. Prove that $\overline{\lim} a_n b_n = a \,\overline{\lim} b_n$.

Proof of Ex. 4. Let $L = \overline{\lim} b_n$.

Case 1. Suppose $0 < L < +\infty$. Let $\epsilon > 0$. Let

$$\delta > 0 \text{ such that } \delta < \min\left\{\sqrt{\frac{\epsilon}{2}}, \; \frac{\epsilon}{2(a+L)}, \; a, \; L\right\}. \tag{4.6}$$

By 4.1.1(b) and 4.2.4(i), let M be such that

$$a - \delta < a_n < a + \delta \quad \text{and} \quad b_n < L + \delta \quad \text{for each } n > M. \tag{4.7}$$

Thus for each $n > M$,

$$\begin{aligned} a_n b_n \;&<\; (a + \delta)(L + \delta) = aL + (a + L)\delta + \delta^2 \\ &<\; aL + \frac{\epsilon}{2} + \frac{\epsilon}{2} = aL + \epsilon \quad \text{by (4.6).} \end{aligned} \tag{4.8}$$

By 4.2.5(b),

$$L - \delta < b_n \quad \text{for infinitely many values of } n. \tag{4.9}$$

By (4.6), we have $a - \delta > 0$ and $L - \delta > 0$. Thus by (4.7) and (4.9), for infinitely many values of n,

$$\begin{aligned} a_n b_n \;&>\; (a - \delta)(L - \delta) \;=\; aL - (a + L)\delta + \delta^2 \\ &>\; aL - (a + L)\delta \;>\; aL - \frac{\epsilon}{2} \quad \text{by (4.6)} \\ &>\; aL - \epsilon. \end{aligned} \tag{4.10}$$

Hence by (4.8), (4.10), and 4.2.4, we have $\overline{\lim} a_n b_n = aL$.

Case 2. Suppose $L = 0$. Then $\overline{\lim} b_n = 0 = \lim b_n$. For by 4.2.4, there is an integer M such that $-\epsilon < b_n < \epsilon$ for each $n > M$. Now

$$\begin{aligned} \lim a_n b_n \;&=\; a \cdot 0 && \text{by 4.1.2} \\ &=\; \overline{\lim} a_n b_n && \text{by Ex. 2.} \end{aligned}$$

Case 3. Suppose $L = +\infty$. Let $m \in N$ such that

$$a_n > \frac{a}{2} \quad \text{for each } n > m. \tag{4.11}$$

Let M be **any** real number and let

$$K = \max\left\{\frac{2M}{a}, b_1, b_2, \cdots, b_m\right\}. \tag{4.12}$$

By 4.2.4(b), let $j \in N$ such that

$$b_j > K. \tag{4.13}$$

Thus $j > m$ by (4.12). By (4.11) and (4.13) with $n = j$,

$$
\begin{aligned}
a_j b_j &> \frac{a}{2} \cdot K \\
&\geq \frac{a}{2} \cdot \frac{2M}{a} \quad \text{by (4.12)} \\
&= M.
\end{aligned}
\tag{4.14}
$$

Thus by (4.14) and 4.2.4(b),

$$\overline{\lim} a_n b_n = +\infty = a(+\infty) = a\,\overline{\lim} b_n.$$

Case 3 (Alternate Proof). Suppose $L = +\infty$. By 4.2.4(b), this means that $\{b_n\}$ is not bounded above. Since $a > 0$, there is an integer m such that if $n > m$, then $a_n > \frac{a}{2}$. Thus $\{a_n b_n\}$ is not bounded above. Hence by 4.2.4(b),

$$\overline{\lim} a_n b_n = +\infty = a(+\infty) = a\,\overline{\lim} b_n.$$

4.3   Power Series

Definition 4.3.1 A series of the form

$$a_0 + \sum_{n=1}^{\infty} a_n(z - z_0)^n$$

is called a power series in $z - z_0$. We denote such a series by

$$\sum_{n=0}^{\infty} a_n (z - z_0)^n \quad \text{or by} \quad \sum a_n (z - z_0)^n.$$

If $z = z_0$, it is clear that a_0 is the sum of the first series given in 4.3.1. This is so since each term, except the first, is zero if $z = z_0$. Thus the sequence of partial sums is the constant sequence given by $S_n = a_0$ for each n. However, without the convention stated in 4.3.1, the first term in the second summation would be $a_0 0^0$ (which is meaningless) if $z = z_0$. To use the second summation as we do, does **not** mean that we now define 0^0 to be 1. We do not! We simply use the brief notation to stand for the first series given in 4.3.1.

Theorem 4.3.2 Given a series $\sum a_n (z - z_0)^n$, let $\rho = \overline{\lim} \sqrt[n]{|a_n|}$ and let

$$r = \begin{cases} 0 & \text{if } \rho = \infty \\ \frac{1}{\rho} & \text{if } 0 < \rho < \infty \\ \infty & \text{if } \rho = 0. \end{cases}$$

Then the series

(a) converges absolutely if $|z - z_0| < r$,

(b) diverges if $|z - z_0| > r$, and

(c) may converge for some values of z for which $|z - z_0| = r$ and may diverge for some values of z for which $|z - z_0| = r$.

Proof. With the Root Test in mind, we observe that

$$\overline{\lim} \, |a_n (z - z_0)^n|^{\frac{1}{n}} \;=\; |z - z_0| \, \overline{\lim} |a_n|^{\frac{1}{n}}$$

$$=\; |z - z_0| \, \rho = \begin{cases} 0 & \text{if } \rho = 0 \\ \infty & \text{if } \rho = \infty \text{ and } z \neq z_0 \end{cases}$$

$$=\; \frac{|z - z_0|}{r} \quad \text{if } 0 < \rho < \infty$$

$$\begin{cases} < 1 & \text{if } |z - z_0| < r \\ > 1 & \text{if } |z - z_0| > r. \end{cases}$$

Hence by 4.2.6, we conclude that $\sum a_n (z - z_0)^n$ converges absolutely if $|z - z_0| < r$ and diverges if $|z - z_0| > r$.

Remarks 4.3.3 (a) The value of r in 4.3.2 is called the **radius of convergence** of the power series and is given by the formula

$$r = \frac{1}{\overline{\lim} \sqrt[n]{|a_n|}} \tag{4.15}$$

where it is understood that $r = +\infty$ or 0 if $\overline{\lim} \sqrt[n]{|a_n|} = 0$ or $+\infty$, respectively. The circle

$$\gamma = \{z : \;\; |z - z_0| = r\}$$

is called the **circle of convergence** of the given series.

(b) Thus the series in 4.3.2

    (i) converges for $z = z_0$ only or

    (ii) converges absolutely for each z in R^2 or

(iii) converges absolutely for each z in $I(\gamma)$ and diverges for each z in $E(\gamma)$.

(c) By (4.15) and Ex. 2 in 4.2.7, the radius of convergence of the given power series is given by

$$r = \frac{1}{\lim \sqrt[n]{|a_n|}} \quad \text{if this limit exists.} \qquad (4.16)$$

However, we often use Part (b) to find r. For example, we may use the ratio test to find the "region of convergence" described in Part (b) for a given series. In particular, consider the series

$$\sum \frac{z^n}{(n+1)2^n}. \qquad (4.17)$$

Using the ratio test,

$$\left| \frac{z^{n+1}}{(n+2)2^{n+1}} \cdot \frac{(n+1)2^n}{z^n} \right| = \frac{n+1}{n+2} \frac{|z|}{2} \rightarrow \frac{|z|}{2} \quad \text{as } n \rightarrow \infty.$$

Thus the series in (4.17) converges if $|z| < 2$ and diverges if $|z| > 2$. Thus $r = 2$ by Part (b).

Theorem 4.3.4 Let r be the radius of convergence of a given series $\sum_{n=0}^{\infty} a_n z^n$. Then r is also the radius of convergence of $\sum_{n=1}^{\infty} n a_n z^{n-1}$ (which is obtained from the given series by termwise differentiation).

Proof. Let r_d be the radius of convergence of $\sum_{n=1}^{\infty} n a_n z^{n-1}$. Now this series converges iff

$$z \sum_{n=1}^{\infty} n a_n z^{n-1} = \sum_{n=1}^{\infty} n a_n z^n$$

converges. From 4.3.3,

$$
\begin{aligned}
r_d &= \frac{1}{\overline{\lim} \sqrt[n]{n|a_n|}} \\
&= \frac{1}{\left(\lim \sqrt[n]{n} \right) \left(\overline{\lim} \sqrt[n]{|a_n|} \right)} \\
&= \frac{1}{\overline{\lim} \sqrt[n]{|a_n|}} \quad \text{by Ex. 3 and Ex. 4 in 4.2.7} \\
&= r \quad \text{by 4.3.3(a).}
\end{aligned}
$$

Theorem 4.3.5 Let f be given by

$$f(z) = \sum_{n=0}^{\infty} a_n z^n \quad \text{for each } z \text{ in } N(0, r) \tag{4.18}$$

where r is the radius of convergence of the series. Then $f'(z)$ exists and is given by

$$f'(z) = \sum_{n=1}^{\infty} n a_n z^{n-1} \quad \text{for each } z \text{ in } N(0, r). \tag{4.19}$$

Thus f is analytic in $N(0, r)$.

Proof. By 4.3.4, let

$$g(z) = \sum_{n=1}^{\infty} n a_n z^{n-1} \quad \text{for each } z \text{ in } N(0, r). \tag{4.20}$$

Let z be fixed in $N(0, r)$ and let ρ be a constant such that $|z| < \rho < r$. Let $w \in N(0, \rho)$ such that $w \neq z$. Then

$$f(w) = a_0 + \sum_{n=1}^{\infty} a_n w^n \quad \text{and} \quad f(z) = a_0 + \sum_{n=1}^{\infty} a_n z^n. \tag{4.21}$$

Thus we have

$$\frac{f(w) - f(z)}{w - z} - g(z) = \sum_{n=1}^{\infty} a_n P_n(w) \tag{4.22}$$

where

$$\begin{aligned}
P_n(w) &= \frac{w^n - z^n}{w - z} - n z^{n-1} \\
&= w^{n-1} + w^{n-2} z + \cdots + w z^{n-2} + z^{n-1} - n z^{n-1} \\
&= (w - z) \sum_{k=1}^{n-1} k z^{k-1} w^{n-k-1}
\end{aligned} \tag{4.23}$$

which can be verified by multiplication. Thus

$$\begin{aligned}
|P_n(w)| &\leq |w - z| \rho^{n-2} \sum_{k=1}^{n-1} k \quad \text{since } |z| < \rho \text{ and } |w| < \rho \\
&\leq |w - z| \frac{n(n-1)}{2} \rho^{n-2}.
\end{aligned} \tag{4.24}$$

[The sum of an arithmetic progression with first term A and last term L is $\frac{n}{2}(A + L)$ where n is the number of terms.] By $(4.22) - (4.24)$,

$$\begin{aligned}
\left| \frac{f(w) - f(z)}{w - z} - g(z) \right| &\leq \frac{|w - z|}{2} \sum_{n=2}^{\infty} n(n-1) |a_n| \rho^{n-2} \\
&\to 0 \quad \text{as } w \to z
\end{aligned} \tag{4.25}$$

since by 4.3.4, the series in (4.25) converges to a constant. But (4.25) implies

$$f'(z) = g(z).$$

Thus (4.20) implies (4.19).

Corollary 4.3.6 A power series represents a function which is analytic in the interior of its circle of convergence.

Theorem 4.3.7 For each z in $N(0,r)$, let

$$f(z) = \sum_{n=0}^{\infty} a_n z^n \quad \text{and} \quad g(z) = \sum_{n=0}^{\infty} \frac{a_n}{n+1} z^{n+1}$$

where r is the radius of convergence of the second series. Then for each z in $N(0,r)$

$$\int_0^z f(t)\, dt = g(z)$$

where the integral is along any contour contained in $N(0,r)$ with initial point 0 and terminal point z. Thus a power series in z may be integrated termwise in the interior of its circle of convergence.

Proof. By 4.3.4, we know that r is the radius of convergence of $\sum_{n=0}^{\infty} a_n z^n$. By 4.3.5,

$$g'(z) = f(z) \quad \text{for each } z \text{ in } N(0,r).$$

Thus by 3.6.3(b),

$$\int_0^z f(t)\, dt = g(z) - g(0) = g(z) \quad \text{for each } z \text{ in } N(0,r).$$

Remark 4.3.8 Theorems relating to functions represented by power series in powers of W and defined in $N(0,r)$ may be readily translated to corresponding theorems on functions represented by power series in powers of $z - z_0 = W$ and defined in $N(z_0, r)$.

Theorem 4.3.9 Let a_n, b_n, and c be complex numbers. If $\sum a_n = A$ and $\sum b_n = B$, then

$$\sum (a_n \pm b_n) = A \pm B \quad \text{and} \quad \sum c\, a_n = c\, A.$$

Proof. Let

$$S_n = \sum_{k=1}^{n} a_k \quad \text{and} \quad T_n = \sum_{k=1}^{n} b_k.$$

Then

$$\sum (a_n \pm b_n) = \lim (S_n \pm T_n) = A \pm B \quad \text{by 4.1.2.}$$

Also

$$\sum c\, a_n = \lim (c\, S_n) = c\, \lim S_n = c\, A \quad \text{by 4.1.2.}$$

Exercises 4.3.10

1. Prove if $\sum a_n$ converges, then $\lim a_n = 0$. **Hint.** $a_n = S_n - S_{n-1}$

2. Prove that if $\{a_n\}$ does not converge to zero, then the series $\sum a_n$ diverges.

In Ex. 3–8, test the series $\sum_{n=1}^{\infty} a_n$ for convergence or divergence where a_n is the given expression.

3. $\dfrac{3^n}{n+1}$ \hfill div.

4. $\dfrac{n^2 + i}{n^3}$ \hfill div.

5. $\dfrac{n - 2i}{3n^3}$ \hfill conv.

6. $\dfrac{3n + 2i + 5}{i^n}$ \hfill div.

7. $\left(\dfrac{n}{n+1}\right)^{4n^2} i^n$ **Hint.** $\lim (1 + \frac{1}{n})^n = e$ \hfill conv.

8. $\dfrac{\operatorname{Log} n}{n!} i^n$ \hfill conv.

9. Find the radius of convergence for each of the following power series (where n varies from 1 to ∞).

 (a) $\sum n! \, z^n$ (b) $\sum \dfrac{z^n}{n^3}$ (c) $\sum \dfrac{z^n}{n}$

 (d) $\sum \dfrac{(z-1)^n}{2^n}$ (e) $\sum n \, z^{n-1}$ (f) $\sum n^{\operatorname{Log} n} z^n$

 (g) $\sum 3^n (z+1)^n$ (h) $\sum \left(\dfrac{n}{n+1}\right)^{n^2} z^n$ (i) $\sum \left(1 - \dfrac{1}{n}\right)^{n^2} z^n$

 (j) $\sum n(n!) \, z^{n-1}$ [Use (a) and 4.3.4.]

 (k) $\sum n(n^{\operatorname{Log} n}) z^{n-1}$ [Use (f) and 4.3.4.]   (l) $\sum \dfrac{n z^{n+1}}{n+1}$

 Ans.   (a) 0 (b) 1 (c) 1 (d) 2 (e) 1 (f) 1 (g) $\dfrac{1}{3}$ (h) e (i) e (j) 0 (k) 1 (l) 1

10. Test $\displaystyle\sum_{n=1}^{\infty} \dfrac{i^n}{n+1}$ for convergence or divergence. [Use 4.2.2.]

11. If f is defined by $f(z) = \sum_{n=0}^{\infty} a_n z^n$ for each z in $N(0,r)$ where r is the radius of convergence, prove that f is continuous on $N(0,r)$. **Hint.** Use 4.3.6.

12. Prove if $\lim z_n = z$, then $\lim |z_n| = |z| = |\lim z_n|$. **Hint.** $||z_n| - |z|| \le |z_n - z|$

13. Prove that if $\sum |z_n|$ converges, then

(a) $\sum z_n$ converges and

(b) $\left| \sum z_n \right| \leq \sum |z_n|$.

Hint for (a). If $z_n = x_n + iy_n$, then $|x_n| \leq |z_n|$ and $|y_n| \leq |z_n|$. Hence by the comparison test in calculus, $\sum |x_n|$ and $\sum |y_n|$ converge; and thus $\sum x_n$ and $\sum y_n$ converge. Now use 4.2.2.

Hint for (b). Since $|S_n| = \left| \sum_{k=1}^{n} z_k \right| \leq \sum_{k=1}^{n} |z_k| = T_n$, we have $\left| \sum z_n \right| = |\lim S_n| = \lim |S_n| \leq \lim T_n = \sum |z_n|$. See Ex. 12 and 4.1.2.

14. Prove that if $\sum_{n=1}^{\infty} z_n = S$, then $\sum_{n=m+1}^{\infty} z_n$ converges and

$$\sum_{n=m+1}^{\infty} z_n = S - \sum_{n=1}^{m} z_n.$$

15. Let f be a function on D into R^2 where $D \subset R^2$, let $z_0 \in D$, and let $\{z_n\}$ be a sequence in D such that $z_n \to z_0$ as $n \to \infty$. Prove if f is continuous at z_0, then $f(z_n) \to f(z_0)$ as $n \to \infty$.

Solution. Let $\epsilon > 0$. Since f is continuous at z_0, let $\delta > 0$ such that $f(z) \in N_\epsilon(f(z_0))$ for each z in $D \cap N_\delta(z_0)$. Now let $m \in N$ such that if $n > m$ then $z_n \in N_\delta(z_0)$. Thus if $n > m$, then $f(z_n) \in N_\epsilon(f(z_0))$.

16. Let $\sum_{n=1}^{\infty} z_n = \sum_{n=1}^{\infty}(a_n + ib_n) = a + ib = S$ and let $\sum z_{r_n}$ be a rearrangement of $\sum z_n$. Prove if $\sum |z_n|$ converges, then $\sum_{n=1}^{\infty} z_{r_n} = S$. [See 6.1.11 for a formal definition of a rearrangement.] Here, it is understood that $a_n = \mathcal{R}(z_n)$, $b = \mathcal{I}(S)$, etc. **Hint.** Recall that $|a_n| \leq |z_n|$ implies $\sum |a_n|$ converges. Thus $\sum_{n=1}^{\infty} a_{r_n} = \sum a_n = a$ by 4.2.2 and by the theorem from the calculus on rearrangement of absolutely convergent series of real numbers.

17. Let $\{a_n\}$ be a bounded sequence in R. Prove each of the following.

(a) If $a_n \leq a_{n+1}$ for each n in N, then $\{a_n\}$ converges.

(b) If $a_n \geq a_{n+1}$ for each n in N, then $\{a_n\}$ converges. **Hint.** Use 1.4.12.

Remarks 4.3.11 (The Geometric Series). A very important power series is the **geometric series** $\sum_{n=0}^{\infty} z^n$. By (4.16), the radius of convergence of the geometric series is 1. The nth partial sum S_n is given by

$$S_n = 1 + z + \cdots + z^{n-1} = \frac{1 - z^n}{1 - z} = \frac{1}{1 - z} - \frac{1}{1 - z} z^n \quad \text{if } z \neq 1.$$

Thus $S_n \to \frac{1}{1-z}$ as $n \to \infty$ for $|z| < 1$. Hence if $f(z)$ is the sum of the geometric series, then

$$f(z) = \frac{1}{1 - z} = \sum_{n=0}^{\infty} z^n \quad \text{for } |z| < 1.$$

If $|z| \geq 1$, then the nth term z^{n-1} does not approach zero as $n \to \infty$. Hence the geometric series diverges if $|z| \geq 1$. See the argument in brackets for the case $\rho > 1$ in the proof of 4.2.3.

4.4 Uniform Convergence of Series

In this section, we use uniform convergence to obtain the Laurent series for a function. However, the same result is given in Chapter 6 without the concept of uniform convergence. For a short course, if desired, the rest of this chapter may be omitted.

Throughout this section, for each n in N, we let f_n be a complex function of a complex variable. If for each z in D (where $D \subset R^2$), the series $\sum f_n(z)$ of complex numbers converges to a number $f(z)$, then the function

$$f(z) = \sum f_n(z) \quad \text{for each } z \text{ in } D$$

is defined; and we say that $\sum f_n$ **converges to f pointwise on D**. We shall let

$$S_n(z) = \sum_{k=1}^{n} f_k(z).$$

Definition 4.4.1 (a) The series $\sum f_n$ **converges to f pointwise on D**, denoted by

$$\sum f_n \xrightarrow{P\,D} f,$$

if and only if

$$\sum f_n(z) = f(z) \quad \text{for each } z \text{ in } D.$$

[This condition means that for each fixed z in D and for each $\epsilon > 0$, there is an integer $M(z, \epsilon)$ such that

$$\text{if } n \geq M(z, \epsilon), \text{ then } |S_n(z) - f(z)| < \epsilon.]$$

(b) The series $\sum f_n$ **converges uniformly on D to f** if and only if for each $\epsilon > 0$, there is an integer M_ϵ such that if $n \geq M_\epsilon$, then

$$|S_n(z) - f(z)| < \epsilon \quad \text{for each } z \text{ in } D.$$

We shall denote this by writing

$$\sum f_n \xrightarrow{U\,D} f,$$

Note that in 4.4.1(a), the number $M(z, \epsilon)$ depends upon z as well as upon ϵ; but in (b), the number M_ϵ depends only on ϵ.

Exercises 4.4.2 Prove each of the following statements.

1. If $\sum f_n \xrightarrow{U D} f$, then $\sum f_n \xrightarrow{P D} f$.

2. If $\sum f_n \xrightarrow{U D} f$ and if f_n is continuous on D for each n in N, then f is continuous on D.

Proof of Ex. 2. Let $z_0 \in D$ and let $\epsilon > 0$. By 4.4.1(b), let $M \in N$ such that if $n \geq M$, then for each z in D,

$$|S_n(z) - f(z)| < \frac{\epsilon}{3}.$$

Now $S_M(z)$ is continuous on D since S_M is the sum of a finite number of continuous functions on D. Thus we let $\delta > 0$ such that for each z in $N_\delta(z_0) \cap D$,

$$|S_M(z) - S_M(z_0)| < \frac{\epsilon}{3}.$$

Then for each z in $N_\delta(z_0) \cap D$,

$$
\begin{aligned}
|f(z) - f(z_0)| &\leq |f(z) - S_M(z)| + |S_M(z) - S_M(z_0)| + |S_M(z_0) - f(z_0)| \\
&< \frac{\epsilon}{3} + \frac{\epsilon}{3} + \frac{\epsilon}{3} = \epsilon.
\end{aligned}
$$

Theorem 4.4.3 Let K be a contour and let $f_n \in C(K)$ for each n in N. If $\sum f_n \xrightarrow{U K} f$, then

$$\int_K f(z)\,dz = \sum \int_K f_n(z)\,dz. \tag{4.26}$$

Proof. Let $\epsilon > 0$ and let

$$R_n(z) = f(z) - S_n(z) \quad \text{for each } z \text{ on } K \quad \text{and} \quad \text{for each } n \text{ in } N. \tag{4.27}$$

By 4.4.1(b) and (4.27), let $M \in N$ such that

$$|R_n(z)| < \frac{\epsilon}{L} \quad \text{for each } z \text{ on } K \quad \text{and} \quad \text{for each } n \geq M \tag{4.28}$$

where L is the length of K. By (4.27) and Ex. 2 in 4.4.2, it follows that $R_n \in C(K)$. Now

$$
\begin{aligned}
\left| \int_K f(z)\,dz - \sum_{j=1}^{n} \int_K f_j(z)\,dz \right| &= \left| \int_K f(z)\,dz - \int_K \left[\sum_{j=1}^{n} f_j(z) \right] dz \right| \\
&= \left| \int_K [f(z) - S_n(z)] dz \right| \\
&= \left| \int_K R_n(z)\,dz \right| \quad \text{by (4.27)} \\
&\leq \frac{\epsilon}{L} \cdot L = \epsilon \quad \text{if } n \geq M \quad \text{by (4.28) and 3.4.10(g).}
\end{aligned}
$$

Thus (4.26) holds (by the definition of the sum of a series of complex numbers).

Theorem 4.4.4 (Weierstrass M-test). Let $D \subset R^2$. If $\sum M_n$ is a convergent series of real numbers and if

$$|f_n(z)| \leq M_n \quad \text{for each } n \text{ in } N \quad \text{and} \quad \text{for each } z \text{ in } D, \qquad (4.29)$$

then

$$\sum f_n \xrightarrow{U\,D} f$$

where $f(z)$ is the sum of the given series for each z in D.

Proof. By (4.29) and the comparison test in calculus, the series

$$\sum |f_n(z)| \quad \text{converges} \quad \text{for each } z \text{ in } D.$$

Hence by Ex. 13 in 4.3.10, the series $\sum f_n(z)$ converges for each z in D, and we denote the sum by $f(z)$. To show that $\sum f_n \xrightarrow{U\,D} f$, let $\epsilon > 0$. By 4.2.1(c), let $M \in N$ such that if $n \geq M$, then

$$\sum_{k=n+1}^{\infty} M_k = \sum_{k=1}^{\infty} M_k - \sum_{k=1}^{n} M_k \quad \text{by Ex. 14 in 4.3.10}$$

$$< \epsilon. \qquad (4.30)$$

Thus for each z in D and for each $n \geq M$,

$$|f(z) - S_n(z)| = \left| \sum_{k=n+1}^{\infty} f_k(z) \right|$$

$$\leq \sum_{k=n+1}^{\infty} |f_k(z)| \quad \text{by Ex. 13 in 4.3.10}$$

$$\leq \sum_{k=n+1}^{\infty} M_k \quad \text{by (4.29)}$$

$$< \epsilon \quad \text{by (4.30).}$$

Therefore,

$$\sum f_n \xrightarrow{U\,D} f \quad \text{by 4.4.1(b).}$$

Theorem 4.4.5 Let K be the cco circle $K = \{w : |w - a| = r\}$. If f is continuous on K, then

(a) $\displaystyle\sum_{n=0}^{\infty} \frac{(z - a)^n f(w)}{(w - a)^{n+1}} \xrightarrow{U\,K} \frac{f(w)}{w - z}$ for each fixed z in $I(K)$, and

(b) $\displaystyle\sum_{n=-1}^{-\infty} \frac{(z - a)^n f(w)}{(w - a)^{n+1}} \xrightarrow{U\,K} -\frac{f(w)}{w - z}$ for each fixed z in $E(K)$.

Proof of (a). Let z be a fixed element of $I(K)$. Then for each w on K,

$$
\begin{aligned}
\frac{f(w)}{w-z} &= \frac{f(w)}{(w-a)-(z-a)} \\
&= \frac{f(w)}{w-a} \cdot \frac{1}{1-\frac{z-a}{w-a}} \quad \text{where } \left|\frac{z-a}{w-a}\right| = \rho < 1 \\
&= \frac{f(w)}{w-a} \sum_{n=0}^{\infty}\left(\frac{z-a}{w-a}\right)^n \quad \text{(geometric series)} \\
&= \sum_{n=0}^{\infty} \frac{(z-a)^n\, f(w)}{(w-a)^{n+1}}.
\end{aligned}
\tag{4.31}
$$

By Ex. 2(b) in 2.4.9, let B be such that

$$
|f(w)| \le B \quad \text{for each } w \text{ on } K.
\tag{4.32}
$$

See Ex. 3 in 3.6.18 or 1.4.15. Thus by (4.32), for each w on K,

$$
\left|\frac{f(w)}{w-a}\left(\frac{z-a}{w-a}\right)^n\right| \le \frac{B}{r}\rho^n \quad \text{where } \rho = \left|\frac{z-a}{w-a}\right| < 1.
$$

Now

$$
\sum_{n=0}^{\infty} \frac{B}{r}\rho^n = \frac{B}{r}\cdot\frac{1}{1-\rho}.
$$

Hence by (4.31) and 4.4.4, the series in (a) converges uniformly on K to $\frac{f(w)}{w-z}$ for a fixed z in $I(K)$.

Proof of (b). Let z be a fixed element of $E(K)$. Then for each w on K,

$$
\begin{aligned}
\frac{f(w)}{z-w} &= \frac{f(w)}{(z-a)-(w-a)} \\
&= \frac{f(w)}{z-a} \cdot \frac{1}{1-\frac{w-a}{z-a}} \quad \text{where } \left|\frac{w-a}{z-a}\right| = \rho_1 < 1 \\
&= \frac{f(w)}{z-a} \sum_{n=0}^{\infty}\left(\frac{w-a}{z-a}\right)^n = \sum_{n=1}^{\infty} \frac{f(w)(w-a)^{n-1}}{(z-a)^n} \\
&= \sum_{n=-1}^{-\infty} f(w)\frac{(z-a)^n}{(w-a)^{n+1}} \quad \text{replacing } n \text{ by } -n.
\end{aligned}
\tag{4.33}
$$

Thus by (4.32), for each w on K,

$$
\left|\frac{f(w)}{z-a}\left(\frac{w-a}{z-a}\right)^n\right| \le \frac{B}{|z-a|}\rho_1^n.
$$

Now

$$\sum_{n=0}^{\infty} \frac{B}{|z-a|} \rho_1^n = \frac{B}{|z-a|} \cdot \frac{1}{1-\rho_1}.$$

Hence by (4.33) and 4.4.4, the series in (b) converges uniformly on K to $-\frac{f(w)}{w-z}$ for a fixed z in $E(K)$.

Theorem 4.4.6 Let K be the cco circle $K = \{w : |w-a| = r\}$. If f is continuous on K, then

(a) $\displaystyle\int_K \frac{f(w)\,dw}{w-z} = \sum_{n=0}^{\infty} b_n (z-a)^n$ for each fixed z in $I(K)$, and

(b) $\displaystyle -\int_K \frac{f(w)\,dw}{w-z} = \sum_{n=-1}^{-\infty} b_n(z-a)^n$ for each fixed z in $E(K)$ where

$$b_n = \int_K \frac{f(w)\,dw}{(w-a)^{n+1}} \quad \text{for each } n \text{ in } J. \tag{4.34}$$

Proof of (a). We shall use 4.4.3 with

$$f_n(w) = \frac{(z-a)^n f(w)}{(w-a)^{n+1}} \quad \text{which is continuous on } K. \tag{4.35}$$

We have for each fixed z in $I(K)$,

$$\begin{aligned}
\int_K \frac{f(w)}{w-z}\,dw &= \int_K \left[\sum_{n=0}^{\infty} \frac{(z-a)^n f(w)}{(w-a)^{n+1}}\right] dw \quad \text{by 4.4.5(a)} \\
&= \sum_{n=0}^{\infty} \left(\int_K \frac{f(w)\,dw}{(w-a)^{n+1}}\right)(z-a)^n \quad \text{by 4.4.3} \\
&= \sum_{n=0}^{\infty} b_n (z-a)^n
\end{aligned}$$

where b_n is given in (4.34).

    Proof of (b). We again use 4.4.3 with $f_n(w)$ given by (4.35). For each fixed z in $E(K)$, we have

$$\begin{aligned}
-\int_K \frac{f(w)\,dw}{w-z} &= \int_K \left[\sum_{n=-1}^{-\infty} \frac{(z-a)^n f(w)}{(w-a)^{n+1}}\right] dw \quad \text{by 4.4.5(b)} \\
&= \sum_{n=-1}^{-\infty} \left(\int_K \frac{f(w)\,dw}{(w-a)^{n+1}}\right)(z-a)^n \quad \text{by 4.4.3} \\
&= \sum_{n=-1}^{-\infty} b_n(z-a)^n
\end{aligned}$$

where b_n is given in (4.34).

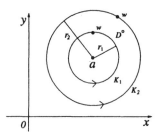

Figure 4.1: $D^\circ = I(K_2) \cap E(K_1)$

Theorem 4.4.7 Let $f \in A(D)$ where D is the **closed annular region** (or **annulus**) given by
$$D = \{z : \; r_1 \le |z - a| \le r_2\} \quad \text{where } 0 < r_1 < r_2.$$
Then for each z in D°,

$$f(z) = \sum_{n=-1}^{-\infty} a_n \, (z-a)^n + \sum_{n=0}^{\infty} a_n \, (z-a)^n \tag{4.36}$$

where

$$a_n = \frac{1}{2\pi i} \int_{K_2} \frac{f(w)\, dw}{(w-a)^{n+1}} \quad \text{for each } n = 0, 1, 2, \cdots \tag{4.37}$$

and

$$a_n = \frac{1}{2\pi i} \int_{K_1} \frac{f(w)\, dw}{(w-a)^{n+1}} \quad \text{for each } n = -1, -2, -3, \cdots, \tag{4.38}$$

and where K_1 and K_2 are the cco circles given by

$$K_1 = \{w : \; |w-a| = r_1\} \quad \text{and} \quad K_2 = \{w : \; |w-a| = r_2\}.$$

Proof. Note that if $z \in D^\circ$, then $z \in I(K_2)$ and $z \in E(K_1)$. See Figure 4.1.
By 3.6.8,

$$\begin{aligned}
f(z) &= \frac{1}{2\pi i} \left(\int_{K_2} \frac{f(w)\, dw}{w-z} - \int_{K_1} \frac{f(w)\, dw}{w-z} \right) \\
&= \frac{1}{2\pi i} \left(\sum_{n=0}^{\infty} b_n \, (z-a)^n + \sum_{n=-1}^{-\infty} b_n \, (z-a)^n \right) \quad \text{by 4.4.6}
\end{aligned} \tag{4.39}$$

where by (4.34),

$$b_n = \begin{cases}
\displaystyle \int_{K_2} \frac{f(w)\, dw}{(w-a)^{n+1}} & \text{if } n \ge 0 \\[3ex]
\displaystyle \int_{K_1} \frac{f(w)\, dw}{(w-a)^{n+1}} & \text{if } n < 0.
\end{cases} \tag{4.40}$$

Now (4.36) – (4.38) follow from (4.39) and (4.40).

Remarks 4.4.8 We use

$$\sum_{n=-\infty}^{\infty} a_n (z - a)^n$$

to denote the right-hand member of (4.36). This doubly infinite series is called the **Laurent series** for f in the annulus D°.

Theorem 4.4.7 has profound consequences which are discussed in detail in Chapter 6.

Theorem 4.4.9 If

$$K = \{z : \ |z - a| = r\} \quad \text{where } r > 0$$

is the circle of convergence of the power series

$$\sum_{n=0}^{\infty} a_n (z - a)^n,$$

then the power series is uniformly convergent on every compact set S which is contained in $I(K)$.

Proof. Let S be some particular nonempty compact subset of $I(K)$. There is a number $r_0 < r$ such that $S \subset N_{r_0}(a) \subset I(K)$.

[For $\mathcal{C} = \{N_\rho(a) : \ 0 < \rho < r\}$ is an **open** cover for S. Since S is compact, there is a finite subcover $\{N_{r_1}(a), N_{r_2}(a), \cdots, N_{r_k}(a)\}$ of $\mathcal{C}$ for S. Let $r_0 = \max\{r_1, r_2, \cdots, r_k\}$.] Thus for each z in S,

$$|a_n(z - a)^n| \leq |a_n| \, r_0^n.$$

Now $z = (a+r_0) \in I(K)$; and hence by 4.3.3(b), $\sum_{n=0}^{\infty} |a_n| |(a+r_0)-a|^n = \sum_{n=0}^{\infty} |a_n| \, r_0^n$ is a convergent series of non-negative real numbers. Thus by 4.4.4, the power series $\sum_{n=0}^{\infty} a_n (z - a)^n$ converges uniformly on S.

Theorem 4.4.10 Let $r \ (> 0)$ be the radius of convergence of the power series

$$\sum_{n=1}^{\infty} a_n (z - a)^n.$$

If

$$K = \{z : \ |z - a| = r_1\} \quad \text{where } r_1 > \frac{1}{r},$$

then

$$\sum_{n=1}^{\infty} a_n (z - a)^{-n} \xrightarrow{U\,D} f(z)$$

where $D = \overline{E(K)}$ (the closure of the exterior of K) and where $f(z)$ is the sum of the latter series.

Proof. Now for each z in D,

$$\left| a_n \left(z - a \right)^{-n} \right| = \frac{|a_n|}{|z - a|^n} \leq \frac{|a_n|}{r_1^n}.$$

By 4.3.2,

$$\sum_{n=1}^{\infty} |a_n| \left(\frac{1}{r_1} \right)^n \quad \text{converges since } \frac{1}{r_1} < r.$$

Thus by 4.4.4,

$$\sum_{n=1}^{\infty} a_n \left(z - a \right)^{-n} \xrightarrow{U\,D} f(z).$$

Exercises 4.4.11

1. Show that the geometric series $\sum_{n=0}^{\infty} z^n$ converges uniformly on any **closed** set contained in the open disk D given by $|z| < 1$.

2. Show that the geometric series $\sum_{n=0}^{\infty} z^n$ does not converge uniformly in the open disk $D = N(0,1)$. **Hint.** If $z = 1 - \frac{1}{n}$, then

$$|S_n(z) - f(z)| = \left| \frac{z^{n+1}}{1 - z} \right| = n \left(1 - \frac{1}{n} \right) \left(1 - \frac{1}{n} \right)^n \rightarrow \infty \quad \text{as } n \rightarrow \infty$$

since

$$\left(1 - \frac{1}{n} \right)^n = \left(\frac{k}{k+1} \right)^{k+1} \quad \text{if } n = k + 1$$

$$= \frac{1}{\left(1 + \frac{1}{k} \right)^k} \cdot \frac{1}{1 + \frac{1}{k}} \rightarrow \frac{1}{e} \quad \text{as } k \rightarrow \infty.$$

If ϵ is a positive number, then the required number M_ϵ in 4.4.1(b), independent of z in D, does not exist.

3. Let $g(w) = \sum_{n=1}^{\infty} a_n w^n$ for $|w| < r$ and let $f(z) = g \left(\frac{1}{z-a} \right) = \sum_{n=1}^{\infty} a_n (z-a)^{-n}$ for each z in $E(C)$ where $C = \left\{ z : \ |z - a| = \frac{1}{r} \right\}$. Prove $f \in A(E(C))$, and $f'(z) = \sum_{n=1}^{\infty} (-n) a_n (z - a)^{-n-1}$ for each z in $E(C)$. **Hint.** Let $w = \frac{1}{z-a}$. By the chain rule, $f'(z) = g'(w) w'(z)$.

4. Prove $\sum_{n=1}^{\infty} a_n (z - a)^{-n}$ converges in $E(K)$ iff $\sum_{n=1}^{\infty} a_n (z - a)^n$ converges in $I(C)$ where $C = \{ z : \ |z - a| = r \}$ and $K = \left\{ z : \ |z - a| = \frac{1}{r} \right\}$. (Draw figures for $r > 1$, $0 < r < 1$, and $r = 1$.)

5. Let $F(z) = \sum_{n=-\infty}^{\infty} b_n (z-a)^n$ for each z in D, where $D = \{ z : \ r_1 < |z-a| < r_2 \}$. Prove $F \in A(D)$, and $F'(z) = \sum_{n=-\infty}^{\infty} n b_n (z - a)^{n-1}$. **Hint.** Let $F(z) = \sum_{n=-1}^{-\infty} b_n (z - a)^n + \sum_{n=0}^{\infty} b_n (z - a)^n$ and use Ex. 3.

6. The function $\zeta(z)$, defined by $\zeta(z) = \sum_{n=1}^{\infty} \frac{1}{n^z}$ for $\mathcal{R}(z) > 1$, (see 5.4.7) is called the **Riemann zeta function** and plays an important role in research in number theory. Prove

$$\sum_{n=1}^{\infty} \frac{1}{n^z} \xrightarrow{U\,H} \zeta(z)$$

where H is the half-plane

$$H = \{z: \ \mathcal{R}(z) \geq p\} \quad \text{with } p > 1.$$

Hint. $\left| \frac{1}{n^z} \right| = \frac{1}{n^{\mathcal{R}(z)}} \leq \frac{1}{n^p}$ for $\mathcal{R}(z) \geq p$ by 5.4.8. Use 4.4.4 where $M_n = \frac{1}{n^p}$ (the nth term of the p-series).

7. Let $g \in C(K)$ where K is a compact set in R^2. Prove if $\sum f_n \xrightarrow{U\,K} f$, then $\sum g f_n \xrightarrow{U\,K} gf$.

Solution. By Ex. 2(b) in 2.4.9, let $M > 0$ such that $|g(z)| \leq M$ on K. Also let S_n and T_n be the nth partial sums of $\sum f_n$ and $\sum g f_n$, respectively. Now let $\epsilon > 0$. By Definition 4.4.1(b) of uniform convergence, let $m \in N$ such that for each $n \geq m$, $|S_n(z) - f(z)| < \frac{\epsilon}{M}$ on K. Then for each $n \geq m$,

$$|T_n(z) - g(z)f(z)| = |g(z)||S_n(z) - f(z)| < \epsilon \quad \text{on } K.$$

Hence $\sum g f_n \xrightarrow{U\,K} gf$.

Theorem 4.4.12 Let $f_n \in A(D)$ for each n in N, where D is a domain, and let

$$\sum_{n=1}^{\infty} f_n \xrightarrow{U\,K} f \qquad \text{for each compact set } K \subset D. \tag{4.41}$$

Then

(a) $f \in A(D)$, and

(b) $\sum_{n=1}^{\infty} f_n' \xrightarrow{U\,K} f'$ for each compact set $K \subset D$.
(Thus $f'(z) = \sum_{n=1}^{\infty} f_n'(z)$ for each z in D; that is, the series may be differentiated termwise in D.)

Proof of (a). Let C be any simple closed contour such that $\overline{I(C)} \subset D$. By Ex. 3 in 3.6.18, we know that C is compact. Thus by 4.4.3,

$$\int_C f(z)\,dz = \sum_{n=1}^{\infty} \int_C f_n(z)\,dz$$
$$= 0 \quad \text{by the Cauchy-Goursat Theorem.}$$

Thus $f \in A(D)$ by Ex. 2 in 3.6.18.

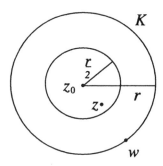

Figure 4.2: $\overline{N_r(z_0)} \subset D$

Proof of (b). Let $z_0 \in D$ and let $r > 0$ such that $\overline{N_r(z_0)} \subset D$. See Figure 4.2. We show that

$$\sum_{n=1}^{\infty} f_n' \xrightarrow{U\ N_{\frac{r}{2}}(z_0)} f'. \tag{4.42}$$

To do this, let $\epsilon > 0$ and let K be the cco boundary of $N_r(z_0)$. Now K is compact. Thus by (4.41), let $M \in N$ such that

$$|f(w) - S_n(w)| < \frac{\epsilon r}{8} \quad \text{for each } w \text{ in } K \quad \text{and} \quad \text{for each } n \geq M,$$

where

$$S_n = \sum_{k=1}^{n} f_k.$$

Now for each z in $N_{\frac{r}{2}}(z_0)$ and for each $n \geq M$,

$$\begin{aligned}
|f'(z) - S_n'(z)| &= \frac{1}{2\pi} \left| \int_K \frac{f(w) - S_n(w)}{(w-z)^2} \, dw \right| \quad \text{by 3.6.9} \\
&\leq \frac{1}{2\pi} \left(\frac{\frac{\epsilon r}{8}}{\frac{r^2}{4}} \right) 2\pi r = \frac{\epsilon}{2} < \epsilon \quad \text{by 3.4.10(g)}.
\end{aligned}$$

Thus (4.42) follows. Hence (b) follows from 4.4.13.

Exercise 4.4.13 Let K be a compact set. Prove if for each z in K, there exists some $r > 0$ such that $\sum_{n=1}^{\infty} f_n \xrightarrow{U\ N_r(z)} f$, then $\sum_{n=1}^{\infty} f_n \xrightarrow{U\ K} f$. **Hints.** Let $D_1, D_2, \cdots, D_n$ be a finite number of open disks which cover K and which are such that $\sum_{n=1}^{\infty} f_n \xrightarrow{U\ D_k} f$ for each $k = 1, 2, \cdots, n$. For a given $\epsilon > 0$, let $M = \max\{M_1, M_2, \cdots, M_n\}$ where $M_k \in N$ such that $|f(z) - \sum_{k=1}^{m} f_k(z)| < \epsilon$ for each z in D_k and for each $m > M_k$.

4.5 Cauchy Condition for Uniform Convergence

Definition 4.5.1 A sequence $\{z_n\}$ in R^2 is called a **Cauchy sequence** iff

$$\text{for each } \epsilon > 0, \text{ there is an integer } M \text{ such that}$$
$$\text{if } m \geq M \text{ and } n \geq M, \text{ then } |z_m - z_n| < \epsilon. \tag{4.43}$$

Theorem 4.5.2 A sequence $\{z_n\}$ in R^2 is convergent (i.e., has a limit in R^2) iff $\{z_n\}$ is a Cauchy sequence.

Proof. First suppose $z_n \to z$ as $n \to \infty$, and let $\epsilon > 0$. Let

$$M \in N \text{ such that } |z_n - z| < \frac{\epsilon}{2} \quad \text{for each } n \geq M.$$

Then for $m \geq M$ and $n \geq M$,

$$|z_m - z_n| \leq |z_m - z| + |z - z_n| < \frac{\epsilon}{2} + \frac{\epsilon}{2} = \epsilon.$$

Hence $\{z_n\}$ is a Cauchy sequence.

Conversely, suppose $\{z_n\}$ is a Cauchy sequence. Then $\{z_n\}$ is bounded. For if it is not, then for each n, there is an integer $m > n$ such that $|z_m| > |z_n| + 1$. Thus

$$|z_m - z_n| \geq ||z_m| - |z_n|| > 1.$$

Hence (4.43) cannot hold.

Now $E = \{z_n : \; n \in N\}$ is either (i) infinite or (ii) finite.

(i) By the Bolzano-Weierstrass Theorem, let z be a limit point of E. Let $\epsilon > 0$. By (4.43), let $M \in N$ such that

$$|z_m - z_n| < \frac{\epsilon}{2} \quad \text{for } m \geq M \text{ and } n \geq M. \tag{4.44}$$

Since z is a limit point of E, the intersection $E \cap N(z, \frac{\epsilon}{2})$ is infinite (by Ex. 4 in 1.4.10). Thus let $k \geq M$ such that

$$|z_k - z| < \frac{\epsilon}{2}. \tag{4.45}$$

By (4.44) and (4.45), for each $m \geq M$,

$$|z_m - z| \leq |z_m - z_k| + |z_k - z| < \frac{\epsilon}{2} + \frac{\epsilon}{2} = \epsilon.$$

Hence $\{z_n\}$ converges to z.

(ii) In this case, the Cauchy condition (4.43) implies that some point is repeated from some index onward, and the sequence will converge to this point.

Theorem 4.5.3 (Cauchy Condition for Uniform Convergence of Series). The series $\sum f_n$ converges uniformly on D iff for each $\epsilon > 0$, there is an integer M such that whenever $m > n > M$,

$$\left| \sum_{k=n+1}^{m} f_k(z) \right| < \epsilon \quad \text{for each } z \text{ in } D. \tag{4.46}$$

Proof. Suppose $\sum f_n \xrightarrow{U D} f$, and let $\epsilon > 0$. By 4.4.1(b), let M be such that whenever $m > n > M$,

$$|S_m(z) - f(z)| < \frac{\epsilon}{2} \quad \text{and} \quad |S_n(z) - f(z)| < \frac{\epsilon}{2} \quad \text{for each } z \text{ in } D.$$

Thus for each z in D,

$$
\begin{aligned}
\left| \sum_{k=n+1}^{m} f_k(z) \right| &= |S_m(z) - S_n(z)| \\
&\leq |S_m(z) - f(z)| + |f(z) - S_n(z)| \\
&< \frac{\epsilon}{2} + \frac{\epsilon}{2} = \epsilon \quad \text{if } m > n > M.
\end{aligned}
$$

Conversely, suppose the Cauchy condition in the theorem holds and let $\epsilon > 0$. Then by this Cauchy condition, we let M be such that (4.46) holds for $m > n > M$. This means that for $m > n > M$,

$$|S_m(z) - S_n(z)| = \left| \sum_{k=n+1}^{m} f_k(z) \right| < \epsilon \quad \text{for each } z \text{ in } D. \tag{4.47}$$

Hence by definition, $\{S_m(z)\}$ is a Cauchy sequence of numbers for each fixed z in D. Thus by 4.5.2,
$$\lim_{m \to \infty} S_m(z) = f(z) \quad \text{for some number } f(z). \tag{4.48}$$

We now show that $\sum f_n \xrightarrow{U D} f$. By the Cauchy condition, we let M^* be such that for each z in D,

$$|S_m(z) - S_n(z)| = \left| \sum_{k=n+1}^{m} f_k(z) \right| < \frac{\epsilon}{2} \quad \text{if } m > n > M^*. \tag{4.49}$$

Thus for a fixed $n > M^*$ and for each z in D,

$$
\begin{aligned}
|f(z) - S_n(z)| &= \left| \lim_{m \to \infty} [S_m(z) - S_n(z)] \right| && \text{by (4.48)} \\
&= \lim_{m \to \infty} |S_m(z) - S_n(z)| && \text{by Ex. 12 in 4.3.10} \\
&\leq \frac{\epsilon}{2} < \epsilon && \text{by (4.49)}.
\end{aligned}
$$

Hence by 4.4.1(b), the series $\sum f_n \xrightarrow{U\,D} f$.

We now use the Cauchy condition to prove the Weierstrass M-test.

Alternate Proof of Theorem 4.4.4 (Weierstrass M-test). Let $\epsilon > 0$. By 4.5.2, the partial sums of $\sum M_n$ form a Cauchy sequence. Thus let M be such that for each $m > n > M$,

$$\sum_{k=n+1}^{m} M_k < \epsilon.$$

By (4.29), for each z in D,

$$\left| \sum_{k=n+1}^{m} f_k(z) \right| \leq \sum_{k=n+1}^{m} |f_k(z)| \leq \sum_{k=n+1}^{m} M_k < \epsilon.$$

Hence $\sum f_n \xrightarrow{U\,D} f$ by 4.5.3.

Exercise 4.5.4 Let h be piecewise continuous on the cco circle $C = \{t : |t| = r_0\}$. Prove $g \in A(I(C))$ if

$$g(z) = \int_C \frac{t+z}{t-z} h(t)\, dt \quad \text{for each } z \text{ in } I(C).$$

Proof. If z is a point in $I(C)$ and t is a point on the circle C, then $\left| \frac{z}{t} \right| < 1$. Thus if z is in $I(C)$ and t is on C, then

$$\begin{aligned}
\frac{t+z}{t-z} &= 1 + \frac{2z}{t-z} = 1 + \frac{2z}{t} \frac{1}{1-\frac{z}{t}} \\
&= 1 + \frac{2z}{t} \sum_{n=0}^{\infty} \frac{z^n}{t^n} \\
&= 1 + 2 \sum_{n=1}^{\infty} \frac{z^n}{t^n}.
\end{aligned}$$

Now let $B > 0$ such that $\left| h(t) \frac{z^n}{t^n} \right| \leq B \left| \frac{z}{t} \right|^n$ where $r = |z| < r_0 = |t|$. For a fixed z in $I(C)$, let $M_n = B \left(\frac{r}{r_0} \right)^n$ and use 4.4.4. Then apply 4.4.3. Thus

$$\begin{aligned}
g(z) &= \int_C h(t)\, dt + 2 \sum_{n=1}^{\infty} z^n \int_C t^{-n} h(t)\, dt \\
&= \sum_{n=0}^{\infty} a_n z^n.
\end{aligned}$$

Hence $g \in A(I(C))$ by 4.3.6.

Definition 4.5.5 Let $\{x_n\}$ be a sequence in R and let $K \in R$. Then K is called the **lower limit** (or **limit inferior**) of $\{x_n\}$ iff for each $\epsilon > 0$,

(i) $x_n > K - \epsilon$ for all but a finite number of values of n, and

(ii) $x_n < K + \epsilon$ for infinitely many values of n. We denote the lower limit of $\{x_n\}$ by $\underline{\lim} \, x_n$. Also, we define

$$\underline{\lim} \, x_n = -\infty \text{ iff for each real number } M, \text{ we have } x_n < M \text{ for some}$$
n.

Finally,

$$\underline{\lim} \, x_n = +\infty \text{ iff for each real number } M, \text{ we have } x_n > M \text{ for all}$$
but a finite number of values of n.

Exercises 4.5.6

1. Find $\underline{\lim} \, x_n$ of each sequence in 4.2.5(d).
   Ans. (i) 0   Ans. (ii) $-\infty$ Ans. (iii) -1 Ans. (iv) $-\infty$ Ans. (v) $-\infty$

2. Give a sequence $\{x_n\}$ such that $\underline{\lim} \, x_n = +\infty$.

In Ex. 3 – 10, let $\{x_n\}$ and $\{y_n\}$ be sequences in R and prove the stated results.

3. $\overline{\lim} \, x_n$ exists and is unique.

4. $\underline{\lim} \, x_n \leq \overline{\lim} \, x_n$. **Hints.** Let $K = \underline{\lim} \, x_n$ and let $L = \overline{\lim} \, x_n$. If $K = -\infty$ or $L = +\infty$, then clearly $K \leq L$. Also, if $L = -\infty$ or $K = +\infty$, then $K = L$. Finally, suppose $K, L \in R$. Assume $K > L$ and let $\epsilon = \frac{K-L}{2}$. By 4.2.4(i), we have $x_n < L + \epsilon = \frac{L+K}{2}$ for all but a finite number of values of n. But by 4.5.5(i), we have $x_n > K - \epsilon = \frac{L+K}{2}$ for all but a finite number of values of n, a contradiction.

5. $\lim x_n = a$ iff $\underline{\lim} \, x_n = a = \overline{\lim} \, x_n$ where $a \in R \cup \{+\infty, -\infty\}$. **Hint.** If $a \in R$, compare (i) and (ii) in 4.2.4, Ex. 1 of 4.2.7, and (i) and (ii) in 4.5.5, and interpret these conditions with the aid of a figure.

6. $\overline{\lim} \, x_n = -\infty$ iff $\lim x_n = -\infty$. (Use Ex. 4 and 5.)

7. $\underline{\lim} \, x_n = +\infty$ iff $\lim x_n = +\infty$. (Use Ex. 4 and 5.)

8. If for each $\epsilon > 0$, there is an integer m such that $x_n < x + \epsilon$ for each $n \geq m$, then $\overline{\lim} \, x_n \leq x$. **Hint.** Let $\overline{\lim} \, x_n = L \in R$ and assume $L > x$. Let $\epsilon = \frac{L-x}{2}$. By 4.2.4(ii), $x_n > L - \epsilon = x + \epsilon$ for infinitely many values of n, which contradicts the hypotheses.

9. If there is an integer m such that $x_n \leq y_n$ for each $n > m$, then $\underline{\lim} \, x_n \leq \underline{\lim} \, y_n$ and $\overline{\lim} \, x_n \leq \overline{\lim} \, y_n$. (Make use of Ex. 8.)

10. If $\{x_n\}$ and $\{y_n\}$ are bounded, then $\overline{\lim}\,(x_n + y_n) \leq \overline{\lim}\,x_n + \overline{\lim}\,y_n$ and $\underline{\lim}\,(x_n + y_n) \geq \underline{\lim}\,x_n + \underline{\lim}\,y_n$.

11. **Definition.** Let $\{z_n\}$ be a sequence in R^2 and let $\{n_k\}$ be a sequence in N such that $n_k < n_{k+1}$ for each k in N. Then the sequence $\{z_{n_k}\}$ is called a subsequence of $\{z_n\}$.
    Prove $\lim z_n = a$ iff $\lim_{k\to\infty} z_{n_k} = a$ for each subsequence $\{z_{n_k}\}$ of $\{z_n\}$.

12. Prove if $\{x_n\}$ is a sequence in R such that $\overline{\lim}\,x_n = L$, then

    (i) there is a subsequence $\{x_{n_k}\}$ of $\{x_n\}$ such that $\lim_{k\to\infty} x_{n_k} = L$, but

    (ii) if $M > L$, then there is no subsequence of $\{x_n\}$ whose limit is M.

    Remark. Thus, for a given sequence $\{x_n\}$ of real numbers, if $A = \{x : x \text{ is the limit of a subsequence of } \{x_n\}\}$, then $\overline{\lim}\,x_n = \max A$, that is, $\overline{\lim}\,x_n$ is the largest member of A.

13. Prove if $c > 0$, then $\lim \sqrt[n]{c} = 1$. **Hint.** $c^{\frac{1}{n}} = e^{\frac{1}{n}\text{Log}\,c}$.

14. Prove if $a_n > 0$ for each n in N, then

$$\underline{\lim}\,\frac{a_{n+1}}{a_n} \leq \underline{\lim}\,\sqrt[n]{a_n} \leq \overline{\lim}\,\sqrt[n]{a_n} \leq \overline{\lim}\,\frac{a_{n+1}}{a_n}.$$

15. Prove the **General Ratio Test**:
    The series $\sum a_n$ of nonzero complex numbers

    (a) converges absolutely if $\overline{\lim}\left|\frac{a_{n+1}}{a_n}\right| < 1$,

    (b) diverges if $\underline{\lim}\left|\frac{a_{n+1}}{a_n}\right| > 1$,

    (c) may converge or may diverge if $\underline{\lim}\left|\frac{a_{n+1}}{a_n}\right| \leq 1 \leq \overline{\lim}\left|\frac{a_{n+1}}{a_n}\right|$.

16. For each n in N, let

$$a_n = \begin{cases} \left(\dfrac{1}{2}\right)^n & \text{if } n \text{ is odd} \\[2mm] \left(\dfrac{1}{3}\right)^n & \text{if } n \text{ is even}. \end{cases}$$

Prove each of the following results.

(a) $\overline{\lim}\,\sqrt[n]{a_n} = \frac{1}{2}$ (Use the remark in Ex. 12.)

(b) $\underline{\lim}\,\sqrt[n]{a_n} = \frac{1}{3}$

(c) $\underline{\lim}\left|\frac{a_{n+1}}{a_n}\right| = 0$ and $\overline{\lim}\left|\frac{a_{n+1}}{a_n}\right| = +\infty$

(d) The series $\sum_{n=1}^{\infty} a_n$ converges.

Remarks. Observe that the root test shows that this series converges, but the ratio test fails. In general, Ex. 14 gives the following two equivalent results.

If the general ratio test shows convergence (or divergence) of a given series, then so does the root test. If the root test is inconclusive, then so is the general ratio test.

Thus (in view of the present example) the root test is stronger than the general ratio test. However, the ratio test is easier to apply in many examples.

17. Prove the series $\sum a_n$ of nonzero complex numbers diverges if $\left|\frac{a_{n+1}}{a_n}\right| \geq 1$ for each $n \geq M$ for some M in N.

   Proof. $\left|\frac{a_{n+1}}{a_n}\right| \geq 1$ for each $n \geq M$ implies a_n does not approach 0 as $n \to \infty$. See the argument in brackets in the proof of 4.2.3.

18. Use the result of Ex. 17 to show that $\sum_{n=1}^{\infty} n$ diverges. Also show that the root test and the general ratio test fail for this series. (See Ex. 5.)

Solution of Ex. 3. If $\{x_n\}$ is not bounded above, then clearly by (a), (b), and (c) of 4.2.4, it follows that $+\infty$ is the unique upper limit of $\{x_n\}$.

Thus suppose $x_n < M$ for each n in N where $M \in R$. Let

$$A = \{x : \quad x_n \leq x \text{ for all but a finite number of values of } n\}. \qquad (4.50)$$

Now $A \neq \emptyset$, since $M \in A$. Hence we let

$$L = \text{glb } A, \qquad (4.51)$$

where $L = -\infty$ if A is not bounded below. (If A is bounded below, then $L \in R$ by 1.4.12.) We now prove $L = \overline{\lim} \, x_n$. (If $L = -\infty$, then $\lim x_n = -\infty$ and $-\infty$ is the unique upper limit of $\{x_n\}$ by (a), (b), and (c) of 4.2.4.) Next suppose $L \neq -\infty$. Then $L \in R$. To verify (i) and (ii) in 4.2.4(a), let $\epsilon > 0$. By (4.51), $L + \epsilon$ is not a lower bound of A. Thus let $y \in A$ such that $y < L + \epsilon$. By (4.50), $x_n \leq y < L + \epsilon$ for all but a finite number of values of n. Thus 4.2.4(i) holds. Also by (4.51), $(L - \epsilon) \notin A$, and hence 4.2.4(ii) holds. Thus L is an upper limit of $\{x_n\}$. Now suppose $L' = \overline{\lim} \, x_n$ and $L' \neq L$. Without loss of generality, we suppose $L' < L$ and let $\epsilon = \frac{L - L'}{2} > 0$. Then $L' + \epsilon = \frac{L' + L}{2} = L - \epsilon$. By 4.2.4(i), $x_n < L' + \epsilon = \frac{L' + L}{2}$ for all but a finite number of values of n. But by 4.2.4(ii), $x_n > L - \epsilon = \frac{L' + L}{2}$ for infinitely many values of n, which is a contradiction. Hence $\overline{\lim} \, x_n$ is unique.

Solution of Ex. 10. Let $A = \overline{\lim} \, x_n$, $B = \overline{\lim} \, y_n$ and let $\epsilon > 0$. By 4.2.4(i), there is an integer m such that for each $n \geq m$, $x_n < A + \frac{\epsilon}{2}$, and $y_n < B + \frac{\epsilon}{2}$. Thus $x_n + y_n < (A + B) + \epsilon$ for each $n \geq m$. Hence $\overline{\lim} \, (x_n + y_n) \leq A + B$ by Ex. 8. The other part is proved similarly.

Solution of Ex. 11. Suppose $\lim_{k\to\infty} z_{n_k} = a$ for each subsequence $\{z_{n_k}\}$ of $\{z_n\}$. Since $\{z_n\}$ is a subsequence of $\{z_n\}$, it follows that $\lim_{n\to\infty} z_n = a$.

Conversely, suppose $\lim z_n = a$ and let $\{z_{n_k}\}$ be a subsequence of $\{z_n\}$. Let $\epsilon > 0$. Let M be such that $|z_n - a| < \epsilon$ for each $n \geq M$. Now $n_1 \geq 1$ and $n_{k+1} > n_k$ implies that $n_k \geq k$ for each k in N. Thus $|z_{n_k} - a| < \epsilon$ for each $k \geq M$. Hence $\lim_{k\to\infty} z_{n_k} = a$.

Solution of Ex. 12. If $L = +\infty$, then $\{x_n\}$ is not bounded above and clearly there is a subsequence $\{x_{n_k}\}$ of $\{x_n\}$ such that $\lim_{k\to\infty} x_{n_k} = +\infty$; and there exists no $M > L$. If $L = -\infty$, see Ex. 6 and observe that $\lim_{k\to\infty} x_{n_k} = -\infty$ for each subsequence.

Now suppose $L \in R$. By (i) and (ii) of 4.2.4, for each k in N, let $n_k \in N$ such that $L - \frac{1}{k} < x_{n_k} < L + \frac{1}{k}$ and such that $n_{k+1} > n_k$. Then $\{x_{n_k}\}$ is a subsequence of $\{x_n\}$ and clearly $\lim_{k\to\infty} x_{n_k} = L$. Finally let $M > L$. If $M \in R$, let $\epsilon = \frac{M-L}{2}$. By 4.2.4(i), $x_n < L + \epsilon = \frac{L+M}{2} = M - \epsilon$ for all but a finite number of values of n. Thus no subsequence of $\{x_n\}$ can converge to M. But if $M = +\infty$, it is obvious that no subsequence of $\{x_n\}$ has M as a limit.

Solution of Ex. 14. Let $L = \overline{\lim} \frac{a_{n+1}}{a_n}$. If $L = +\infty$, then $\overline{\lim} \sqrt[n]{a_n} \leq L$. Now suppose $L \in R$, and let $p \in R$ such that $L < p$. By 4.2.4(i), let $M \in N$ such that $\frac{a_{n+1}}{a_n} < p$ for each $n \geq M$. Then

$$\frac{a_{M+1}}{a_M} \cdot \frac{a_{M+2}}{a_{M+1}} \cdots \frac{a_{M+k}}{a_{M+k-1}} < p^k.$$

That is,

$$a_{M+k} < a_M p^k \quad \text{for each } k \text{ in } N.$$

Putting $n = M + k$ for each k in N, we have

$$a_n < a_M \, p^{n-M} = a_M \, p^{-M} \, p^n \quad \text{for each } n > M.$$

Thus for each $n > M$, $\sqrt[n]{a_n} < p \sqrt[n]{a_M \, p^{-M}}$. Hence by Ex. 13, 4.1.2, and Ex. 5 and 9,

$$\overline{\lim} \sqrt[n]{a_n} \leq p. \tag{4.52}$$

Since (4.52) holds for each $p > L$, it follows that $\overline{\lim} \sqrt[n]{a_n} \leq L$.

The proof of the inequality

$$\underline{\lim} \frac{a_{n+1}}{a_n} \leq \underline{\lim} \sqrt[n]{a_n}$$

is similiar.

Solution of Ex. 15.

(a) If $\overline{\lim} \left|\frac{a_{n+1}}{a_n}\right| < 1$, then $\overline{\lim} \sqrt[n]{|a_n|} < 1$ by Ex. 14. Hence the series converges absolutely by 4.2.6.

(b) Suppose $\varlimsup \left|\frac{a_{n+1}}{a_n}\right| > 1$. Then by Ex. 14, $\varlimsup \sqrt[n]{|a_n|} > 1$. Hence the series diverges by 4.2.6.

(c) See the proof of the third part of 4.2.6.

Remarks 4.5.7 If, in the general ratio test in Ex. 15, we replace Part (b) by the result in Ex. 17, then we obtain a test, which we may call the "strong ratio test." The hypothesis in Part (b) implies the hypothesis in Ex. 17, but not conversely. (For suppose $\varlimsup \left|\frac{a_{n+1}}{a_n}\right| = K > 1$, and take $\epsilon = \frac{K-1}{2}$. By the definition of $\varlimsup x_n$ in 4.5.5, we have $\left|\frac{a_{n+1}}{a_n}\right| > K - \epsilon \geq 1$ for each $n \geq M$ for some M.) Now Ex. 14 implies

(i) if the general ratio test shows convergence for a given series, then so does the root test.

Exercise 16 shows that

(ii) the root test verifies convergence for at least one series when the general ratio test fails.

α. For convergence, the root test is stronger than the general ratio test.

β. For divergence, the root test fails for one series when the strong ratio test is conclusive.

Chapter 5

Some Elementary Functions

We have considered such elementary functions as polynomials and rational functions. In this chapter, we study such elementary functions as the exponential, logarithmic, trigonometric, and hyperbolic functions.

5.1 The Exponential Function

Theorem 5.1.1 The radius of convergence of the power series

$$\sum_{n=0}^{\infty} \frac{z^n}{n!}$$

is ∞.

Proof. To apply the ratio test, we observe that

$$
\begin{aligned}
\lim \left| \frac{a_{n+1}}{a_n} \right| &= \lim \frac{|z|^{n+1}}{(n+1)!} \cdot \frac{n!}{|z|^n} \\
&= \lim \frac{|z|}{n+1} \\
&= 0.
\end{aligned}
$$

Thus by the ratio test, $r = \infty$. Hence the power series in 5.1.1 defines a function for each z in R^2. By 4.3.6, the function is analytic in R^2. By 3.6.9, all of its derivatives are analytic in R^2. This function reduces to

$$\sum_{n=0}^{\infty} \frac{x^n}{n!} \quad \text{when } z = x \in R$$

which is the Maclaurin's series representation of e^x. Also, e^z will be seen to satisfy the same basic laws as e^x. Thus it seems reasonable to adopt the following definition.

127

Definition 5.1.2

$$e^z = \sum_{n=0}^{\infty} \frac{z^n}{n!} \quad \text{for each } z \text{ in } R^2. \tag{5.1}$$

Theorem 5.1.3 The function e^z is an entire function, and

$$\frac{d}{dz} e^z = e^z \quad \text{on } R^2. \tag{5.2}$$

Proof. By 4.3.5,

$$\frac{d}{dz} e^z = \frac{d}{dz}\left(1 + \sum_{n=1}^{\infty} \frac{z^n}{n!}\right) = \sum_{n=1}^{\infty} \frac{nz^{n-1}}{n!}$$

$$= \sum_{n=1}^{\infty} \frac{z^{n-1}}{(n-1)!} = \sum_{n=0}^{\infty} \frac{z^n}{n!} = e^z.$$

Lemma 5.1.4 If z and z^* are complex numbers, then

$$e^z e^{z^*-z} = e^{z^*}. \tag{5.3}$$

Proof. For a fixed z^* in R^2, let

$$f(z) = e^z e^{z^*-z} \quad \text{for each } z \text{ in } R^2.$$

By (5.2),

$$f'(z) = e^z(-e^{z^*-z}) + e^{z^*-z} e^z = 0 \quad \text{for each } z \text{ in } R^2.$$

Therefore by 2.3.15(a),

$$f(z) = e^z e^{z^*-z} = C \quad \text{for each } z \text{ in } R^2 \text{ where } C \text{ is a constant.} \tag{5.4}$$

For $z = 0$,

$$f(0) = e^{z^*} = C. \tag{5.5}$$

By (5.4) and (5.5), we obtain (5.3).

Theorem 5.1.5 (The Addition Theorem). If z_1 and z_2 are complex numbers, then

$$e^{z_1} e^{z_2} = e^{z_1+z_2}. \tag{5.6}$$

Proof. In (5.3), let $z = z_1$ and $z^* = z_1 + z_2$.

Remark 5.1.6 $e^z e^{-z} = e^{z-z} = e^0 = 1$ which implies

$$e^{-z} = \frac{1}{e^z} \quad \text{for each } z \text{ in } R^2.$$

5.2 The Trigonometric Functions

Theorem 5.2.1 For each of the following series, the radius of convergence is ∞ :

$$\sum_{m=0}^{\infty}(-1)^m \frac{z^{2m}}{(2m)!} \quad \text{and} \quad \sum_{m=0}^{\infty}(-1)^m \frac{z^{2m+1}}{(2m+1)!}. \tag{5.7}$$

Proof. If $z \in R^2$ such that $z \neq 0$, then

$$\lim \left| \frac{z^{2m+2}}{(2m+2)!} \cdot \frac{(2m)!}{z^{2m}} \right| = \lim \frac{|z^2|}{(2m+2)(2m+1)} = 0.$$

Thus by the ratio test, 4.2.3, the first series converges for each $z \neq 0$ in R^2. But clearly the series converges for $z = 0$.

The proof for the second series is similar.

Remark 5.2.2 Thus the two power series in 5.2.1 define functions on R^2. By 4.3.6 and by 3.6.9, these functions and all of their derivatives are analytic in R^2. Since the power series in 5.2.1 reduce, respectively, to the Maclaurin series for $\cos x$ and $\sin x$ when $z = x \in R$, it seems reasonable to adopt the following definition.

Definition 5.2.3 For each z in R^2,

$$\cos z = \sum_{m=0}^{\infty}(-1)^m \frac{z^{2m}}{(2m)!} \quad \text{and} \quad \sin z = \sum_{m=0}^{\infty}(-1)^m \frac{z^{2m+1}}{(2m+1)!}. \tag{5.8}$$

Remark 5.2.4 By (5.8), for each z in R^2,

$$\cos(-z) = \cos z \quad \text{and} \quad \sin(-z) = -\sin z.$$

Thus $\cos z$ is an **even** function and $\sin z$ is an **odd** function.

Euler's Formulas 5.2.5 For each z in R^2,

$$e^{iz} = \cos z + i \sin z \quad \text{and} \quad e^{-iz} = \cos z - i \sin z. \tag{5.9}$$

Proof. By 5.1.2, we have

$$
\begin{aligned}
e^{iz} &= \sum_{n=0}^{\infty} \frac{(iz)^n}{n!} = 1 + iz + \frac{(iz)^2}{2!} + \frac{(iz)^3}{3!} + \cdots \\
&= [1 + iz] + \left[\frac{(iz)^2}{2!} + \frac{(iz)^3}{3!} \right] + \left[\frac{(iz)^4}{4!} + \frac{(iz)^5}{5!} \right] + \cdots \quad \text{by Ex. 16 in 5.2.11} \\
&= \sum_{m=0}^{\infty} \frac{(iz)^{2m}}{(2m)!} + \sum_{m=0}^{\infty} \frac{(iz)^{2m+1}}{(2m+1)!} \quad \text{by 4.3.9} \\
&= \sum_{m=0}^{\infty} (i^2)^m \frac{z^{2m}}{(2m)!} + \sum_{m=0}^{\infty} i(i^2)^m \frac{z^{2m+1}}{(2m+1)!} \\
&= \sum_{m=0}^{\infty} (-1)^m \frac{z^{2m}}{(2m)!} + i \sum_{m=0}^{\infty} (-1)^m \frac{z^{2m+1}}{(2m+1)!} \\
&= \cos z + i \sin z \quad \text{by 5.2.3.}
\end{aligned}
$$

This proves the first part of (5.9). To prove the second part, we have

$$e^{-iz} = \cos(-z) + i\sin(-z) \quad \text{by the first part of (5.9)}$$
$$= \cos z - i\sin z \qquad\qquad \text{by 5.2.4.}$$

Remarks 5.2.6 (a) If we take $z = x$ in 5.2.5 where x is real, we obtain the result that $e^{ix} = \cos x + i\sin x = \operatorname{cis} x$.

(b) The function e^z, for each $z = (x + iy)$ in R^2, can be expressed as

$$e^z = e^x(\cos y + i\sin y) = e^x e^{iy}.$$

For

$$e^z = e^{x+iy} = e^x e^{iy} \qquad \text{by 5.1.5}$$
$$= e^x(\cos y + i\sin y) \quad \text{by (a).}$$

Remarks 5.2.7 (a) From 5.2.6(a), we obtain the following important results which involve the special numbers e, π, $\pm i$, and ± 1.

$$e^{2\pi i} = 1 \qquad e^{\pi i} = -1 \qquad e^{\frac{\pi}{2}i} = i \qquad e^{-\frac{\pi}{2}i} = -i$$

(b) Also by 5.2.6(a), for each real number y,

$$|e^{iy}| = |\cos y + i\sin y| = \sqrt{\cos^2 y + \sin^2 y} = 1.$$

(c) For each $z = x + iy$ where x and y are real, we have

$$|e^z| = |e^x e^{iy}| \quad \text{by 5.2.6(b)}$$
$$= |e^x||e^{iy}|$$
$$= e^x \cdot 1 \quad \text{by (b)}$$
$$= e^{\mathcal{R}(z)}.$$

(d) For each $z = x + iy$ where x and y are real, we have

$$\arg e^z = \arg\left[e^x(\cos y + i\sin y)\right] = y = \mathcal{I}(z).$$

(e) For each integer k,
$$e^{z+2k\pi i} = e^z \cdot e^{2k\pi i} = e^z.$$

This means that the function e^z is periodic with period $2\pi i$.

(f) We next prove

$$\text{if } e^{z_1} = e^{z_2}, \text{ then } z_2 = z_1 + 2k\pi i \quad \text{for some integer } k.$$

To prove this, suppose that

$$e^{z_1} = e^{z_2}.$$

Multiplying both sides by e^{-z_1}, we have

$$\begin{aligned}
1 = e^0 = e^{z_1-z_1} &= e^{z_2-z_1} \\
&= e^{x_2-x_1}\left[\cos(y_2 - y_1) + i\sin(y_2 - y_1)\right].
\end{aligned}$$

Thus $x_2 - x_1 = 0$ and $y_2 - y_1 = 2k\pi$ for some integer k. Hence

$$z_2 - z_1 = (x_2 - x_1) + i(y_2 - y_1) = 0 + 2k\pi i = 2k\pi i.$$

Therefore, $z_2 = z_1 + 2k\pi i$.

(g) Solving the two equations in 5.2.5 simultaneously for $\cos z$, we obtain

$$\cos z = \frac{e^{iz} + e^{-iz}}{2}.$$

Solving the same two equations for the $\sin z$, we have

$$\sin z = \frac{e^{iz} - e^{-iz}}{2i}.$$

Theorem 5.2.8 (Addition Theorems). The following identities hold for each pair of complex numbers z_1 and z_2.

$$\begin{aligned}
\cos(z_1 + z_2) &= \cos z_1 \cos z_2 - \sin z_1 \sin z_2 \\
\sin(z_1 + z_2) &= \sin z_1 \cos z_2 + \cos z_1 \sin z_2
\end{aligned} \tag{5.10}$$

Proof. From 5.2.7(g) and 5.1.5, we obtain the following two identities.

$$\cos(z_1 + z_2) = \frac{e^{iz_1}e^{iz_2} + e^{-iz_1}e^{-iz_2}}{2}$$

$$\sin(z_1 + z_2) = \frac{e^{iz_1}e^{iz_2} - e^{-iz_1}e^{-iz_2}}{2i} \tag{5.11}$$

By 5.2.5,

$$e^{iz_1}e^{iz_2} = (\cos z_1 + i\sin z_1)(\cos z_2 + i\sin z_2).$$

Hence we have the next two identities.

$$\begin{aligned}
e^{iz_1}e^{iz_2} &= (\cos z_1 \cos z_2 - \sin z_1 \sin z_2) + i(\sin z_1 \cos z_2 + \cos z_1 \sin z_2) \\
e^{-iz_1}e^{-iz_2} &= (\cos z_1 \cos z_2 - \sin z_1 \sin z_2) - i(\sin z_1 \cos z_2 + \cos z_1 \sin z_2)
\end{aligned}$$

Adding and subtracting in the last two equations and using (5.11), we obtain the formulas for $\cos(z_1 + z_2)$ and $\sin(z_1 + z_2)$.

Theorem 5.2.9 For each $z = (x, y)$ in R^2,

$$\begin{aligned}
\cos z &= \cos(x + iy) &= \cos x \ \cosh y - i \sin x \ \sinh y \\
\sin z &= \sin(x + iy) &= \sin x \ \cosh y + i \cos x \ \sinh y.
\end{aligned} \tag{5.12}$$

Formulas (5.12) separate the real and imaginary parts of $\cos z$ and $\sin z$ so that they may be evaluated by tables of trigonometric and hyperbolic functions.

Proof. Taking $z_1 = x$ and $z_2 = iy$ in 5.2.8, we obtain for real numbers x and y,

$$\begin{aligned}
\cos(x + iy) &= \cos x \ \cos iy - \sin x \ \sin iy \\
\sin(x + iy) &= \sin x \ \cos iy + \cos x \ \sin iy.
\end{aligned} \tag{5.13}$$

By the definition of $\cosh y$ and $\sinh y$, and by taking $z = iy$ in 5.2.7(g), we obtain

$$\cos iy = \frac{e^y + e^{-y}}{2} = \cosh y, \quad \text{and}$$

$$\sin iy = \frac{e^{-y} - e^y}{2i} = i \frac{e^y - e^{-y}}{2} = i \sinh y. \tag{5.14}$$

Thus (5.12) follows from (5.13) and (5.14).

Theorem 5.2.10 The functions $\cos z$ and $\sin z$ are entire functions. Their derivatives are given by

$$\frac{d}{dz} \cos z = -\sin z \quad \text{and} \quad \frac{d}{dz} \sin z = \cos z \quad \text{on } R^2.$$

Proof. By 4.3.5 and 5.2.3, the derivative of $\cos z$ exists for each z in R^2 and is given by the following formula.

$$\begin{aligned}
\frac{d}{dz} \cos z &= \frac{d}{dz} \left(\sum_{m=0}^{\infty} (-1)^m \frac{z^{2m}}{(2m)!} \right) = \sum_{m=1}^{\infty} (-1)^m \frac{2m \, z^{2m-1}}{(2m)!} \\
&= \sum_{m=1}^{\infty} (-1)^m \frac{z^{2m-1}}{(2m-1)!} = -\sum_{m=0}^{\infty} (-1)^m \frac{z^{2m+1}}{(2m+1)!} \\
&= -\sin z
\end{aligned}$$

Likewise, the derivative of $\sin z$ exists for each z in R^2 and is given by our next formula.

$$\begin{aligned}
\frac{d}{dz} \sin z &= \frac{d}{dz} \left(\sum_{m=0}^{\infty} (-1)^m \frac{z^{2m+1}}{(2m+1)!} \right) = e \sum_{m=0}^{\infty} (-1)^m \frac{2m+1}{(2m+1)!} z^{2m} \\
&= \sum_{m=0}^{\infty} (-1)^m \frac{z^{2m}}{(2m)!} = \cos z
\end{aligned}$$

Exercises 5.2.11

1. Prove each of the following identities holds for all values of z in R^2.
   (a) $\sin^2 z + \cos^2 z = 1$ (b) $\sin(-z) = -\sin z$
   (c) $\cos(-z) = \cos z$ (d) $\sin 2z = 2 \sin z \cos z$
   (e) $\cos 2z = \cos^2 z - \sin^2 z$

2. Prove all of the zeros of $\sin z$ and $\cos z$ are real, and they are given respectively by
$$z = n\pi \quad \text{and by} \quad z = (2n+1)\frac{\pi}{2} \quad \text{for each } n \text{ in } J.$$
   Hint. Use 5.2.9 and the definitions of $\sinh y$ and $\cosh y$.

3. Let E be the set of all complex numbers which are not **odd** integral multiples of $\frac{\pi}{2}$, and let G be the set of all complex numbers which are not integral multiples of π. We define $\tan z$, $\cot z$, $\sec z$, and $\csc z$ as follows

$$\tan z = \frac{\sin z}{\cos z} \qquad \sec z = \frac{1}{\cos z} \qquad \text{for each } z \text{ in } E.$$

$$\cot z = \frac{\cos z}{\sin z} \qquad \csc z = \frac{1}{\sin z} \qquad \text{for each } z \text{ in } G.$$

   Prove each of the following formulas.

   (a) $1 + \tan^2 z = \sec^2 z$ on E

   (b) $1 + \cot^2 z = \csc^2 z$ on G

   (c) $\dfrac{d}{dz} \tan z = \sec^2 z$ on E

   (d) $\dfrac{d}{dz} \sec z = \sec z \, \tan z$ on E

   (e) $\dfrac{d}{dz} \cot z = -\csc^2 z$ on G

   (f) $\dfrac{d}{dz} \csc z = -\csc z \, \cot z$ on G

4. Prove each of the following holds for each z in R^2.

   (a) $|\sin z|^2 = \sin^2 x + \sinh^2 y = \cosh^2 y - \cos^2 x$

   (b) $|\cos z|^2 = \cos^2 x + \sinh^2 y$

   Hint. For Parts (a) and (b), use 5.2.9 and $\cosh^2 y - \sinh^2 y = 1$.

In Ex. 5 and 6, prove each inequality holds for all complex numbers $z = x + iy$, where x and y are real.

5. $|\sinh y| \le |\sin z| \le \cosh y$

6. $|\sinh y| \le |\cos z| \le \cosh y$

7. Find all values of z such that $\sin z = 4$.

   Hint. Use 5.2.9. Ans. $z = \dfrac{4n+1}{2}\pi + i\cosh^{-1}4 = \dfrac{4n+1}{2}\pi \pm i\operatorname{Log}\left(4+\sqrt{15}\right)$
   where $\operatorname{Log}\left(4+\sqrt{15}\right)$ is the real logarithm from the calculus

8. Show that for each n in J, each of the following identities holds.
   (a) $\cos(z+2n\pi) = \cos z$ (b) $\sin(z+2n\pi) = \sin z$
   (c) $\cos[z+(2n+1)\pi] = -\cos z$ (d) $\sin[z+(2n+1)\pi] = -\sin z$

9. Show that $e^z = i$ iff $z = \dfrac{(4n+1)\pi i}{2}$ where $n \in J$.

10. Solve the equation $e^z = a + ib$ for z where a and b are real and $ab \ne 0$.
    Ans. $z = \dfrac{1}{2}\operatorname{Log}(a^2+b^2) + iy$ where $\sin y = \dfrac{b}{\sqrt{a^2+b^2}}$ and $\cos y = \dfrac{a}{\sqrt{a^2+b^2}}$

11. State appropriate theorems showing why the following functions are entire.

    $$\text{(a)} \ \ 3iz^2 + 4 + \frac{1}{e^z} \qquad\qquad \text{(b)} \ \ e^{\sin z}$$

12. Show that $\sin \bar{z}$ and $\cos \bar{z}$ are not analytic anywhere.
    Hint. $\sin \bar{z} = \overline{\sin z}$. Use Ex. 17 in 2.3.16.

13. Show that $e^z P(z)$ is an entire function where P is a polynomial in z.
    Hint. Recall 2.3.5, 2.3.14, and 5.1.3.

14. Evaluate $\lim\limits_{x \to -\infty} e^{2x+3iy}$. [See Ex. 5 in 2.2.7.]

15. From 5.1.1 and 5.1.2, we know that $e = \lim\limits_{n \to \infty} S_n = \lim\limits_{n \to \infty} \sum\limits_{k=0}^{n} \dfrac{1}{k!}$. Show that
    $\lim\limits_{n \to \infty} T_n = e$ where $T_n = \left(1 + \frac{1}{n}\right)^n$.

Hints.

(i) Observe $S_n < 1 + \sum\limits_{n=0}^{\infty} \dfrac{1}{2^n} = 3$.

(ii) By the binomial theorem,

$$
\begin{aligned}
T_n &= 1 + 1 + \frac{1}{2!}\left(1 - \frac{1}{n}\right) + \cdots + \frac{1}{n!}\left(1 - \frac{1}{n}\right)\left(1 - \frac{2}{n}\right)\cdots\left(1 - \frac{n-1}{n}\right) \\
&\le S_n < 3.
\end{aligned}
$$

(iii) Now $T_n \le T_{n+1}$ by (ii). Thus $\lim_{n \to \infty} T_n = T \le e$.

Next we show that $T = e$.

(iv) Observe that if $m > n$, then

$$T_m > \left[1 + 1 + \frac{1}{2!}\left(1 - \frac{1}{m}\right) + \cdots + \frac{1}{n!}\left(1 - \frac{1}{m}\right) \cdots \left(1 - \frac{n-1}{m}\right)\right]$$
$$\to S_n \text{ as } m \to \infty.$$

Thus for a fixed n, we have $T = \lim_{m \to \infty} T_m \ge S_n$.

(v) Therefore $T \ge e$.

Thus $T = e$ by (iii) and (v).

We observe that the series $1 - 1 + 1 - 1 + 1 - \cdots$ diverges. However the series $(1 - 1) + (1 - 1) + (1 - 1) + \cdots$ converges to zero since each term is zero. This shows that we may group terms in some divergent series and obtain a convergent series. But Ex. 16 shows that if a series converges, then the insertion of parentheses does not alter its sum.

16. Let $\{n_k\}$ be a sequence in N such that $n_1 < n_2 < n_3 < \cdots$. Prove if

$$s = \sum_{n=1}^{\infty} z_n \quad \text{where } z_n \in R^2,$$

then

$$\begin{aligned} s &= (z_1 + z_2 + \cdots + z_{n_1}) + (z_{n_1+1} + z_{n_1+2} + \cdots + z_{n_2}) \\ &\quad + (z_{n_2+1} + z_{n_2+2} + \cdots + z_{n_3}) + \cdots \\ &= A_1 + A_2 + A_3 + \cdots \end{aligned}$$

where A_k is the sum in the kth pair of parentheses.

Proof. Let S_n be the nth partial sum of $\sum_{m=1}^{\infty} z_m$ and let T_k be the kth partial sum of $\sum_{m=1}^{\infty} A_m$. Now $T_k = S_{n_k}$. Thus $\{T_k\}$ is a subsequence of $\{S_n\}$ and hence has the same limit as $\{S_n\}$ by Ex. 11 in 4.5.6.

17. Review the proof of the first part of (5.9) and observe how Ex. 16 is used in that proof.

5.3 The Hyperbolic Functions

Definition 5.3.1 For each z in R^2,

$$\cosh z = \frac{e^z + e^{-z}}{2} \quad \text{and} \quad \sinh z = \frac{e^z - e^{-z}}{2}.$$

We see that for the case in which $z = x$ where x is real, the above formulas become the familiar formulas for $\cosh x$ and $\sinh x$ from the calculus of functions of a real variable.

Remarks 5.3.2 For each z in R^2,

$$\cosh z = \sum_{m=0}^{\infty} \frac{z^{2m}}{(2m)!} \quad \text{and} \quad \sinh z = \sum_{m=0}^{\infty} \frac{z^{2m+1}}{(2m+1)!}.$$

To obtain the above formula for $\cosh z$, we observe that for each z in R^2,

$$\begin{aligned} \cosh z &= \frac{e^z + e^{-z}}{2} \\ &= \frac{1}{2}\left[\sum_{n=0}^{\infty} \frac{z^n}{n!} + \sum_{n=0}^{\infty} \frac{(-1)^n z^n}{n!}\right] \quad \text{by 5.1.2} \\ &= \sum_{m=0}^{\infty} \frac{z^{2m}}{(2m)!} \quad \text{by 4.3.9.} \end{aligned}$$

The formula for $\sinh z$ is obtained in the same way.

Basic Identities 5.3.3 The following identities hold for all complex values of z, z_1, and z_2 and for all real values of x and y.

1. $\sinh(-z) = -\sinh z$

2. $\cosh(-z) = \cosh z$

3. $\sinh iz = i \sin z$

4. $\cosh iz = \cos z$

5. $\sin iz = i \sinh z$

6. $\cos iz = \cosh z$

7. $\cosh^2 z - \sinh^2 z = 1$

8. $\sinh(z_1 + z_2) = \sinh z_1 \cosh z_2 + \cosh z_1 \sinh z_2$

9. $\cosh(z_1 + z_2) = \cosh z_1 \cosh z_2 + \sinh z_1 \sinh z_2$

10. $\sinh(x + iy) = \sinh x \cos y + i \cosh x \sin y$

11. $\cosh(x + iy) = \cosh x \cos y + i \sinh x \sin y$

12. $\sinh 2z = 2 \sinh z \cosh z$

13. $|\sinh z|^2 = \sinh^2 x + \sin^2 y = \cosh^2 x - \cos^2 y$

14. $|\cosh z|^2 = \sinh^2 x + \cos^2 y = \cosh^2 x - \sin^2 y$

15. $\sinh z = \sum\limits_{m=0}^{\infty} \dfrac{z^{2m+1}}{(2m+1)!}$ and $\cosh z = \sum\limits_{m=0}^{\infty} \dfrac{z^{2m}}{(2m)!}$

Proofs.

(1) $\sinh(-z) = \dfrac{e^{-z} - e^{z}}{2} = -\sinh z$ by 5.3.1

(3) $\sinh iz = \dfrac{e^{iz} - e^{-iz}}{2} = i \sin z$ by 5.3.1 and 5.2.7(g)

(5) $\sin iz = \dfrac{e^{-z} - e^{z}}{2i} = i \sinh z$ by 5.2.7(g) and 5.3.1

(7) $\cosh^2 z - \sinh^2 z = \dfrac{e^{2z} + 2 + e^{-2z}}{4} - \dfrac{e^{2z} - 2 + e^{-2z}}{4} = 1$ by 5.3.1

(9) $\begin{aligned}
\cosh(z_1 + z_2) &= \cos(iz_1 + iz_2) & \text{by 6} \\
&= \cos iz_1 \cos iz_2 - \sin iz_1 \sin iz_2 & \text{by 5.2.8} \\
&= \cosh z_1 \cosh z_2 - i^2 \sinh z_1 \sinh z_2 & \text{by 5} \\
&= \cosh z_1 \cosh z_2 + \sinh z_1 \sinh z_2 &
\end{aligned}$

(11) $\begin{aligned}
\cosh(x + iy) &= \cosh x \cosh iy + \sinh x \sinh iy & \text{by 9} \\
&= \cosh x \cos y + i \sinh x \sin y & \text{by 3 and 4}
\end{aligned}$

(13) $\begin{aligned}
|\sinh z|^2 &= \sinh^2 x \cos^2 y + \cosh^2 x \sin^2 y & \text{by 10} \\
&= \sinh^2 x \cos^2 y + (1 + \sinh^2 x) \sin^2 y & \text{by 7} \\
&= \sinh^2 x + \sin^2 y &
\end{aligned}$

Definition 5.3.4 For each complex number z which is not of the form $\frac{2n+1}{2}\pi i$ where $n \in J$,

$$\tanh z = \frac{\sinh z}{\cosh z} \quad \text{and} \quad \operatorname{sech} z = \frac{1}{\cosh z}.$$

Also, for each z in R^2 such that $z \neq n\pi i$ where n is an integer,

$$\coth z = \frac{\cosh z}{\sinh z} \quad \text{and} \quad \operatorname{csch} z = \frac{1}{\sinh z}.$$

(**Hint.** See the fifth and sixth identities in 5.3.3 and Ex. 2 in 5.2.11.)

Theorem 5.3.5 The functions $\cosh z$ and $\sinh z$ are entire functions. Their derivatives are given by

$$\frac{d}{dz} \cosh z = \sinh z \quad \text{and} \quad \frac{d}{dz} \sinh z = \cosh z \quad \text{on } R^2.$$

Partial Proof. $\frac{d}{dz}\cosh z = \frac{1}{2}\frac{d}{dz}(e^z + e^{-z}) = \frac{e^z - e^{-z}}{2} = \sinh z$ by 5.3.1, (5.2), and the chain rule.

Exercises 5.3.6

1. Prove the even numbered identities and the last one in 5.3.3.

In Ex. 2–5, prove the stated result.

2. $\sinh(z + \pi i) = -\sinh z$ for each z in R^2

3. $\cosh(z + \pi i) = -\cosh z$ for each z in R^2

4. $\tanh(z + \pi i) = \tanh z$ for each $z \neq \dfrac{2n+1}{2}\pi i$ where $n \in J$

5. $|\sinh x| \leq |\cosh z| \leq \cosh x$. **Hint.** Use the 14th identity in 5.3.3.

6. Find all values of z such that $\cosh z = -2$. **Hint.** Use the 11th identity in 5.3.3. Ans. $z = \text{Log}\left(2 \pm \sqrt{3}\right) + (2n+1)\pi i$ for each n in J where $\text{Log } x$ is the real logarithm from the calculus

7. Find all of the zeros of the given function.

    (a) $\sinh z$ $n\pi i$ where $n \in J$

    (b) $\cosh z$ $\dfrac{2n+1}{2}\pi i$ where $n \in J$

8. Show that the function $\cosh e^z$ is entire.

5.4   The Logarithmic Function

Definition 5.4.1 If $z = x + iy$ is a given nonzero complex number, then the unique real number θ in the interval $(-\pi, \pi]$ such that

$$x = |z|\cos\theta \quad \text{and} \quad y = |z|\sin\theta$$

is called the **principal argument** of z and is denoted by $\text{Arg } z$.

Theorem 5.4.2 Every complex number $z \neq 0$ can be written in the exponential form $z = re^{i\theta}$ where $r = |z|$ and $\theta = \text{Arg } z + 2n\pi$ for each n in J.

Proof. Let $z \neq 0$ and let $r = |z|$. Then

$$\begin{aligned} z &= r(\cos\theta + i\sin\theta) \quad \text{by (1.15)} \\ &= re^{i\theta} \quad\quad\quad\quad\quad \text{by 5.2.5.} \end{aligned}$$

Theorem 5.4.3 If $z_1 z_2 \neq 0$, then

$$\operatorname{Arg} z_1 z_2 = \operatorname{Arg} z_1 + \operatorname{Arg} z_2 + 2n\pi \tag{5.15}$$

where n has one of the values $0, 1$, or -1.

Proof. For $j = 1$ and $j = 2$, we have $z_j = |z_j|(\cos\theta_j + i\sin\theta_j)$ where

$$-\pi < \theta_j = \operatorname{Arg} z_j \leq \pi. \tag{5.16}$$

Then by Theorem 1.2.2,

$$\begin{aligned} z_1 z_2 &= |z_1||z_2| \operatorname{cis}(\theta_1 + \theta_2) \\ &= |z_1||z_2| \operatorname{cis}(\theta_1 + \theta_2 + 2k\pi) \quad \text{for each } k \text{ in } J. \end{aligned} \tag{5.17}$$

By (5.16), we see that

$$-2\pi < \theta_1 + \theta_2 \leq 2\pi. \tag{5.18}$$

Thus $\theta_1 + \theta_2$ is in one of the intervals $(-2\pi, -\pi]$, $(-\pi, \pi]$, or $(\pi, 2\pi]$.
If $\theta_1 + \theta_2$ is in $(-\pi, \pi]$, then

$$\begin{aligned} \operatorname{Arg} z_1 z_2 &= \theta_1 + \theta_2 \quad \text{by (5.17)} \\ &= \operatorname{Arg} z_1 + \operatorname{Arg} z_2. \end{aligned}$$

Thus in this case, (5.15) holds with $n = 0$.
Next suppose $\theta_1 + \theta_2$ is in $(\pi, 2\pi]$. Then

$$\pi < \theta_1 + \theta_2 \leq 2\pi.$$

Adding -2π to each member of this inequality, we have

$$-\pi < \theta_1 + \theta_2 - 2\pi \leq 0 \leq \pi.$$

Thus in this case, by (5.17) with $k = -1$ and by (5.16), we see that (5.15) holds for $n = -1$.

Similarly, in the remaining case where $\theta_1 + \theta_2$ is in $(-2\pi, -\pi]$, we can see from (5.17) that (5.15) holds with $n = 1$.

Theorem 5.4.4 If z is any nonzero complex number, then there is a complex number w such that

$$e^w = z. \tag{5.19}$$

One such value of w is given by

$$w_1 = \operatorname{Log}|z| + i\operatorname{Arg} z.$$

Moreover, all values of w are given by

$$w = \operatorname{Log}|z| + i(\operatorname{Arg} z + 2n\pi) \quad \text{where } n \in J \tag{5.20}$$

and where $\operatorname{Log}|z|$ denotes the real logarithm of $|z|$.

Proof. By 5.1.5 and 5.4.2, we have $e^{w_1} = e^{\text{Log}\,|z| + i\,\text{Arg}\,z} = e^{\text{Log}\,|z|}e^{i\,\text{Arg}\,z} = |z|e^{i\,\text{Arg}\,z} = z$. Now if w is any solution of (5.19), then $e^w = z = e^{w_1}$. Thus by 5.2.7(f), we have $w = w_1 + 2n\pi i$ where $n \in J$. Hence (5.20) follows from the last equation since $w_1 = \text{Log}\,|z| + i\,\text{Arg}\,z$.

Definition 5.4.5 Let $z \in R^2$ such that $z \neq 0$. If $w \in R^2$ such that $e^w = z$, then w is called **a logarithm** of z and we write $w = \log z$. The **principal logarithm** of z (denoted by $\text{Log}\,z$) is the unique number given by

$$\text{Log}\,z = \text{Log}\,|z| + i\,\text{Arg}\,z \tag{5.21}$$

where $\text{Log}\,|z|$ is the real logarithm of the real number $|z|$.

From (5.20), we see that all values of $\log z$ are given by

$$\begin{aligned}
\log z &= \text{Log}\,|z| + i\,(\text{Arg}\,z + 2n\pi) \\
&= \text{Log}\,z + 2n\pi i \quad \text{where } n \in J.
\end{aligned}$$

Theorem 5.4.6 If $z_1 \in R^2$, $z_2 \in R^2$, and $z_1 z_2 \neq 0$, then

$$\text{Log}\,z_1 z_2 = \text{Log}\,z_1 + \text{Log}\,z_2 + 2n\pi i \tag{5.22}$$

where n has one of the values 0, 1, or -1.

Proof. By (5.15) and (5.21) with $z = z_1 z_2$,

$$\begin{aligned}
\text{Log}\,z_1 z_2 &= \text{Log}\,|z_1 z_2| + i\,\text{Arg}\,(z_1 z_2) \\
&= \text{Log}\,|z_1| + \text{Log}\,|z_2| + i\,[\text{Arg}\,z_1 + \text{Arg}\,z_2 + 2n\pi] \\
&= (\text{Log}\,|z_1| + i\,\text{Arg}\,z_1) + (\text{Log}\,|z_2| + i\,\text{Arg}\,z_2) + 2n\pi i \\
&= \text{Log}\,z_1 + \text{Log}\,z_2 + 2n\pi i \quad \text{by (5.21).}
\end{aligned}$$

Definition 5.4.7 If z and w are given complex numbers such that $z \neq 0$, then z^w is defined by

$$z^w = e^{w\log z}.$$

Note that $z^w = |z|^w e^{iw\,\text{arg}\,z}$ where $\text{arg}\,z = \text{Arg}\,z + 2n\pi$. Observe that if $w \notin J$, then z^w is not single-valued. The **principal value** of z^w is defined to be the unique number

$$e^{w\,\text{Log}\,z}.$$

In this book, we shall use z^w to denote $e^{w\,\text{Log}\,z}$

unless the contrary is stated. $\hspace{4cm}$ (5.23)

Theorem 5.4.8 If z, w_1, and w_2 are complex numbers and $z \neq 0$, then

$$z^{w_1} z^{w_2} = z^{w_1 + w_2}.$$

If z_1, z_2, and w are complex numbers and $z_1 z_2 \neq 0$, then

$$(z_1 z_2)^w = z_1^w z_2^w e^{2n\pi i w} \quad \text{for some } n = 0, \pm 1.$$

Proof. By (5.23) and 5.1.5,

$$z^{w_1 + w_2} = e^{(w_1 + w_2) \operatorname{Log} z} = e^{w_1 \operatorname{Log} z} e^{w_2 \operatorname{Log} z} = z^{w_1} z^{w_2}.$$

By (5.22) and (5.23),

$$
\begin{aligned}
(z_1 z_2)^w &= e^{w \operatorname{Log} z_1 z_2} = e^{w (\operatorname{Log} z_1 + \operatorname{Log} z_2 + 2n\pi i)} \\
&= e^{w \operatorname{Log} z_1} e^{w \operatorname{Log} z_2} e^{2n\pi i w} = z_1^w z_2^w e^{2n\pi i w}.
\end{aligned}
$$

We now express the Cauchy-Riemann equations in polar coordinates and apply them to show that the function $\operatorname{Log} z$ is analytic in D where D is given in 5.4.10.

Theorem 5.4.9 Let $f = u + iv$ be defined on an open set $D \subset R^2$. Let $z = re^{i\theta}$ be a fixed point in D such that $z \neq 0$. Suppose u_r, v_r, u_θ, and v_θ exist at each point in some neighborhood of z and are continuous at z. Then $f'(z)$ exists iff

$$u_r = \frac{1}{r} v_\theta \quad \text{and} \quad \frac{1}{r} u_\theta = -v_r \quad \text{at } z. \tag{5.24}$$

Furthermore, if conditions (5.24) hold, then

$$f'(z) = e^{-i\theta}[u_r(r, \theta) + iv_r(r, \theta)]. \tag{5.25}$$

Proof of "Only If Part." Suppose $f'(z)$ exists. Then by 2.3.7,

$$u_x = v_y \quad \text{and} \quad u_y = -v_x \quad \text{at } z. \tag{5.26}$$

Now

$$z = x + iy = re^{i\theta} = r(\cos\theta + i\sin\theta).$$

Thus

$$u = u(x, y) \text{ and } v = v(x, y) \text{ where } x = r\cos\theta \text{ and } y = r\sin\theta. \tag{5.27}$$

Hence

$$
\begin{aligned}
u_r &= u_x x_r + u_y y_r = v_y \cos\theta - v_x \sin\theta \\
&= \frac{1}{r}(v_y r \cos\theta - v_x r \sin\theta) = \frac{1}{r}(v_y y_\theta + v_x x_\theta) = \frac{1}{r} v_\theta.
\end{aligned}
$$

The proof of $\frac{1}{r} u_\theta = -v_r$ is similiar.

Proof of "If Part." Let (5.24) hold. Then by (5.27), the following equations hold.

$$
\begin{aligned}
u_r &= u_x \cos\theta + u_y \sin\theta & v_\theta &= r(-v_x \sin\theta + v_y \cos\theta) \\
v_r &= v_x \cos\theta + v_y \sin\theta & u_\theta &= r(-u_x \sin\theta + u_y \cos\theta)
\end{aligned}
\tag{5.28}
$$

By (5.28), the equations in (5.24) become the following two equations,

$$
\begin{aligned}
u_x \cos\theta + u_y \sin\theta &= -v_x \sin\theta + v_y \cos\theta \\
-u_x \sin\theta + u_y \cos\theta &= -v_x \cos\theta - v_y \sin\theta
\end{aligned}
\tag{5.29}
$$

Solving (5.29) (by Cramer's rule) for u_x and u_y in terms of v_x and v_y, we obtain

$$
u_x = \frac{\begin{vmatrix} -v_x \sin\theta + v_y \cos\theta & \sin\theta \\ -v_x \cos\theta - v_y \sin\theta & \cos\theta \end{vmatrix}}{\begin{vmatrix} \cos\theta & \sin\theta \\ -\sin\theta & \cos\theta \end{vmatrix}} = \frac{v_y}{1} = v_y.
$$

Similarly, using Cramer's rule in (5.29) to solve for u_y, we obtain $u_y = -v_x$. Hence $f'(z)$ exists by 2.3.11.

Proof of (5.25). By (5.26) and (5.28), the next two equations hold at the point z.

$$
\begin{aligned}
u_r &= u_x \cos\theta - v_x \sin\theta \\
v_r &= u_x \sin\theta + v_x \cos\theta
\end{aligned}
\tag{5.30}
$$

Solving (5.30) for u_x and v_x, we obtain formulas for u_x and v_x which must hold at the point $z = re^{i\theta}$.

$$
\begin{aligned}
u_x &= u_r \cos\theta + v_r \sin\theta \\
v_x &= -u_r \sin\theta + v_r \cos\theta
\end{aligned}
\tag{5.31}
$$

Thus by (5.31) and (2.26),

$$
\begin{aligned}
f'(z) &= u_x + iv_x = u_r(\cos\theta - i\sin\theta) + iv_r(\cos\theta - i\sin\theta) \\
&= e^{-i\theta}(u_r + iv_r),
\end{aligned}
$$

which is (5.25).

The conditions stated in (5.24) are called the **Cauchy-Riemann equations in polar coordinates**.

Theorem 5.4.10 Let

$$
D = \{z : \quad z \neq 0 \quad \text{and} \quad -\pi < \operatorname{Arg} z < \pi \}.
$$

Then the function $\operatorname{Log} z$ is analytic in D, and

$$
\frac{d}{dz}(\operatorname{Log} z) = \frac{1}{z} \quad \text{on } D.
$$

Proof. $\text{Log}\, z = \text{Log}\, r + i\theta$ where $r = |z|$ and $\theta = \text{Arg}\, z$. Thus $u(r,\theta) = \text{Log}\, r$ and $v(r,\theta) = \theta$. Hence $u_r = \frac{1}{r}$, $v_r = 0$, $u_\theta = 0$, and $v_\theta = 1$ on D. Hence (5.24) is satisfied, and $\frac{d}{dz}(\text{Log}\, z)$ exists on D. By (5.25),

$$\frac{d}{dz}(\text{Log}\, z) \;=\; e^{-i\theta}\left(\frac{1}{r}\right) \;=\; \frac{1}{z}.$$

Exercises 5.4.11

1. Evaluate each of the following principal logarithms.

   (a) $\text{Log}\, i$ (b) $\text{Log}\,(-1)$ (c) $\text{Log}\,(-1 - i)$
   (d) $\text{Log}\,(-1 + \sqrt{3}i)$ (e) $\text{Log}\,(1 - i)$

   Ans. (a) $\dfrac{\pi i}{2}$ (b) πi (c) $\dfrac{1}{2}\text{Log}\, 2 - \dfrac{3\pi i}{4}$
        (d) $\text{Log}\, 2 + \dfrac{2\pi i}{3}$ (e) $\dfrac{1}{2}\text{Log}\, 2 - \dfrac{\pi i}{4}$

2. In (a) and (b), find all values of z which satisfy the given equation.

   (a) $\text{Log}\, z = -\dfrac{\pi}{2}i$ (b) $e^z = a$ where $a < 0$

          Ans. (a) $-i$ (b) $\text{Log}\,|a| + (2n + 1)\pi i$ where $n \in J$.

3. Find the principal value of each of the following exponential expressions.

   (a) $(1 - i)^i$ (b) $\left(1 + \sqrt{3}i\right)^{-i}$ (c) i^{-i}

        Ans. (a) $e^{\frac{\pi}{4}}\left[\cos\left(\dfrac{1}{2}\text{Log}\, 2\right) + i\sin\left(\dfrac{1}{2}\text{Log}\, 2\right)\right]$
          (b) $e^{\frac{\pi}{3}}[\cos(\text{Log}\, 2) - i\sin(\text{Log}\, 2)]$ (c) $e^{\frac{\pi}{2}}$

4. Let a and b be positive numbers. In (a) through (c), find the value of n such that the given equation holds.

   (a) $\text{Log}\,[(a + bi)(a - bi)] = \text{Log}\,(a + bi) + \text{Log}\,(a - bi) + 2n\pi i$
   (b) $\text{Log}\,(-ai) = \text{Log}\,(-a) + \text{Log}\, i + 2n\pi i$
   (c) $\text{Log}\,(ab) = \text{Log}\,(-a)(-b) = \text{Log}\,(-a) + \text{Log}\,(-b) + 2n\pi i$

          Ans. (a) 0 (b) -1 (c) -1

5. Show that $\text{Log}\, z$ is not continuous at any point on the nonpositive real axis.
   Hint. Let $z_0 = x_0 < 0$. $\text{Log}\, z = u + iv$ where $v = \text{Arg}\, z$. Now $v(x_0, 0) = \pi$. But $v(x, y) = \text{Arg}\, z$ is near $-\pi$ in the "strict" lower half of each small neighborhood of z_0. Thus v is not continuous at z_0. Hence $\text{Log}\, z$ is not continuous at z_0 by 2.2.11. Why is $\text{Log}\, z$ not continuous at $z = 0$?

6. Show that $\operatorname{Log} z$ is not analytic at any point on the nonpositive real axis.
   Hint. See Ex. 5 and 2.3.3.

7. Let $a \in (R - J)$ and let $f(z) = z^a$. Prove each of the following statements.

   (a) $z^a = |z|^a e^{ia\operatorname{Arg} z}$

   (b) f is not continuous at any point on the nonpositive real axis and hence
   is not analytic at any such point. **Hint.** $f(z) \to |x|^a e^{i\pi a}$ if $z \to x$ where
   $x < 0$ from above the x-axis. But $f(z) \to |x|^a e^{-i\pi a}$ if $z \to x < 0$ from
   below the x-axis. See 5.2.7(f).

   (c) $f \in A(R^2 - H)$ where H is the nonpositive real axis. **Hint.** Note that f
   is the composite of $(a \operatorname{Log} z)$ and e^w and use 2.3.6 and 2.3.12.

8. Let $g(x) = \operatorname{Log}(-z)$. Prove $g \in A(R^2 - K)$ where K is the non-negative real
   axis. **Hint.** Let $h(z) = -z$ and $f(z) = \operatorname{Log} z$. Then $g(z) = f(h(z))$.

9. Let $a \in (R - J)$ and let $h(z) = (-z)^a$.

   (a) Prove $h \in A(R^2 - K)$ where K is the non-negative real axis.

   (b) Prove h is not continuous at any point of K.

10. Prove if $z_1 z_2 \neq 0$, then $\operatorname{Log} \dfrac{z_1}{z_2} = \operatorname{Log} z_1 - \operatorname{Log} z_2 + 2n\pi i$ for some $n = 0, \pm 1$.

11. Let

$$g(z) = \operatorname{Log} r + i\theta$$

where $r = |z| > 0$ and where $\theta_0 < \theta = \arg z < \theta_0 + 2\pi$ for some fixed θ_0. Prove
$g \in A(D)$ and

$$g'(z) = \frac{1}{z} \quad \text{for each } z \text{ in } D$$

where D is the domain of g. **Hint.** Apply (5.24) and (5.25). Note that if $z \in D$,
then $g(z)$ is one of the values of $\log z$.

12. Let z_1, z_2, and w be complex numbers such that $z_1 z_2 \neq 0$. Prove that

$$\left(\frac{z_1}{z_2}\right)^w = \frac{z_1^w}{z_2^w} e^{2n\pi i w} \quad \text{for some } n = 0, \pm 1.$$

13. Let z^a be the function given in Ex. 7. Prove

$$\frac{d}{dz} z^a = a z^{a-1} \quad \text{for each } z \text{ in } (R^2 - H)$$

where H is the nonpositive real axis.

14. Let z, w, and a be complex numbers such that $z \neq 0$. Prove each of the following results.

(i) $\operatorname{Log} z^w = w \operatorname{Log} z + 2n\pi i$ where

$$n \in J \text{ such that } -\pi < \mathcal{I}(w \operatorname{Log} z) + 2n\pi \leq \pi \qquad (5.32)$$

(ii) $(z^w)^a = z^{aw} e^{2n\pi i a}$ where n satisfies (5.32)

Proof of (i). If $q \in R^2$, then $e^q = e^{q + 2m\pi i}$ for each m in J. Thus

$$\operatorname{Log} e^q = q + 2k\pi i \qquad (5.33)$$

where $k \in J$ such that

$$-\pi < \mathcal{I}(q) + 2k\pi \leq \pi \quad \text{by 5.4.5 and 5.2.7(d).}$$

Now

$$
\begin{aligned}
\operatorname{Log} z^w &= \operatorname{Log} e^{w \operatorname{Log} z} &&\text{by (5.23)} \\
&= w \operatorname{Log} z + 2n\pi i &&\text{by (5.33)}
\end{aligned}
$$

where n satisfies (5.32).

Remarks 5.4.12 By 5.4.5 and the proof of 5.4.6, the usual law of logarithms

$$\log z_1 z_2 = \log z_1 + \log z_2 \qquad (5.34)$$

holds in the following sense. If any two terms in (5.34) are given, then the third term can be chosen so that (5.34) holds. Similarly,

$$\log \frac{z_1}{z_2} = \log z_1 - \log z_2.$$

Also

$$(z_1 z_2)^w = z_1^w z_2^w \qquad (5.35)$$

holds in the sense that if two powers in (5.34) are given, then the third power may be chosen so that (5.35) is true (where here we are not using principal values). [For

$$
\begin{aligned}
e^{w \log z_1 z_2} &= e^{w \log z_1 + w \log z_2} &&\text{by (5.34)} \\
&= e^{w \log z_1} e^{w \log z_2} &&\text{by 5.1.5} \\
&= z_1^w z_2^w &&\text{by 5.4.7.}]
\end{aligned}
$$

Chapter 6

Laurent Series, Poles and Residues

6.1 Laurent Series

Theorem 6.1.1 Let f be analytic in the closed annular region S given by

$$S = \{t : \ r_1 \leq |t - a| \leq r_2\}$$

where a is a fixed complex number and r_1 and r_2 are fixed real numbers with $0 < r_1 < r_2$. Then for each z in S°,

$$f(z) = \sum_{n=1}^{\infty} a_{-n}(z - a)^{-n} + \sum_{n=0}^{\infty} a_n(z - a)^n \qquad (6.1)$$

where

$$a_{-n} = \frac{1}{2\pi i} \int_{\gamma_1} \frac{f(t)}{(t - a)^{-n+1}} \, dt \quad \text{and} \quad a_n = \frac{1}{2\pi i} \int_{\gamma_2} \frac{f(t)}{(t - a)^{n+1}} \, dt \qquad (6.2)$$

and where γ_1, γ_2 are the cco circles given by

$$\gamma_1 = \{t : \ |t - a| = r_1\} \quad \text{and} \quad \gamma_2 = \{t : \ |t - a| = r_2\}.$$

Proof. See Figure 6.1. For each z in S°

$$f(z) = \frac{1}{2\pi i} \int_{\gamma_2} \frac{f(t)}{t - z} \, dt - \frac{1}{2\pi i} \int_{\gamma_1} \frac{f(t)}{t - z} \, dt \qquad \text{by 3.6.8.} \qquad (6.3)$$

For each t on γ_2,

$$\frac{f(t)}{t - z} = \frac{f(t)}{(t - a) - (z - a)} = \frac{f(t)}{t - a} \frac{1}{1 - \frac{z-a}{t-a}}. \qquad (6.4)$$

147

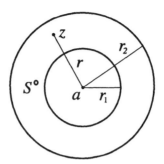

Figure 6.1: $S^{\circ} = I(\gamma_2) \cap E(\gamma_1)$

For each t on γ_1,

$$\frac{f(t)}{z - t} = \frac{f(t)}{(z - a) - (t - a)} = \frac{f(t)}{z - a} \frac{1}{1 - \frac{t-a}{z-a}}. \tag{6.5}$$

We observe that if $\alpha \neq 1$, then

$$\frac{1}{1 - \alpha} = \sum_{k=0}^{n-1} \alpha^k + \frac{\alpha^n}{1 - \alpha}. \tag{6.6}$$

[We may verify (6.6) by expressing the right member as a single fraction.] We now apply (6.6) to (6.4) with

$$\alpha = \frac{z - a}{t - a}$$

and obtain for each t on γ_2,

$$\frac{f(t)}{t - z} = \frac{f(t)}{t - a} \sum_{k=0}^{n-1} \left(\frac{z - a}{t - a}\right)^k + \frac{f(t)}{t - z} \left(\frac{z - a}{t - a}\right)^n. \tag{6.7}$$

Again we apply (6.6) to (6.5) with $\alpha = \frac{t-a}{z-a}$. Thus for each t on γ_1

$$\frac{f(t)}{z - t} = \frac{f(t)}{z - a} \sum_{k=0}^{n-1} \left(\frac{t - a}{z - a}\right)^k + \frac{f(t)}{z - t} \left(\frac{t - a}{z - a}\right)^n. \tag{6.8}$$

Hence from (6.7),

$$\frac{1}{2\pi i} \int_{\gamma_2} \frac{f(t)}{t - z} \, dt = \sum_{k=0}^{n-1} (z - a)^k \frac{1}{2\pi i} \int_{\gamma_2} \frac{f(t)}{(t - a)^{k+1}} \, dt + R_2 \tag{6.9}$$

where

$$R_2 = \frac{(z - a)^n}{2\pi i} \int_{\gamma_2} \frac{f(t)}{(t - z)(t - a)^n} \, dt. \tag{6.10}$$

From (6.8) and replacing k by $k - 1$, we obtain

$$\frac{1}{2\pi i}\int_{\gamma_1}\frac{f(t)}{z-t}\,dt = \sum_{k=1}^{n}(z-a)^{-k}\frac{1}{2\pi i}\int_{\gamma_1}\frac{f(t)}{(t-a)^{-k+1}}\,dt + R_1 \qquad (6.11)$$

where

$$R_1 = \frac{1}{2\pi i(z-a)^n}\int_{\gamma_1}\frac{f(t)(t-a)^n}{z-t}\,dt. \qquad (6.12)$$

But $f \in A(S)$ implies $f \in C(\gamma_1 \cup \gamma_2)$. Thus by Ex. 2(b) in 2.4.9,

there is some $M > 0$ such that $|f(t)| < M$ for each $t \in (\gamma_1 \cup \gamma_2)$. \qquad (6.13)

For each z in S°, we have $r_1 < r < r_2$ where $r = |z - a|$. \qquad (6.14)

For each t on γ_2, $|t - a| = r_2$ and for each t on γ_1, $|t - a| = r_1$. \qquad (6.15)

Now

$$|t - z| = |(t-a)-(z-a)| \geq ||t-a|-|z-a|| = \begin{cases} r_2 - r & \text{if } t \in \gamma_2 \\ r - r_1 & \text{if } t \in \gamma_1. \end{cases} \qquad (6.16)$$

Hence by (6.10) and (6.14)–(6.16),

$$|R_2| \leq \frac{r^n}{2\pi}\frac{M2\pi r_2}{(r_2-r)r_2^n} = \frac{r_2 M}{r_2 - r}\left(\frac{r}{r_2}\right)^n \quad \text{where } \frac{r}{r_2} < 1. \qquad (6.17)$$

By (6.12) and (6.14)–(6.16),

$$|R_1| \leq \frac{1}{2\pi r^n}\frac{M2\pi r_1 r_1^n}{r - r_1} = \frac{r_1 M}{r - r_1}\left(\frac{r_1}{r}\right)^n \quad \text{where } \frac{r_1}{r} < 1. \qquad (6.18)$$

Thus by (6.17) and (6.18),

$$\lim_{n\to\infty} R_1 = 0 = \lim_{n\to\infty} R_2.$$

Hence by (6.3), (6.9), and (6.11) we obtain (6.1) and (6.2).

Remarks 6.1.2 (a) The sum on the right in 6.1.1 is denoted by

$$\sum_{n=-\infty}^{\infty} a_n(z-a)^n$$

and is called the **Laurent** series for f in the **annulus** S°.

(b) The series

$$\sum_{n=1}^{\infty} a_{-n}(z-a)^{-n}$$

in 6.1.1 is called the **principal part** of f in S°.

(c) Laurent's Theorem is 6.1.1 with γ_1 and γ_2 in (6.2) replaced by any cco simple closed contour $C \subset S$ such that $a \in I(C)$. [See 3.5.11.]

(d) The formulas in (6.2) may be combined into a single formula as

$$a_n = \frac{1}{2\pi i} \int_C \frac{f(t)}{(t-a)^{n+1}} \, dt \quad \text{for each } n \text{ in } J$$

where C is given in Part (c).

Theorem 6.1.3 If in the closed annulus S of 6.1.1, the function f is represented by

$$f(z) = \sum_{n=-\infty}^{\infty} b_n (z-a)^n, \tag{6.19}$$

then

$$b_n = \frac{1}{2\pi i} \int_C \frac{f(t)}{(t-a)^{n+1}} \, dt \quad \text{for each } n \text{ in } J \tag{6.20}$$

where C is any simple closed contour in S such that $a \in I(C)$; that is, the series (6.19) is the Laurent series representation of f in S°.

Proof. By (6.19),

$$\frac{f(z)}{(z-a)^{k+1}} = \sum_{n=k-1}^{-\infty} b_n (z-a)^{n-k-1} + \frac{b_k}{z-a} + \sum_{n=k+1}^{\infty} b_n (z-a)^{n-k-1} \quad \text{for each } z \text{ in } S^\circ.$$

By 4.4.3, 4.4.9, and 4.4.10,

$$\int_C \frac{f(t)\,dt}{(t-a)^{k+1}} = \sum_{n=k-1}^{-\infty} b_n \int_C (t-a)^{n-k-1} dt + b_k \int_C \frac{dt}{t-a} + \sum_{n=k+1}^{\infty} b_n \int_C (t-a)^{n-k-1} dt$$

$$= 2\pi i b_k \quad \text{by 3.6.6 and 3.6.11,}$$

and thus (6.20) holds.

Remark. Theorem 6.1.3 shows that a Laurent series for f in a **given** annulus is unique.

We now obtain the following corollary of 6.1.1.

Corollary 6.1.4 If $f \in A(D)$ where

$$D = \{z : \ |z-a| < r\},$$

then f is represented by

$$f(z) = \sum_{n=0}^{\infty} a_n (z-a)^n \quad \text{for each } z \text{ in } D \tag{6.21}$$

where

$$a_n = \frac{f^{(n)}(a)}{n!}. \tag{6.22}$$

Proof. Let $z \in (D - \{a\})$. Choose r_1 and r_2 such that $0 < r_1 < |z - a| < r_2 < r$. Thus there is determined an annular region

$$S = \{t : \ r_1 \leq |t - a| \leq r_2\}$$

to which we apply Laurent's Theorem. Hence

$$f(z) = \sum_{n=-\infty}^{\infty} a_n(z - a)^n \quad \text{where } a_n \text{ is given in 6.1.2(d).} \tag{6.23}$$

Thus

$$a_n = \begin{cases} 0 & \text{if } n \leq -1 \quad \text{by 3.5.10} \\ \dfrac{1}{2\pi i} \displaystyle\int_C \dfrac{f(t)}{(t - a)^{n+1}} \, dt = \dfrac{1}{n!} f^{(n)}(a) & \text{if } n \geq 0 \quad \text{by 3.6.9.} \end{cases} \tag{6.24}$$

Thus (6.23) reduces to (6.21) where (6.22) holds for each z in $(D - \{a\})$, since z was an arbitrary point in $D - \{a\}$. If $z = a$, then (6.21) is obvious.

Definition 6.1.5 The series given by (6.21) is called the **Taylor's series for f** in powers of $(z - a)$, or the Taylor's series for f about a. If $a = 0$, then (6.21) is called the Maclaurin's series for f.

Combining 6.1.4 with 4.3.6 and 4.3.8, we have the following theorem.

Theorem 6.1.6 A function f is analytic at the point $z = a$ iff, in some neighborhood of a, f can be represented by a power series in powers of $z - a$.

Remark 6.1.7 Since the Laurent series for f in a **given** annulus is unique (and so is the Taylor series for f about a), the coefficients a_n and b_n may be computed by the simplest method available.

Examples 6.1.8 Let

$$f(z) = \frac{A}{az + b} + \frac{B}{cz + d} \tag{6.25}$$

where $A, B, a, b, c,$ and d are nonzero complex numbers, and let

$$r_1 = \left| \frac{b}{a} \right| < \left| \frac{d}{c} \right| = r_2.$$

Find the Laurent series for f in

(i) $S = \{z : \ r_1 < |z| < r_2\}$,

(ii) $G = \{z : \ |z| > r_2\}$, and

(iii) $D = N_{r_1}(0)$.

Solutions. We shall use the geometric series of Ex. 3 in 6.1.9.

(i) Now for each z in S,

$$\left|\frac{b}{az}\right| < 1 \quad \text{and} \quad \left|\frac{cz}{d}\right| < 1. \tag{6.26}$$

By (6.25) for each z in S,

$$
\begin{aligned}
f(z) &= \frac{A}{az\left(1 + \frac{b}{az}\right)} + \frac{B}{d\left(1 + \frac{cz}{d}\right)} \\
&= A\sum_{n=0}^{\infty}(-1)^n\frac{b^n}{(az)^{n+1}} + B\sum_{n=0}^{\infty}(-1)^n\frac{c^n}{d^{n+1}}z^n \quad \left\{ \begin{array}{l} \text{by (6.26) and} \\ \text{Ex. 3 in 6.1.9} \end{array} \right. \\
&= \sum_{n=0}^{\infty}(-1)^n\frac{Ab^n}{a^{n+1}}\frac{1}{z^{n+1}} + \sum_{n=0}^{\infty}(-1)^n\frac{Bc^n}{d^{n+1}}z^n.
\end{aligned}
$$

Thus this series represents f in S and hence by 6.1.7, it must be the Laurent series for f in any annulus centered at $z = 0$ and contained in S.

(ii) If $z \in G$, then

$$\left|\frac{b}{az}\right| < 1 \quad \text{and} \quad \left|\frac{d}{cz}\right| < 1. \tag{6.27}$$

Now for each z in G,

$$
\begin{aligned}
f(z) &= \frac{1}{z}\left(\frac{A}{a\left(1 + \frac{b}{az}\right)} + \frac{B}{c\left(1 + \frac{d}{cz}\right)}\right) \quad \text{by (6.25)} \\
&= \frac{1}{z}\left(\frac{A}{a}\sum_{n=0}^{\infty}(-1)^n\frac{b^n}{(az)^n} + \frac{B}{c}\sum_{n=0}^{\infty}(-1)^n\frac{d^n}{(cz)^n}\right) \quad \left\{ \begin{array}{l} \text{by (6.27) and} \\ \text{Ex. 3 in 6.1.9} \end{array} \right. \\
&= \sum_{n=0}^{\infty}(-1)^n\left(\frac{Ab^n}{a^{n+1}} + \frac{Bd^n}{c^{n+1}}\right)\frac{1}{z^{n+1}}.
\end{aligned}
$$

Hence this series is the Laurent series for f in G, in the sense that it is the Laurent series for f in any annulus centered at $z = 0$, which is contained in G.

(iii) If $z \in D$, then

$$\left|\frac{az}{b}\right| < 1 \quad \text{and} \quad \left|\frac{cz}{d}\right| < 1. \tag{6.28}$$

Thus for each z in D,

$$
\begin{aligned}
f(z) &= \frac{A}{b\left(1 + \frac{az}{b}\right)} + \frac{B}{d\left(1 + \frac{cz}{d}\right)} \quad \text{by (6.25)} \\
&= \frac{A}{b}\sum_{n=0}^{\infty}(-1)^n\left(\frac{az}{b}\right)^n + \frac{B}{d}\sum_{n=0}^{\infty}(-1)^n\left(\frac{cz}{d}\right)^n \\
&\qquad \text{by (6.28) and Ex. 3 in 6.1.9} \\
&= \sum_{n=0}^{\infty}(-1)^n\left(\frac{Aa^n}{b^{n+1}} + \frac{Bc^n}{d^{n+1}}\right)z^n,
\end{aligned}
$$

which is the Maclaurin's series for f. Since this representation is unique, $\frac{f^{(n)}(0)}{n!}$ is the coefficient of z^n in this series.

Exercises 6.1.9

1. Find the Maclaurin series for $f(z) = \dfrac{1}{1-z}$ for each z such that $|z| < 1$.

   Hint. Use 4.3.11 and 6.1.7 or use 6.1.4.
   $$\sum_{n=0}^{\infty} z^n \text{ for each } z \text{ such that } |z| < 1.$$

2. Find the Laurent series which represents $f(z) = \dfrac{1}{1-z}$ in the region $\{z : |z| > 1\}$.

   Hint. $\frac{1}{1-z} = -\frac{1}{z} \cdot \frac{1}{1-\frac{1}{z}}$ and use the result of Ex. 1.

3. Find the Maclaurin series for $f(z) = \dfrac{1}{1+z}$ for each z such that $|z| < 1$.
   Hint. Use the result of Ex. 1.
   $$\sum_{n=0}^{\infty} (-1)^n z^n \text{ for each } z \text{ such that } |z| < 1.$$

4. Show that for each z such that $|z| < 1$,

   $$\frac{1}{(1+z)^2} = \sum_{n=1}^{\infty} (-1)^{n-1} n\, z^{n-1} = \sum_{n=0}^{\infty} (-1)^n (n+1) z^n.$$

   Hint. Use Ex. 3 and 4.3.5.

5. Find the Maclaurin series for $f(z) = \dfrac{z}{1-3z}$ for each z such that $|z| < \dfrac{1}{3}$.
   Hint. Use the result of Ex. 1.

6. Find the Taylor series for $f(z) = \dfrac{1}{z-b}$ in powers of $z - a$, where a, $b \in R^2$, and $a \neq b$ for $|z - a| < |a - b|$. **Hint.** $\frac{1}{z-b} = \frac{1}{(a-b)[1+\frac{z-a}{a-b}]}$ and see Ex. 1 or 3.
   $$\sum_{n=0}^{\infty} \frac{(-1)^n}{(a-b)^{n+1}} (z - a)^n \text{ for each } z \text{ such that } |z - a| < |a - b|.$$

7. Show that $\mathrm{Log}\,(z+1) = \displaystyle\sum_{n=1}^{\infty} \frac{(-1)^{n+1} z^n}{n}$ for each z such that $|z| < 1$.
   Hint. Use Ex. 3 and 4.3.7.

8. Find the Laurent series which represents $f(z) = \dfrac{6z + 8}{(2z + 3)(4z + 5)}$ in the given region.

(a) $\left\{ z : \dfrac{5}{4} < |z| < \dfrac{3}{2} \right\}$ **Hint.** See 6.1.8.

(b) $\left\{ z : |z| > \dfrac{3}{2} \right\}$

(c) $\left\{ z : |z| < \dfrac{5}{4} \right\}$

Ans. (a) $\displaystyle\sum_{n=0}^{\infty}(-1)^n \frac{2^n z^n}{3^{n+1}} + \sum_{n=0}^{\infty}(-1)^n \frac{5^n}{4^{n+1}} z^{-n-1}, \quad \frac{5}{4} < |z| < \frac{3}{2}$

(b) $\displaystyle\sum_{n=0}^{\infty}(-1)^n \left(\frac{3^n}{2^{n+1}} + \frac{5^n}{4^{n+1}} \right) z^{-n-1}, \quad |z| > \frac{3}{2}$

(c) $\displaystyle\sum_{n=0}^{\infty}(-1)^n \left(\frac{2^n}{3^{n+1}} + \frac{4^n}{5^{n+1}} \right) z^{n}, \quad |z| < \frac{5}{4}$

9. (a) Find the Taylor series for $f(z) = \dfrac{1}{z^2}$ in powers of $z - a$ where $a \in R^2$ and $a \neq 0$.
    Hint. $\frac{1}{z^2} = \frac{1}{(z-a+a)^2} = \frac{1}{a^2(1+\frac{z-a}{a})^2}$.   Now use Ex. 4.

   (b) Find the Laurent series for $f(z) = \dfrac{1}{z^2}$ for $|z - a| > |a| \neq 0$.
    Hint. $\frac{1}{z^2} = \frac{1}{(z-a+a)^2} = \frac{1}{(z-a)^2} \frac{1}{[1+\frac{a}{z-a}]^2}$. Use Ex. 4.

Ans. (a) $\displaystyle\sum_{n=1}^{\infty}(-1)^{n-1} \frac{n(z-a)^{n-1}}{a^{n+1}}$ for each z such that $|z - a| < |a|$

(b) $\displaystyle\sum_{n=1}^{\infty}(-1)^{n-1} \frac{na^{n-1}}{(z-a)^{n+1}}$ for each z such that $|z - a| > |a|$

10. Represent $z \sinh z^2$ by a Maclaurin series which is valid on R^2.
    Hint. Use 5.3.2. $\displaystyle\sum_{n=1}^{\infty} \frac{z^{4n-1}}{(2n-1)!}$ for each z in R^2

11. Find the Laurent series for $e^{\frac{1}{z}}$ for $|z| > 0$.
    Hint. Use 5.1.2. $\displaystyle\sum_{n=0}^{\infty} \frac{1}{n!z^n}$ for $|z| > 0$

12. Prove that $\dfrac{\cos z^2}{z^3} = \dfrac{1}{z^3} + \displaystyle\sum_{n=1}^{\infty}(-1)^n \frac{z^{4n-3}}{(2n)!}$ for each z such that $|z| > 0$.

13. Find the Laurent series for $f(z) = \dfrac{1}{cz - z^2}$ where $c \neq 0$

    (a) in the region $0 < |z| < |c|$, and

    (b) in the region $|z| > |c|$.

14. Find the Laurent series for $f(z) = \dfrac{1}{(z+1)(z-3)}$

    (a) in powers of $z + 1$ which is valid in $\{z : \ 0 < |z+1| < 4\}$,

    (b) in powers of $z + 1$ which is valid in $\{z : \ |z+1| > 4\}$,

    (c) in powers of $z - 2$ which is valid in $\{z : \ 1 < |z-2| < 3\}$.

    Ans. (a) $\ -\displaystyle\sum_{n=0}^{\infty} \dfrac{(z+1)^{n-1}}{4^{n+1}}, \ \ 0 < |z+1| < 4$

    (b) $\displaystyle\sum_{n=0}^{\infty} \dfrac{4^n}{(z+1)^{n+2}}, \ \ 4 < |z+1|$

    (c) $\dfrac{1}{12}\displaystyle\sum_{n=0}^{\infty} \dfrac{(-1)^{n+1}(z-2)^n}{3^n} + \dfrac{1}{4}\displaystyle\sum_{n=0}^{\infty} \dfrac{1}{(z-2)^{n+1}}, \ \ 1 < |z-2| < 3$

15. Verify (6.13). (See Ex. 3 in 3.6.18. But $\gamma_1 \cup \gamma_2$ is not a curve.)

    Solution 1. Now γ_1 is a curve. Thus by Ex. 3 in 3.6.18, γ_1 is compact. By Ex. 2(b) in 2.4.9, let M_1 be such that $|f(z)| < M_1$ on γ_1. Similarly let M_2 be such that $|f(z)| < M_2$ on γ_2. Now let $M = \max\{M_1, M_2\}$.

    Solution 2. Since γ_1 and γ_2 are closed by Ex. 1 in 1.4.16 it follows that $\gamma_1 \cup \gamma_2$ is closed by Ex. 2(b) in 1.4.20. Clearly $\gamma_1 \cup \gamma_2$ is bounded. Thus $\gamma_1 \cup \gamma_2$ is compact by the Heine-Borel Theorem.

    Solution 3. Since γ_1 and γ_2 are compact, $\gamma_1 \cup \gamma_2$ is compact. (For let C be any open cover for $\gamma_1 \cup \gamma_2$. Let D be a finite subfamily of C which covers γ_1 and let $\mathcal{E}$ be a finite subcover of C for γ_2. Then $D \cup \mathcal{E}$ is a finite subcover of C for $\gamma_1 \cup \gamma_2$.) Now we apply Ex. 2(b) in 2.4.9 to obtain M.

16. Prove Laurent's Theorem as stated in 6.1.2(c).

    Proof. If $C \subset S^\circ$, the conclusion follows from 3.5.11. But if $C \cap (\gamma_1 \cup \gamma_2) \neq \emptyset$, this theorem does not apply directly and we proceed as follows. For each z on γ_2, let $N(z)$ be a neighborhood of z such that $f \in A[N(z)]$. Since γ_2 is compact, let $\{N_{\rho_1}(z_1), N_{\rho_2}(z_2), \cdots, N_{\rho_k}(z_k)\}$ be a finite subcover of $\{N(z) : \quad z \in \gamma_2\}$ for γ_2. Now let Γ be a circle centered at $z = a$ with radius δ such that

$$r_2 < \delta \quad \text{and} \quad \Gamma \subset N_{\rho_1}(z_1) \cup N_{\rho_2}(z_2) \cup \cdots \cup N_{\rho_k}(z_k).$$

Thus $f \in A(D)$ where $D = \{z : \quad r_1 \le |z - a| \le \delta\}$.

Now by (6.2),

$$
\begin{aligned}
a_n &= k \int_{\gamma_2} g(t)\, dt = k \int_{\Gamma} g(t)\, dt \quad \text{by 3.5.11} \\
&= k \int_C g(t)\, dt \quad \text{by 3.5.11}
\end{aligned}
$$

where k and $g(t)$ are obvious from (6.2).

Also (where h is obvious), by (6.2) and 3.5.11,

$$a_{-n} = k \int_{\gamma_1} h(t)\, dt = k \int_{\Gamma} h(t)\, dt = k \int_C h(t)\, dt.$$

Theorem 6.1.10 If

$$f(z) = \sum_{n=0}^{\infty} a_n\, z^n \quad \text{and} \quad g(z) = \sum_{n=0}^{\infty} b_n\, z^n \quad \text{for each } z \text{ in } N_r(0), \tag{6.29}$$

then

$$(fg)(z) = f(z)g(z) = \sum_{n=0}^{\infty} c_n\, z^n \quad \text{for each } z \text{ in } N_r(0) \tag{6.30}$$

where

$$c_n = \sum_{k=0}^{n} a_k\, b_{n-k}. \tag{6.31}$$

[The series (6.30) is called the **Cauchy product** of the two series in (6.29).]

Proof. By 6.1.6, we see that f and g are analytic on $N_r(0)$.
By 6.1.4 and 6.1.7,

$$a_n = \frac{f^{(n)}(0)}{n!} \quad \text{and} \quad b_n = \frac{g^{(n)}(0)}{n!}. \tag{6.32}$$

Observe by 2.3.5 that $fg \in A[N_r(0)]$. Thus by 6.1.4,

$$(fg)(z) = \sum_{n=0}^{\infty} \frac{(fg)^{(n)}(0)}{n!}\, z^n \quad \text{for each } z \text{ in } N_r(0). \tag{6.33}$$

But by Leibniz's rule for the nth derivative of the product of two functions (Ex. 10 in 6.1.14),

$$(fg)^{(n)}(z) = \sum_{k=0}^{n} \frac{n!}{k!\, (n-k)!}\, f^{(k)}(z)\, g^{(n-k)}(z)$$

and hence

$$
\begin{aligned}
\frac{(fg)^{(n)}(0)}{n!} &= \frac{1}{n!} \sum_{k=0}^{n} \frac{n!}{k! \, (n-k)!} \, f^{(k)}(0) \, g^{(n-k)}(0) \\
&= \sum_{k=0}^{n} \frac{f^{(k)}(0) g^{(n-k)}(0)}{k! \, (n-k)!} \\
&= \sum_{k=0}^{n} a_k b_{n-k} \quad \text{by (6.32).} \tag{6.34}
\end{aligned}
$$

Now (6.30) and (6.31) follow from (6.33) and (6.34).

Definition 6.1.11 Let $\sum_{n=1}^{\infty} a_n$ be a series in R^2 and let $r(n)$ be a one-to-one mapping of N onto N. Then the series

$$
\sum_{n=1}^{\infty} a_{r_n} \quad \text{where } r_n = r(n) \quad \text{for each } n \text{ in } N
$$

is called a **rearrangement** of the series $\sum_{n=1}^{\infty} a_n$.

Theorem 6.1.12 Let $\sum_{n=1}^{\infty} a_n = S$ and let $\sum b_n$ be any rearrangement of $\sum a_n$. If $\sum a_n$ converges absolutely, then $\sum b_n$ converges absolutely and $\sum_{n=1}^{\infty} b_n = S$.

Proof. Since $\sum b_n$ is a rearrangement of $\sum a_n$ by 6.1.11,

$$
b_n = a_{r_n} \quad \text{for each } n \text{ in } N, \tag{6.35}
$$

where r is a one-to-one mapping of N onto N. Clearly,

$$
T_n = \sum_{k=1}^{n} |b_k| \le \sum_{n=1}^{\infty} |a_n|.
$$

Thus $\{T_n\}$ is a nondecreasing sequence which is bounded above. Hence by Ex. 28 in 4.3.10, $\lim T_n$ exists so that the series

$$
\sum_{n=1}^{\infty} b_n \text{ converges absolutely.}
$$

Let $\epsilon > 0$. Since $\sum |a_n|$ converges, let $M \in N$ such that

$$
\sum_{k=M+1}^{\infty} |a_k| < \frac{\epsilon}{2}. \tag{6.36}
$$

Then

$$
\left| \sum_{k=1}^{M} a_k - S \right| = \left| \sum_{k=M+1}^{\infty} a_k \right| \le \sum_{k=M+1}^{\infty} |a_k| < \frac{\epsilon}{2}. \tag{6.37}
$$

Choose L sufficiently large so that

$$\{1, 2, \cdots, M\} \subset \{r_1, r_2, \cdots, r_L\}. \tag{6.38}$$

Then for each n in N such that $n > L$,

$$
\begin{aligned}
\left| \sum_{k=1}^{n} b_k - S \right| &= \left| \sum_{k=1}^{n} b_k - \sum_{k=1}^{M} a_k + \sum_{k=1}^{M} a_k - S \right| \\
&\leq \left| \sum_{k=1}^{n} b_k - \sum_{k=1}^{M} a_k \right| + \left| \sum_{k=1}^{M} a_k - S \right| \\
&< \left| \sum_{k=1}^{n} a_{r_k} - \sum_{k=1}^{M} a_k \right| + \frac{\epsilon}{2} \quad \text{by (6.35) and (6.37)} \\
&= \left| \sum_{j=1}^{n-M} a_{t_j} \right| + \frac{\epsilon}{2} \quad \text{by (6.38) where } t_j > M \text{ for each } j \leq n - M \\
&\leq \sum_{j=1}^{n-M} \left| a_{t_j} \right| + \frac{\epsilon}{2} \\
&\leq \sum_{k=M+1}^{\infty} |a_k| + \frac{\epsilon}{2} \quad \text{since } t_j > M \\
&< \frac{\epsilon}{2} + \frac{\epsilon}{2} = \epsilon \quad \text{by (6.36).}
\end{aligned}
$$

Thus $\displaystyle \sum_{n=1}^{\infty} b_n = S$.

Remarks 6.1.13 If f and g are given by (6.29) and if $g(x)$ is never 0 in $N_r(0)$, then $\frac{f}{g} \in A[N_r(0)]$. Hence by 6.1.4,

$$\left(\frac{f}{g} \right)(z) = \frac{f(z)}{g(z)} = \sum_{n=0}^{\infty} q_n z^n \quad \text{for each } z \text{ in } N_r(0)$$

where

$$q_n = \frac{1}{n!} \left(\frac{f}{g} \right)^{(n)} (0)$$

may be determined from 6.1.10 as follows. Let

$$h(z) = \frac{f(z)}{g(z)} = \sum_{n=0}^{\infty} q_n z^n.$$

Then $f(z) = h(z) g(z)$ so that by (6.31),

$$a_n = q_0 b_n + q_1 b_{n-1} + \cdots + q_n b_0.$$

Hence
$$q_n = \frac{a_n - q_0\, b_n - q_1\, b_{n-1} - \cdots - q_{n-1}\, b_1}{b_0} \quad (b_0 = g(0) \neq 0).$$

Thus
$$q_0 = \frac{a_0}{b_0}, \quad q_1 = \frac{(a_1 - q_0\, b_1)}{b_0}, \quad q_2 = \frac{(a_2 - q_0\, b_2 - q_1\, b_1)}{b_0}, \text{ etc.}$$

But these coefficients are exactly those obtained by the long division process as in the quotient of two polynomials, and in practice the long division process is more convenient.

Exercises 6.1.14

In Ex. 1–4, use long division to find the first three nonzero terms in the Maclaurin series for the given function f and determine the open disk in which the series represents f.

1. $f(z) = \dfrac{4 + z^2}{2 - z + 2z^2 - z^3}$ $\qquad\qquad 2 + z - z^2, \; |z| < 1$

2. $f(z) = \sec z = \dfrac{1}{\cos z} = \dfrac{1}{1 - \frac{z^2}{2} + \frac{z^4}{24} - \cdots}$ $\qquad 1 + \dfrac{z^2}{2} + \dfrac{5}{24}z^4, \; |z| < \dfrac{\pi}{2}$

3. $f(z) = \tan z = \dfrac{\sin z}{\cos z}$ $\qquad\qquad z + \dfrac{1}{3}z^3 + \dfrac{2}{15}z^5, \; |z| < \dfrac{\pi}{2}$

4. $f(z) = \dfrac{1}{1 + e^z}$ $\qquad\qquad \dfrac{1}{2} - \dfrac{z}{4} + \dfrac{z^3}{48}, \; |z| < \pi$

In Ex. 5–9, use long division to find the first three nonzero terms in the Laurent series for the given function g and find the open annulus in which the series represents g.

5. $g(z) = \dfrac{4 + z^2}{2z^2 - z^3 + 2z^4 - z^5} = \dfrac{1}{z^2}f(z)$ where $f(z)$ is given in Ex. 1, $0 < |z| < 1$

6. $g(z) = \csc z = \dfrac{1}{\sin z} = \dfrac{1}{z}\dfrac{1}{1 - \frac{z^2}{6} + \frac{z^4}{120} - \cdots}$ $\qquad \dfrac{1}{z} + \dfrac{z}{6} + \dfrac{7z^3}{360}, \; 0 < |z| < \pi$

7. $g(z) = \dfrac{1}{1 - e^z}$ $\qquad\qquad -\dfrac{1}{z} + \dfrac{1}{2} - \dfrac{z}{12}, \; 0 < |z| < 2\pi$

8. $g(z) = \csc 2z^2$ **Hint.** Use Ex. 6. $\quad 0 < |z| < \sqrt{\frac{\pi}{2}}$

9. $g(z) = \cot z$ $\qquad\qquad \dfrac{1}{z} - \dfrac{z}{3} - \dfrac{z^3}{45}, \; 0 < |z| < \pi$

10. Prove **Leibniz's rule** for the nth derivative of a product. If $f^{(n)}(z)$ and $g^{(n)}(z)$ exist, then

$$(fg)^{(n)}(z) = \sum_{k=0}^{n} \binom{n}{k} f^{(n-k)}(z)g^{(k)}(z) \quad \text{where} \quad \binom{n}{k} = \frac{n!}{k!(n-k)!}.$$

Hint. Observe that

(I) $\quad (fg)^{(n)}(z) = \sum_{k=0}^{n} C_k f^{(n-k)}(z)g^{(k)}(z)$

for some $C_k \in N$ and where the coefficients C_k are independent of the choice of f and g. Hence take $f(z) = z^{n-k}$ and $g(z) = z^k$. Note that

$$(fg)^{(n)}(z) = (z^{n-k}z^k)^{(n)} = n!. \tag{6.39}$$

$$\frac{d^{n-j}}{dz^{n-j}} z^{n-k} = \begin{cases} (n-k)! & \text{if } j = k \\ 0 & \text{if } j < k \end{cases} \quad \text{and} \quad \frac{d^j}{dz^j} z^k = \begin{cases} 0 & \text{if } j > k \\ k! & \text{if } j = k. \end{cases}$$

Thus all of the terms on the right in (I), for our choice of f and g, are zero except the term involving C_k which is $C_k k!(n-k)!$. Hence $C_k = \frac{n!}{k!(n-k)!}$ by (6.39).

Example 6.1.15 Let

$$f(z) = \begin{cases} \dfrac{z}{e^z - 1} & \text{if } 0 < |z| < 2\pi \\[2mm] 1 & \text{if } z = 0. \end{cases}$$

Find the Maclaurin series for f in $N_{2\pi}(0)$.

Solution.

$$\frac{z}{e^z - 1} = \frac{z}{z + \frac{z^2}{2!} + \frac{z^3}{3!} + \cdots} = \frac{1}{1 + \frac{z}{2!} + \frac{z^2}{3!} + \cdots + \frac{z^n}{(n+1)!} + \cdots}$$

$$= 1 - \frac{1}{2}z + \frac{1}{12}z^2 - \frac{1}{720}z^4 + \cdots \quad \text{by 6.1.13}. \tag{6.40}$$

Remarks 6.1.16 (Bernoulli Numbers). The Bernoulli numbers $B_n(n = 0, 1, 2, \cdots)$ are defined by

$$B_0 = 1 \quad \text{and} \quad B_n = n!q_n \tag{6.41}$$

where q_n is the coefficient of z^n in the Maclaurin series (6.40) for $\frac{z}{e^z - 1}$. Thus

$$\frac{z}{e^z - 1} = \sum_{n=0}^{\infty} \frac{B_n}{n!} z^n. \tag{6.42}$$

One may use long division to obtain the first several Bernoulli numbers. However, we may use 6.1.13 to obtain a recursive formula for B_n. By 6.1.13, if

$$f(z) = \sum_{n=0}^{\infty} a_n z^n \quad \text{and} \quad g(z) = \sum_{n=0}^{\infty} b_n z^n \quad \text{for each } z \text{ in } N_r(0)$$

then

$$\frac{f}{g}(z) = \frac{f(z)}{g(z)} = \sum_{n=0}^{\infty} q_n z^n \quad \text{for each } z \text{ in } N_r(0)$$

where

$$a_n = q_0 b_n + q_1 b_{n-1} + \cdots + q_n b_0. \tag{6.43}$$

In (6.40), taking $f(z) = 1$ and $g(z) = 1 + \frac{z}{2!} + \cdots + \frac{z^n}{(n+1)!} + \cdots$, we observe that

$$a_n = \begin{cases} 0 & \text{if } n \in N \\ 1 & \text{if } n = 0 \end{cases} \quad \text{and} \quad b_n = \frac{1}{(n+1)!}, \quad n = 0, 1, 2, \cdots. \tag{6.44}$$

Thus by (6.43) and (6.44),

$$\frac{1}{(n+1)!} q_0 + \frac{1}{n!} q_1 + \cdots + q_n = 0. \tag{6.45}$$

Hence by (6.41) and (6.45),

$$\frac{1}{0!(n+1)!} + \frac{B_1}{1!n!} + \cdots + \frac{B_n}{n!1!} = 0. \tag{6.46}$$

Multiplying each member of (6.46) by $(n+1)!$, we have

$$\frac{1}{0!} + \frac{(n+1)B_1}{1!} + \frac{(n+1)nB_2}{2!} + \cdots + \frac{(n+1)B_n}{1!} = 0. \tag{6.47}$$

Now using the binomial theorem on $(1+\beta)^n$, we see that (6.47) is obtained from

$$(1+\beta)^{n+1} - \beta^{n+1} = 0 \tag{6.48}$$

by replacing β^n by B_n. By (6.48), if

$$n = 1, \quad 1 + 2B_1 = 0, \quad B_1 = -\frac{1}{2}$$

$$n = 2, \quad 1 + 3B_1 + 3B_2 = 0, \quad B_2 = \frac{1}{6}$$

$$n = 3, \quad 1 + 4B_1 + 6B_2 + 4B_3 = 0, \quad B_3 = 0$$

$$n = 4, \quad 1 + 5B_1 + 10B_2 + 10B_3 + 5B_4 = 0, \quad B_4 = -\frac{1}{30}$$

$$n = 5, \quad 1 + 6B_1 + 15B_2 + 20B_3 + 15B_4 + 6B_5 = 0, \quad B_5 = 0.$$

We now show that

$$B_{2n+1} = 0 \text{ for } n > 0. \tag{6.49}$$

Let

$$f(z) = \begin{cases} \dfrac{z}{e^z - 1} & \text{if } 0 < |z| < 2\pi \\ 1 & \text{if } z = 0. \end{cases}$$

By (6.42), we have $f(-z) = \sum\limits_{n=0}^{\infty} (-1)^n \dfrac{B_n}{n!} z^n$. Thus

$$f(z) - f(-z) = \sum_{m=0}^{\infty} \frac{2B_{2m+1}\, z^{2m+1}}{(2m+1)!}. \tag{6.50}$$

But

$$f(z) - f(-z) = f(z) + \frac{z}{e^{-z} - 1} = \frac{z}{e^z - 1} - \frac{ze^z}{e^z - 1} = -z. \tag{6.51}$$

Now (6.49) follows by equating coefficients of like powers in (6.50) and (6.51). Thus by (6.42) and (6.49),

$$f(z) = \frac{z}{e^z - 1} = -\frac{z}{2} + \sum_{n=0}^{\infty} \frac{B_{2n}\, z^{2n}}{(2n)!} \quad \text{for } |z| < 2\pi. \tag{6.52}$$

Example 6.1.17 Let

$$g(z) = \begin{cases} z \cot z & \text{if } 0 < |z| < \pi \\ 1 & \text{if } z = 0. \end{cases}$$

Find the Maclaurin series for $g(z)$.

Solution. Now

$$
\begin{aligned}
z \cot z &= iz \frac{e^{iz} + e^{-iz}}{e^{iz} - e^{-iz}} = iz \frac{e^{2iz} + 1}{e^{2iz} - 1} = iz \left(1 + \frac{2}{e^{2iz} - 1}\right) = iz + \frac{2iz}{e^{2iz} - 1} \\
&= \sum_{n=0}^{\infty} \frac{(-1)^n 4^n B_{2n}\, z^{2n}}{(2n)!} \quad \text{for } 0 < |z| < \pi \quad \text{by (6.52)}.
\end{aligned}
$$

Exercises 6.1.18

Verify each of the following.

1. $\tan z = \cot z - 2 \cot 2z = \sum\limits_{n=1}^{\infty} \dfrac{(-1)^{n+1}\, 2^{2n}(2^{2n} - 1)B_{2n}\, z^{2n-1}}{(2n)!}$ for $|z| < \dfrac{\pi}{2}$

2. $\csc z = \cot z + \tan \dfrac{z}{2} = \sum\limits_{n=0}^{\infty} \dfrac{(-1)^{n+1}(2^{2n} - 2)B_{2n}\, z^{2n-1}}{(2n)!}$ for $0 < |z| < \pi$

3. Let $f(z) = \begin{cases} \dfrac{z}{\sin z} & \text{if } z \neq 0 \\ 1 & \text{if } z = 0. \end{cases}$

   Then $f(z) = \displaystyle\sum_{n=0}^{\infty} \frac{(-1)^{n+1}(2^{2n} - 2)B_{2n}\, z^{2n}}{(2n)!}$ for $|z| < \pi$.

4. $z \coth z = \displaystyle\sum_{n=0}^{\infty} \frac{4^n B_{2n}\, z^{2n}}{(2n)!}$ for $0 < |z| < \pi$.

5. $\dfrac{1}{e^z - 1} = -\dfrac{1}{2} + \displaystyle\sum_{n=0}^{\infty} \frac{B_{2n}\, z^{2n-1}}{(2n)!}$ for $0 < |z| < 2\pi$. **Hint.** See (6.52).

6.2 Zeros of Analytic Functions

Let f be analytic at z_0 and let $f(z_0) = 0$. By 6.1.5, we know that f is represented by its Taylor series near z_0. Thus,

$$f(z) = \sum_{n=1}^{\infty} a_n(z - z_0)^n \quad \text{for each } z \text{ in } N(z_0, r) \quad \text{where } a_n = \frac{f^{(n)}(z_0)}{n!}.$$

Thus if there is no neighborhood of z_0 in which $f \equiv 0$, we are led to the following definition.

Definition 6.2.1 Let f be analytic at $z_0 \in R^2$. Then z_0 is a **zero of f of order m** ≥ 1 iff

$$f(z_0) = f'(z_0) = \cdots = f^{(m-1)}(z_0) = 0 \quad \text{and} \quad f^{(m)}(z_0) \neq 0.$$

Remark 6.2.2 Let f be analytic at z_0. Then z_0 is a zero of order $m\, (\geq 1)$ of f iff

$$f(z) = (z - z_0)^m p(z) \quad \text{for each } z \text{ in } N(z_0, r) \text{ for some } r > 0 \qquad (6.53)$$

where

$$p \text{ is analytic at } z_0 \quad \text{and} \quad p(z_0) \neq 0.$$

[To see this, note that

$$\begin{aligned} f(z) &= \sum_{k=m}^{\infty} a_k(z - z_0)^k = (z - z_0)^m \sum_{k=0}^{\infty} a_{m+k}(z - z_0)^k \\ &= (z - z_0)^m p(z) \end{aligned}$$

where

$$p(z) = \sum_{k=0}^{\infty} a_{m+k}(z - z_0)^k.]$$

Theorem 6.2.3 Let $f \in A(D)$ where D is a domain contained in R^2 and let $z_0 \in D$ such that $f(z_0) = 0$.

If there is no neighborhood of z_0 on which $f \equiv 0$, then there is some $r > 0$ such that $f(z) \neq 0$ if $0 < |(z - z_0)| < r$.

Proof. Now z_0 is a zero of f of some order say $m \geq 1$ (since $f \not\equiv 0$ in any neighborhood of z_0). Thus by 6.2.2, we know that f is given by (6.53) where p is analytic at z_0 and $p(z_0) \neq 0$. By Remark α in the proof of 2.2.6(c), let $r > 0$ such that $p(z)$ is never zero in $N_r(z_0)$. Thus by (6.53),

$$\text{if } 0 < |z - z_0| < r, \text{ then } f(z) \neq 0$$

since $(z - z_0)^m \neq 0$ if $z \neq z_0$.

Definition 6.2.4 A zero z_0 of an analytic function f is **isolated** iff there exists an $r > 0$ such that f has no zero in $N'(z_0, r)$.

Remark 6.2.5 Let f be analytic at z_0 and let $f(z_0) = 0$. By 6.2.3, either z_0 is an isolated zero of f or else f is identically zero in some neighborhood of z_0.

Theorem 6.2.6 Let $f \in A(D)$ where D is a domain contained in R^2. If $f \equiv 0$ on a nonempty open set $E \subset D$, then $f \equiv 0$ on D.

Proof. Let

$$G = \{z : \ z \in D \ \text{ and } \ f \equiv 0 \ \text{ on some neighborhood of } z\} \qquad (6.54)$$

and let $H = D - G$. Clearly

$$G \text{ is open, } D = G \cup H, \ G \cap H = \emptyset, \text{ and } \ G \supset E \neq \emptyset. \qquad (6.55)$$

(Indeed G is the largest open subset of D on which $f \equiv 0$.) To show that H is open, assume $z_0 \in H$. **Suppose** $f(z_0) \neq 0$. Then by Remark α in the proof of 2.2.6(c), there is a neighborhood $N(z_0) \subset D$ such that f is never zero in $N(z_0)$. Thus by (6.54), we have $N(z_0) \subset H$ in this case.

Now **suppose** $f(z_0) = 0$. Then (since $z_0 \notin G$) there is by 6.2.3, some neighborhood $N(z_0, r)$ such that f is never zero in $N'(z_0, r)$. Hence $N(z_0, r) \subset H$ in this case.

Thus in either case, there exists some neighborhood of z_0 contained in H so that H is open.

If $H \neq \emptyset$, then D is not connected by (6.55) and Ex. 2 in 1.4.7. But D must be connected since D is a domain. Thus we must have $H = \emptyset$. By (6.55), this means $G = D$. Hence $f \equiv 0$ on D by (6.54).

Theorem 6.2.7 Let $f \in A(D)$ where D is a domain in R^2. Let $T \subset D$ such that T has a limit point z_0 in D. If $f \equiv 0$ on T, then $f \equiv 0$ on D.

Proof. $f \equiv 0$ on T implies $f(z_0) = 0$. [For suppose $f(z_0) \neq 0$. Since f is continuous at z_0, there exists $\delta > 0$ such that f is never zero in $N_\delta(z_0)$ by Remark α in the proof of 2.2.6(c). But this is impossible since $T \cap N_\delta(z_0) \neq \emptyset$.]

Since z_0 is a limit point of T, each neighborhood of z_0 contains points of T different from z_0. Thus $f \equiv 0$ on some neighborhood of z_0 by 6.2.3. Now $f \equiv 0$ on D by 6.2.6.

Theorem 6.2.8 (Identity Theorem for Analytic Functions). Let f and g be analytic in D where D is a domain contained in R^2, and let

$$T \subset D \text{ such that } T^* \cap D \neq \emptyset.$$

If $f \equiv g$ on T, then $f \equiv g$ on D.

Proof. Now $f - g \equiv 0$ on D by 6.2.7. Hence $f \equiv g$ on D.

Definition 6.2.9 Let $f \in A(T)$ and $g \in A(D)$ where $T \subset D$ and $T^* \cap D \neq \emptyset$. Then g is the **analytic continuation of f into D** iff

$$g(z) = f(z) \quad \text{for each } z \text{ in } T.$$

Examples 6.2.10 (a) Show that $\sin^2 z + \cos^2 z = 1$ for each z in R^2.

Solution. Let $f(z) = \sin^2 z + \cos^2 z$ and let $g(z) = 1$. Now f and g are analytic in R^2 and $f(z) = g(z)$ for each z in R. Also, R has a limit point in R^2. Thus $f \equiv g$ on R^2 by 6.2.8.

(b) Let $f(z) = \frac{1}{3-z}$ for z in $D_1 = (R^2 - \{3\})$,

$$g(z) = \sum_{n=0}^{\infty} \frac{z^n}{3^{n+1}} \quad \text{for } z \text{ in } D_2 = \{z : \ |z| < 3\},$$

and

$$h(z) = \sum_{n=0}^{\infty} \frac{(z-i)^n}{(3-i)^{n+1}} \quad \text{for } z \text{ in } D_3 = \{z : \ |z - i| < \sqrt{10}\}.$$

Then by 6.2.8 and 6.2.9,

h is the analytic continuation of g into D_3 [since $h(z) = g(z)$ on $D_2 \cap D_3 = T$ and $T^* \cap D_3 \neq \emptyset$],

g is the analytic continuation of h into D_2, and

f is the analytic continuation of g into D_1.

(c) Show that

$$\sin(z + w) = \sin z \cos w + \cos z \sin w. \tag{6.56}$$

Solution. Let $f(z, w) = \sin(z + w)$ and let $g(z, w)$ be the right hand member of (6.56). Then for a fixed w in R, we have $f(z, w) = g(z, w)$ for each z in R.

Then by 6.2.8 (for the fixed w in R), it follows that $f(z, w) = g(z, w)$ for each z in R^2.

Now (since w was arbitrary in R) for a fixed z in R^2, we have $f(z, w) = g(z, w)$ for each w in R. Thus by 6.2.8 (for a fixed z), it follows that $f(z, w) = g(z, w)$ for each w in R^2. Hence $f(z, w) = g(z, w)$ for all complex numbers z and w.

Theorem 6.2.11 Let $f \in A(D)$ where D is a domain. Then $|f|$ cannot assume a maximum in D unless f is constant on D.

Proof. Suppose there is some z_0 in D such that

$$|f(z)| \leq |f(z_0)| = M \quad \text{for each } z \text{ in } D. \tag{6.57}$$

Since D is open, there exists $D_0 \subset D$ where D_0 is a closed disk given by

$$\{z : \ |z - z_0| \leq r\} \quad \text{where } r > 0.$$

Then

$$|f(z)| \leq |f(z_0)| \quad \text{for each } z \text{ in } D_0 \quad \text{by (6.57).} \tag{6.58}$$

Hence f is constant on D_0 by 3.6.17. Thus f is constant on D by 6.2.8.

Theorem 6.2.12 (Maximum Modulus Theorem). Let $f \in A(D)$ where D is a bounded domain, let $f \in C(\overline{D})$, and let $M = \max\{|f(z)| : \ z \in \overline{D}\}$. If f is not constant on D, then

(i) $f(z_0) = M$ for some point z_0 on the boundary of D and

(ii) $|f(z)| < M$ for each z in D (that is , $|f(z)|$ assumes its maximum on $B(D)$ but never in D).

Proof. By Ex. 2(b) in 2.4.9, we know that $|f|$ assumes a maximum on $\overline{D}$. Furthermore $|f|$ does not assume its maximum in D by 6.2.11. Thus (i) and (ii) follow.

Lemma 6.2.13 (Schwarz's Lemma). Let D be the unit disk $\{z : \ |z| < 1\}$ and let $f \in A(D)$ such that $f(D) \subset \overline{D}$ and $f(0) = 0$.

(a) Then

$$|f(z)| \leq |z| \quad \text{for each } z \text{ in } D \tag{6.59}$$

and

$$|f'(0)| \leq 1. \tag{6.60}$$

(b) If equality holds in (6.59) for at least one z in $(D - \{0\})$ or if equality holds in (6.60), then $f(z) = az$ on D for some a in R^2 such that $|a| = 1$.

Proof. Let

$$g(z) = \begin{cases} \dfrac{f(z)}{z} & \text{if } z \in (D - \{0\}) \\[2mm] f'(0) & \text{if } z = 0. \end{cases} \tag{6.61}$$

Then

$$g \in A(D - \{0\}). \tag{6.62}$$

To see that $g \in A(0)$, we notice by 6.1.4 that

$$g(z) = \frac{f'(0)z + a_2 z^2 + \cdots}{z} = f'(0) + a_2 z + \cdots \quad \text{on } N_1(0) - \{0\} \tag{6.63}$$

since $f(0) = 0$. Also by (6.61), we see that $g(0) = f'(0)$ which is the sum of the last series in (6.63) for $z = 0$. Hence g is represented by this power series in $N_1(0)$ and thus $g \in A(0)$. Hence $g \in A(D)$ by (6.62).

Since $|f(z)| \leq 1$ on D we see that for $z \neq 0$, we have

$$|g(z)| \leq \frac{1}{|z|} \quad \text{by (6.61)}. \tag{6.64}$$

Now g is analytic in the closed disk $D_r = \overline{N_r(0)}$ if $0 < r < 1$. Thus by (6.64) and 6.2.12,

$$|g(z)| \leq \frac{1}{r} \quad \text{on } D_r. \tag{6.65}$$

Letting $r \to 1$, this inequality gives

$$|g(z)| \leq 1 \quad \text{on } D. \tag{6.66}$$

Now Part (a) follows from (6.61) and (6.66).

Finally, if the hypothesis in Part (b) holds, then $|g(z)|$ assumes its maximum of 1 in the **domain** D by (6.61) and (6.66). Hence $g(z) = a$ on D by 6.2.11 where $|a| = 1$. Thus $f(z) = az$ on D by (6.61), and Part (b) is proved.

Remark 6.2.14 From Schwarz's Lemma, we have the following result. If f is an analytic mapping of the unit disk into the unit disk which leaves the origin fixed, then either f is a rotation or f moves each nonzero point closer to the origin.

Exercises 6.2.15

1. Let $f \in A(D)$ where D is a bounded domain, let f be continuous on $\overline{D}$, and let m be the minimum of $|f|$ on $\overline{D}$. If f is never zero in D and f is not constant on D show that $|f(z)|$ assumes its minimum on $B(D)$, but never in D. **Hint.** If $m \neq 0$, consider $\frac{1}{f}$ and use 6.2.12.

2. Let $f(z) = 0$ for each z in R^2 and let $z_0 \in R^2$. We observe that z_0 is a "zero of f," but is not a zero of f of order m for any m in N. **Hint.** Use 2.3.13(c) and 6.2.1.

3. Verify the first statement given in the proof of 6.2.3. **Hint.** Use 6.2.1 and 6.1.4.

6.3   Singular Points (Singularities)

Recall that a function f is said to be analytic at z_0 iff $f'(z)$ exists for each z in some neighborhood of z_0.

Definitions 6.3.1 **(a)** A point z_0 is called a **singular point** of f iff f is not analytic at z_0, but is analytic at some point in each neighborhood of z_0.

(b) A point z_0 is an **isolated singular point** of f iff $f'(z_0)$ does not exist, but f is analytic in $N'(z_0)$ where $N(z_0)$ is some neighborhood of z_0.

Examples 6.3.2 **(a)** The function f given by

$$f(z) = \frac{3z + 2}{z^2(z^2 + 1)}$$

has three isolated singularities; namely, $z = 0$, $z = i$, and $z = -i$.

(b) The function f given by

$$f(z) = \frac{1}{\sin \frac{\pi}{z}}$$

has infinitely many isolated singularities in the interval $(0, 1]$. The origin is a singular point but not an isolated singular point.

Exercise 6.3.3 Let

$$f(z) = \begin{cases} z^2 & \text{if } z \neq \dfrac{1}{n} \\ 0 & \text{if } \dfrac{1}{z} \in N. \end{cases} \quad \text{where } n \in N$$

Prove each of the following statements.

(a) $z = \frac{1}{m}$ is an isolated singularity of f for each m in N.

(b) $z = 0$ is a singularity of f but is not an isolated singularity.

(c) $f'(0) = 0$.

Solution of (a). Let $m \in N$. Then

$$f\left(\frac{1}{m}\right) = 0 \neq \frac{1}{m^2} = \lim_{z \to \frac{1}{m}} f(z).$$

Thus f is not continuous at $\frac{1}{m}$ and hence f is not analytic at $\frac{1}{m}$. But for each $r > 0$, $f \in A(z)$ for some z in $N_r\left(\frac{1}{m}\right)$. Thus $\frac{1}{m}$ is a singular point of f. Also, $\frac{1}{m}$ is an isolated singular point since

$$f \in A\left(N'\left(\frac{1}{m}, r\right)\right) \quad \text{where } r = \frac{1}{m} - \frac{1}{m+1} = \frac{1}{m(m+1)}.$$

Solution of (b). For each $r > 0$, the neighborhood $N_r(0)$ contains points of the form $\frac{1}{m}$. Now $f \notin C\left(\frac{1}{m}\right)$ implies that $f'\left(\frac{1}{m}\right)$ does not exist. Thus $f \notin A(0)$. But for each $r > 0$, the function f is analytic at some point z in $N_r(0)$. Hence 0 is a singular point of f. Now 0 is not an isolated singularity since there are singular points of the form $\frac{1}{m}$ in each neighborhood of 0.

Solution of (c). For $z \neq 0$,

$$\frac{f(z) - f(0)}{z - 0} = \begin{cases} z & \text{if } z \neq \dfrac{1}{m} \\ 0 & \text{if } z = \dfrac{1}{m}. \end{cases}$$

Thus $f'(0) = 0$.

Remarks 6.3.4 If z_0 is an isolated singular point of f, then by 6.1.2, there is some $r > 0$ such that $f(z)$ is given by the sum of the Laurent series in powers of $z - z_0$ for each z in $N_r'(z_0)$. Thus

$$f(z) = \sum_{k=-\infty}^{\infty} a_k(z - z_0)^k \quad \text{for each } z \text{ in } N'(z_0, r).$$

The form of the Laurent series serves to classify isolated singularities into three main types.

Case I. Suppose $a_k = 0$ for each $k < 0$, then we have

$$f(z) = \sum_{k=0}^{\infty} a_k(z - z_0)^k \quad \text{for each } z \text{ in } N_r'(z_0).$$

In this case, we refer to the singularity z_0 as a **removable singularity.** For if we "define or redefine $f(z_0)$ to be a_0," then we obtain a function (different from f at z_0) which is analytic at z_0, since this new function is the sum of its Taylor series about z_0. For an example of a removable singularity, we let $f(z) = \frac{\sin z}{z}$. Now $z_0 = 0$ is the only singularity of f, and we have

$$f(z) = \frac{\sin z}{z} = 1 - \frac{z^2}{3!} + \frac{z^4}{5!} - \cdots z \neq 0.$$

If we extend the domain of f to include $z_0 = 0$ and define $f(0)$ to be 1, then we obtain a new function (which we may continue to denote by f) given by

$$f(z) = \begin{cases} \dfrac{\sin z}{z} & \text{if } z \neq 0 \\ 1 & \text{if } z = 0. \end{cases}$$

Now this new function is analytic at $z_0 = 0$ (and indeed is an entire function), and we may say that we have "removed the singularity of the original function f."

Case II. Suppose the Laurent series for f contains only a finite number ($\neq 0$) of negative powers of $z - z_0$. Then for some $r > 0$, we have for each z in $N'_r(z_0)$,

$$f(z) = \frac{a_{-m}}{(z - z_0)^m} + \cdots + \frac{a_{-1}}{z - z_0} + \sum_{k=0}^{\infty} a_k(z - z_0)^k \qquad (6.67)$$

where $a_{-m} \neq 0$.

In this case, we say that f has a **pole of order m at z_0**. If $m = 1$, then z_0 is called a **simple pole** of f.

Case III. Suppose infinitely many negative powers of $z - z_0$ appear in the Laurent series for f. Then for some $r > 0$,

$$f(z) = \sum_{k=-1}^{-\infty} a_k(z - z_0)^k + \sum_{k=0}^{\infty} a_k(z - z_0)^k \qquad \text{for each } z \text{ in } N'_r(z_0).$$

In this case, f is said to have an **essential singularity** at z_0. Thus an isolated singularity which is not removable and which is not a pole, is an **essential singularity**. For example,

$$f(z) = e^{\frac{1}{z}} = \sum_{n=0}^{\infty} \frac{z^{-n}}{n!}$$

has an essential singularity at $z = 0$.

Remark 6.3.5 From 6.3.4 (Case II), a function h has a pole of order m at z_0 iff for some $r > 0$,

$$h(z) = (z - z_0)^{-m} q(z) \qquad \text{for each } z \text{ in } N'_r(z_0) \qquad (6.68)$$

where by 6.1.6, the function q is analytic at z_0 and $q(z_0) \neq 0$. (To see this, note that $q(z) = a_{-m} + a_{-m+1}(z - z_0) + \cdots$, obtained from (6.67) by replacing f by h.)

Theorem 6.3.6 Let p and q be analytic in a domain $D \subset R^2$ and let

$$p(z_0) \neq 0 \qquad (6.69)$$

where $z_0 \in D$. Then

$$f(z) = \frac{p(z)}{q(z)} \qquad (6.70)$$

has a pole of order m at z_0 iff z_0 is a zero of order m of q.

Proof of "If Part." Let q have a zero of order m at z_0. Then by 6.2.2,

$$q(z) = (z - z_0)^m h(z) \qquad \text{for each } z \text{ in } N(z_0, r_1) \qquad (6.71)$$

where

$$h \text{ is analytic at } z_0 \text{ and } h(z_0) \neq 0.$$

By (6.70) and (6.71),

$$f(z) = (z - z_0)^{-m} g(z) \quad \text{for each } z \text{ in } N'(z_0, r_2) \tag{6.72}$$

where

$$g = \frac{p}{h} \text{ is analytic at } z_0 \quad \text{and by (6.69)}, \quad g(z_0) \neq 0.$$

Thus f has a pole of order m at z_0 by 6.3.5 and (6.72).

Proof of "Only If Part." Let f have a pole of order m at z_0. Then by (6.70) and 6.3.5, there is a function g and a neighborhood $N(z_0)$ of z_0 such that

$$(z - z_0)^{-m} g(z) = f(z) = \frac{p(z)}{q(z)} \quad \text{for each } z \text{ in } N'(z_0) \tag{6.73}$$

where

$$g \text{ is analytic at } z_0 \quad \text{and} \quad g(z_0) \neq 0. \tag{6.74}$$

Thus by (6.73), for some $r_3 > 0$,

$$q(z) = (z - z_0)^m h(z) \quad \text{for each } z \text{ in } N'(z_0, r_3)$$

where

$$h(z) = \frac{p(z)}{g(z)} \text{ is analytic at } z_0 \quad \text{by (6.74)}$$

(since p is analytic at z_0) and $h(z_0) \neq 0$ by (6.69). Hence q has a zero of order m at z_0 by 6.2.2.

Example 6.3.7 Let $P(z)$ and $Q(z)$ be polynomials which have no zero in common. Then the poles of the rational function $\frac{P(z)}{Q(z)}$ are the zeros of $\frac{Q(z)}{P(z)}$. In particular, the rational function

$$\frac{3z + 2}{(z^2 + 4)(z + 1)^3}$$

has poles at $\pm 2i$ of order 1 and a pole at -1 of order 3.

Remarks 6.3.8 Let z_0 be an isolated singular point of f. If $f \in A[N'_r(z_0)]$, then by 6.1.2(a), (c), and (d), the value of $f(z)$ is given by

$$f(z) = \sum_{n=-\infty}^{\infty} a_n(z - z_0)^n \quad \text{for each } z \text{ in } N'_r(z_0),$$

where

$$a_n = \frac{1}{2\pi i} \int_C \frac{f(z)\,dz}{(z - z_0)^{n+1}}$$

and where C is any cco simple closed contour contained in $N'(z_0)$ such that $z_0 \in I(C)$. In particular, the coefficient a_{-1} of $\frac{1}{z - z_0}$ is called the **residue** of f at the isolated singular point z_0 and is denoted by $\text{Res}(f, z_0)$. Thus

$$\text{Res}(f, z_0) = a_{-1} = \frac{1}{2\pi i} \int_C f(z)\, dz. \tag{6.75}$$

Theorem 6.3.9 Let γ be a cco circle with center at z_0, an isolated singularity of f. If $f \in A\left(\overline{I(\gamma)} - \{z_0\}\right)$, then

$$\int_\gamma f(z)\, dz = 2\pi i \, \text{Res}(f, z_0). \tag{6.76}$$

Proof. Formula (6.76) follows from (6.75).

Theorem 6.3.10 (Cauchy's Residue Theorem). Let C be a cco simple closed contour and let

$$f \in A\left(\overline{I(C)} - \{z_1, z_2, \cdots, z_n\}\right)$$

where $z_1, z_2, \cdots, z_n$ are distinct isolated singularities of f in $I(C)$. Then

$$\int_C f(z)\, dz = 2\pi i \sum_{k=1}^n \text{Res}(f, z_k).$$

Proof. For each $k = 1, 2, \cdots, n$, let γ_k be a cco circle with center at z_k such that

$$\gamma_k \subset I(C) \quad \text{and} \quad \gamma_j \subset E(\gamma_k) \ \text{ if } j \neq k.$$

Then

$$f \in A\left(\overline{I(C)} - \cup_{k=1}^n I(\gamma_k)\right).$$

By 3.5.11,

$$\begin{aligned}
\int_C f(z)\, dz &= \sum_{k=1}^n \int_{\gamma_k} f(z)\, dz \\
&= 2\pi i \sum_{k=1}^n \text{Res}(f, z_k) \quad \text{by (6.76).}
\end{aligned}$$

Theorem 6.3.11 If z_0 is a simple pole of f, then

$$\text{Res}(f, z_0) = \lim_{z \to z_0} (z - z_0) f(z). \tag{6.77}$$

Proof. By the definition of a simple pole, for some $r > 0$,

$$f(z) = a_{-1}(z - z_0)^{-1} + \sum_{k=0}^\infty a_k (z - z_0)^k \quad \text{for each } z \text{ in } N'(z_0, r).$$

Hence

$$(z - z_0)f(z) = a_{-1} + \sum_{k=0}^{\infty} a_k(z - z_0)^{k+1} \quad \text{for each } z \text{ in } N'(z_0, r).$$

Thus

$$\lim_{z \to z_0} (z - z_0)f(z) = a_{-1} = \text{Res}(f, z_0).$$

Examples 6.3.12 (a) Find the residue of f at each isolated singularity if $f(z)$ is given by

$$f(z) = \frac{e^z}{z^2 - 1}.$$

Solution. By 6.3.6, the poles of f are $+1$ and -1. By 6.3.11,

$$\text{Res}(f, 1) = \lim_{z \to 1} (z - 1)f(z) = \lim_{z \to 1} \frac{e^z}{z + 1} = \frac{e}{2}.$$

$$\text{Res}(f, -1) = \lim_{z \to -1} (z + 1)f(z) = \lim_{z \to -1} \frac{e^z}{z - 1} = -\frac{e^{-1}}{2}.$$

(b) Evaluate $\int_C \frac{e^z}{z^2 - 1} dz$ where C is the cco circle $|z| = 2$.

Solution. By Cauchy's Residue Theorem,

$$\int_C \frac{e^z}{z^2 - 1} dz = 2\pi i \left[\text{Res}(f, 1) + \text{Res}(f, -1)\right]$$

$$= 2\pi i \left(\frac{e - e^{-1}}{2}\right) = 2\pi i \sinh 1 \quad \text{by Part (a)}.$$

Theorem 6.3.13 Let $f(z) = \frac{g(z)}{(z - z_0)^m}$ where $m \geq 1$, where g is analytic at z_0, and where $g(z_0) \neq 0$. Then f has a pole of order m at $z = z_0$, and

$$\text{Res}(f, z_0) = \frac{g^{(m-1)}(z_0)}{(m - 1)!} \quad \text{where } g^{(0)} = g. \tag{6.78}$$

Proof. By 6.3.6, it follows that f has a pole of order m at z_0. By hypothesis, g is analytic at z_0. Thus by 6.1.4, we know that g is represented by its Taylor series about z_0. Therefore,

$$g(z) = a_{-m} + a_{-m+1}(z - z_0) + \cdots + a_{-1}(z - z_0)^{m-1}$$

$$+ \sum_{k=0}^{\infty} a_k(z - z_0)^{m+k} \quad \text{where } a_{-m} \neq 0. \tag{6.79}$$

Hence by (6.79),

$$
\begin{aligned}
f(z) &= \frac{g(z)}{(z - z_0)^m} \\
&= \frac{a_{-m}}{(z - z_0)^m} + \frac{a_{-m+1}}{(z - z_0)^{m-1}} + \cdots + \frac{a_{-1}}{z - z_0} \\
&\quad + \sum_{k=0}^{\infty} a_k (z - z_0)^k.
\end{aligned}
\tag{6.80}
$$

Thus

$$
a_{-1} = \operatorname{Res}(f, z_0).
\tag{6.81}
$$

But by (6.79), the coefficient of $(z - z_0)^{m-1}$ in the Taylor series for $g(z)$ is given by

$$
a_{-1} = \frac{g^{(m-1)}(z_0)}{(m - 1)!}.
\tag{6.82}
$$

Hence (6.78) follows from (6.81) and (6.82).

Example 6.3.14 Find $\operatorname{Res}(f, 0)$ if $f(z) = -\dfrac{e^{2z}}{z^4}$.

Solution. By 6.3.6, we know that $z = 0$ is a pole of order 4 of the function f. Thus by (6.78),

$$
\operatorname{Res}(f, 0) = \frac{1}{3!} \left. \frac{d^3}{dz^3} \left(-e^{2z} \right) \right|_{z=0} = -\left. \frac{8e^{2z}}{6} \right|_{z=0} = -\frac{4}{3}.
$$

Theorem 6.3.15 Let g and h be analytic at z_0 where $g(z_0) \neq 0$. If $f = \frac{g}{h}$ has a simple pole at z_0, then

$$
\operatorname{Res}(f, z_0) = \frac{g(z_0)}{h'(z_0)}.
$$

Proof. By 6.3.11,

$$
\begin{aligned}
\operatorname{Res}(f, z_0) &= \lim_{z \to z_0} (z - z_0) f(z) \\
&= \lim_{z \to z_0} \left(\frac{z - z_0}{h(z) - h(z_0)} \, g(z) \right) \quad \text{since } h(z_0) = 0 \\
&= \lim_{z \to z_0} \frac{g(z)}{\frac{h(z) - h(z_0)}{z - z_0}} = \frac{g(z_0)}{h'(z_0)}.
\end{aligned}
$$

Exercises 6.3.16

In Ex. $1 - 9$, find

(a) the poles of the given function,

(b) the orders of these poles, and

(c) the residues of the function at these poles.

1. $\dfrac{3z}{z^2 + iz + 2}$ $\qquad\qquad$ (a) $i, -2i$ (b) 1,1 (c) 1,2

2. $\dfrac{e^{iz}}{z^2 + 6iz - 9}$ $\qquad\qquad$ (a) $-3i$ (b) 2 (c) $e^3 i$

3. $\dfrac{\sin z}{z^4}$ **Hint.** 6.3.6 and 6.3.13 do not apply since $\sin 0 = 0$. Use 5.2.3 and 6.3.8.

$\qquad\qquad\qquad\qquad\qquad\qquad\qquad\qquad$ (a) 0   (b) 3   (c) $-\dfrac{1}{6}$

4. $\tan z$ **Hint.** $\tan z = \frac{\sin z}{\cos z}$. See Ex. 2 and 3 of 5.2.11. Apply 6.3.6 and 6.3.15.

$\qquad\qquad\qquad\qquad$ (a) $z_0 = \dfrac{(2n+1)\pi}{2}$, $n \in J$ (b) 1   (c) -1

5. $\dfrac{e^{iz}}{z^2 - \pi^2}$ $\qquad\qquad$ (a) $\pi, -\pi$ (b) 1,1   (c) $-\dfrac{1}{2\pi}, \dfrac{1}{2\pi}$.

6. $\dfrac{e^{-\frac{z}{i}}}{z^4 + 2a^2 z^2 + a^4}$ $\qquad$ (a) $ai, -ai$ (b) 2,2   (c) $\dfrac{e^{-i}}{4a^3}(1 - i)$, $\dfrac{e^i}{4a^3}(1 + i)$

7. $\dfrac{(z + i)^4}{z^3 + 9iz^2 - 27z - 27i}$ $\qquad$ (a) $-3i$ (b) 3   (c) -24

8. $\dfrac{z + i}{\sin z}$ $\qquad\qquad$ (a) $n\pi$, $n \in J$ (b) 1   (c) $(-1)^n(n\pi + 1)$, $n \in J$

9. $\csc z$ $\qquad\qquad$ (a) $n\pi$, $n \in J$ (b) 1   (c) $(-1)^n$, $n \in J$

10. If $f(z) = z^{-4} \csc z^3$, find $\mathrm{Res}\,(f, 0)$.   **Hint.** Use Ex. 6 in 6.1.14 and the definition of a residue. $\qquad\qquad\qquad\qquad\qquad\qquad\qquad \dfrac{1}{6}.$

11. Evaluate $\displaystyle\int_C \dfrac{z^3 - 1}{(z - 3)(z^2 + 4)}\, dz$ where C is the indicated cco circle.

$\qquad$ (a) $|z - 3| = 1$ $\qquad\qquad\qquad\qquad\qquad\qquad$ $4\pi i$

$\qquad$ (b) $|z| = 4$ $\qquad\qquad\qquad\qquad\qquad\qquad\qquad$ $6\pi i$

12. Evaluate $\displaystyle\int_C \tan nz\, dz$ where C is the cco circle $|z| = \dfrac{\pi}{n}$ and $n \in N$.

$\qquad\qquad\qquad\qquad\qquad\qquad\qquad\qquad\qquad\qquad$ $-\dfrac{4\pi i}{n}$

13. Evaluate $\int_C \operatorname{csch} nz\, dz$ where C is the cco circle $|z| = \dfrac{3\pi}{2n}$ and $n \in N$.

$$-\frac{2\pi i}{n}$$

14. Evaluate $\int_C \dfrac{\sin z\, dz}{z^2}$ where C is the cco circle $|z| = 1$. $\qquad\qquad 2\pi i$

15. Evaluate $\int_C \dfrac{\cos(z - a)\, dz}{(z - a)^3}$ where $a \in R^2$ and C is the cco circle $|z - a| = \dfrac{|a|}{2}$.

$$-\pi i$$

16. Evaluate $\int_C \dfrac{z^2 + az + b}{(z - z_1)(z - z_2)(z - z_3)}\, dz$ where $a, b \in R^2$, where C is a cco circle such that z_1, z_2, z_3 are distinct points in $I(C)$ and where $z_k^2 + az_k + b \neq 0$ for $k = 1, 2, 3$.

$$\frac{2\pi i [z_1^2(z_2 - z_3) + z_2^2(z_3 - z_1) + z_3^2(z_1 - z_2)]}{(z_1 - z_2)(z_1 - z_3)(z_2 - z_3)}$$

17. Prove if $f \in A(z_0)$ and

$$q(z) = \begin{cases} \dfrac{f(z) - f(z_0)}{z - z_0} & \text{if } z \neq z_0 \\[2mm] f'(z_0) & \text{if } z = z_0, \end{cases}$$

then $q \in A(z_0)$. (That is,

$$\frac{f(z) - f(z_0)}{z - z_0}$$

has a removeable singularity at z_0.)

Proof. By 6.1.4, there is some $r > 0$ such that

$$f(z) = \sum_{n=0}^{\infty} \frac{f^{(n)}(z_0)}{n!}(z - z_0)^n \quad \text{for each } z \text{ in } N_r(z_0).$$

Hence

$$\frac{f(z) - f(z_0)}{z - z_0} = \sum_{n=1}^{\infty} \frac{f^{(n)}(z_0)}{n!}(z - z_0)^{n-1} \quad \text{for each } z \text{ in } N_r'(z_0).$$

Thus

$$q(z) = \sum_{n=1}^{\infty} \frac{f^{(n)}(z_0)}{n!}(z - z_0)^{n-1} \quad \text{for each } z \text{ in } N_r(z_0).$$

Therefore, $q \in A(z_0)$ by 6.1.6.

Chapter 7

Residue Calculus

7.1 Integrals of Rational Functions of $\sin\theta$ and $\cos\theta$

Let $F(\sin\theta, \cos\theta)$ be a rational function of $\sin\theta$ and $\cos\theta$ with real coefficients which is defined for each θ in $[0, 2\pi]$. Let γ be the cco unit circle given by $z = e^{i\theta}$ for θ in $[0, 2\pi]$. Then for $z \in \gamma$, we have

$$z = e^{i\theta} \quad \text{and} \quad dz = e^{i\theta} i\, d\theta = iz\, d\theta \quad \text{for } \theta \text{ in } [0, 2\pi]. \tag{7.1}$$

From 5.2.7(g),

$$\cos\theta = \frac{e^{i\theta} + e^{-i\theta}}{2} = \frac{1}{2}\left(z + \frac{1}{z}\right) \quad \text{and} \quad \sin\theta = \frac{e^{i\theta} - e^{-i\theta}}{2i} = \frac{1}{2i}\left(z - \frac{1}{z}\right). \tag{7.2}$$

Hence by 3.4.4 and by (7.1) and (7.2),

$$
\begin{aligned}
\int_0^{2\pi} F(\sin\theta, \cos\theta)\, d\theta &= \int_\gamma F\left(\frac{z^2 - 1}{2iz}, \frac{z^2 + 1}{2z}\right) \frac{dz}{iz} \\
&= \begin{cases} 0 \text{ if } G \in A(\overline{I(\gamma)} - \{0\}) \text{ and } 0 \\ \quad \text{is a removable singularity of } G \\[2mm] 2\pi i \sum_{k=1}^{n} \operatorname{Res}(G, z_k) \end{cases}
\end{aligned}
\tag{7.3}
$$

where $z_1, z_2, \cdots, z_n$ are the poles (if any) of G in $I(\gamma)$ and where

$$G(z) = \frac{1}{iz} F\left(\frac{z^2 - 1}{2iz}, \frac{z^2 + 1}{2z}\right). \tag{7.4}$$

Remark 7.1.1 To evaluate the integral in (7.3) for a given F, we determine G in (7.4) by replacing $\sin\theta$ and $\cos\theta$ in F by $\frac{z^2-1}{2iz}$ and $\frac{z^2+1}{2z}$, respectively, and multiplying the result by $\frac{1}{iz}$. Then we use (7.3).

Examples 7.1.2 (a) Evaluate $I = \int_0^{2\pi} \dfrac{d\theta}{\cos\theta + 2}$.

Solution.

$$G(z) = \frac{1}{iz} \frac{1}{\frac{z^2+1}{2z} + 2} = -\frac{i}{\frac{z^2+1}{2} + 2z} = -\frac{2i}{z^2 + 4z + 1}.$$

Now $z_0 = -2 + \sqrt{3}$ is the only root of $z^2 + 4z + 1 = 0$ which is in $\overline{I(\gamma)}$. By 6.3.15, with $g(x) = -2i$ and $h(z) = z^2 + 4z + 1$,

$$\operatorname{Res}(G, z_0) = \frac{g(z_0)}{h'(z_0)} = -\frac{2i}{2z_0 + 4} = -\frac{2i}{2\sqrt{3}} = -\frac{i}{\sqrt{3}}. \tag{7.5}$$

Thus by (7.3), (7.5), and 6.3.10,

$$I = 2\pi i \left(-\frac{i}{\sqrt{3}} \right) = \frac{2\pi\sqrt{3}}{3}.$$

(b) Evaluate $I_1 = \int_0^\pi \dfrac{d\theta}{\cos\theta + 2}$.

Solution. We observe that

$$
\begin{aligned}
\int_0^\pi \frac{d\theta}{\cos\theta + 2} &= \frac{1}{2} \int_{-\pi}^\pi \frac{d\theta}{\cos\theta + 2} \qquad \text{since } \cos(-\theta) = \cos\theta \\
&= \frac{1}{2} \int_\gamma G(z)\, dz \qquad \text{where } G \text{ is given in (7.4)} \\
&= \frac{1}{2} \int_0^{2\pi} \frac{d\theta}{\cos\theta + 2} \\
&= \frac{\pi\sqrt{3}}{3} \qquad \text{by Part (a).}
\end{aligned}
$$

Exercises 7.1.3

In Ex. 1–8, use Cauchy's Residue Theorem to verify each result.

1. $\displaystyle \int_0^\pi \frac{d\theta}{1 + c\cos\theta} = \frac{\pi}{\sqrt{1 - c^2}}$ if $0 < c^2 < 1$

2. $\displaystyle \int_0^{2\pi} \frac{d\theta}{1 + c\sin\theta} = \frac{2\pi}{\sqrt{1 - c^2}}$ if $0 < c^2 < 1$

3. $\displaystyle \int_0^\pi \frac{d\theta}{3 + \cos\theta} = \frac{\pi}{2\sqrt{2}}$

4. $\displaystyle \int_0^{2\pi} \frac{3\, d\theta}{3 - \sin\theta} = \frac{3\pi}{\sqrt{2}}$

5. $\displaystyle\int_0^\pi \frac{\cos\theta\,d\theta}{a+b\cos\theta} = \frac{\pi}{b}\left[1 - \frac{a}{\sqrt{a^2-b^2}}\right]$ if $a,b \in R$ and $a > |b| > 0$

6. $\displaystyle\int_0^{2\pi} \frac{\sin\theta\,d\theta}{a+b\sin\theta} = \frac{2\pi}{b}\left[1 - \frac{a}{\sqrt{a^2-b^2}}\right]$ if $a,b \in R$ and $a > |b| > 0$

7. $\displaystyle\int_0^\pi \frac{\cos\theta\,d\theta}{2-\cos\theta} = \pi\left[\frac{2}{\sqrt{3}} - 1\right]$

8. $\displaystyle\int_0^{2\pi} \frac{\sin\theta}{3+\sin\theta}\,d\theta = 2\pi\left[1 - \frac{3\sqrt{2}}{4}\right]$

We now state Leibniz's rule for differentiating a function which is given by a definite integral involving a parameter in the integrand.

Theorem. If f and f_x are continuous on the closed region $E = \{(x,y):\ x \in [a,b]$ and $y \in [c,d]\}$, then the function $g(x) = \int_c^d f(x,y)\,dy$ is differentiable on $[a,b]$ and $g'(x) = \int_c^d f_x(x,y)\,dy$.

Use Leibniz's rule (stated above) and Ex. 1 and 2 to verify the results in Ex. 9 and 10.

9. $\displaystyle\int_0^\pi \frac{\cos\theta\,d\theta}{(1+c\cos\theta)^2} = \frac{-\pi c}{(1-c^2)^{\frac{3}{2}}}$ if $0 < c^2 < 1$

10. $\displaystyle\int_0^{2\pi} \frac{\sin\theta\,d\theta}{(1+c\sin\theta)^2} = \frac{-2\pi c}{(1-c^2)^{\frac{3}{2}}}$ if $0 < c^2 < 1$

11. Show that $\displaystyle\int_0^\pi \frac{\cos 2\theta\,d\theta}{1+c^2-2c\cos\theta} = \frac{\pi c^2}{|1-c^2|}$ if $0 < c^2 < 1$.

   Hint. $\frac{\cos 2\theta}{1+c^2-2c\cos\theta} = -\frac{1}{c}\frac{2\cos^2\theta-1}{2\cos\theta-\frac{c^2+1}{c}} = -\frac{1}{c}\cos\theta - \frac{c^2+1}{2c^2} + \frac{c^4+1}{2c^2(1+c^2-2c\cos\theta)}$.

   The last step is accomplished by long division.

7.2   Improper Integrals

Definition 7.2.1 (The Cauchy Principal Value Integral). Let f be a real function of a real variable which is Riemann integrable on each closed interval $[a,b]$.

(a) The **improper integral** $\displaystyle\int_{-\infty}^\infty f(x)\,dx$ is defined by

$$\int_{-\infty}^\infty f(x)\,dx = \lim_{a\to-\infty}\int_a^c f(x)\,dx + \lim_{b\to\infty}\int_c^b f(x)\,dx \quad \text{where } c \in R, \qquad (7.6)$$

if each limit in the right member of (7.6) exists.

(b) The Cauchy principal value integral $P \int_{-\infty}^{\infty} f(x)\,dx$ is defined by

$$P \int_{-\infty}^{\infty} f(x)\,dx = \lim_{\rho \to \infty} \int_{-\rho}^{\rho} f(x)\,dx$$

if this limit exists.

Remarks 7.2.2 If $\int_{-\infty}^{\infty} f(x)\,dx$ exists, then $P \int_{-\infty}^{\infty} f(x)\,dx$ exists and the two are

equal. [To see this, put $a = -\rho$ and $b = \rho$ in (7.6).]

However for some f, $P \int_{-\infty}^{\infty} f(x)\,dx$ may exist while $\int_{-\infty}^{\infty} f(x)\,dx$ does not exist. For

example, let $f(x) = x$. Then

$$P \int_{-\infty}^{\infty} f(x)\,dx = \lim_{\rho \to \infty} \int_{-\rho}^{\rho} x\,dx = \lim_{\rho \to \infty} \left[\frac{\rho^2}{2} - \frac{\rho^2}{2} \right] = \lim_{\rho \to \infty} 0 = 0.$$

But

$$\begin{aligned}
\int_{-\infty}^{\infty} x\,dx &= \lim_{a \to -\infty} \int_{a}^{0} x\,dx + \lim_{b \to \infty} \int_{0}^{b} x\,dx \\
&= \lim_{a \to -\infty} \left[\frac{x^2}{2} \right]_{a}^{0} + \lim_{b \to \infty} \left[\frac{x^2}{2} \right]_{0}^{b} \\
&= \lim_{a \to -\infty} -\frac{a^2}{2} + \lim_{b \to \infty} \frac{b^2}{2},
\end{aligned}$$

which is indeterminate.

Theorem 7.2.3 Let $f \in A(H - S)$ where H is the upper half-plane $y \geq 0$, where $S = \{z_1, z_2, \cdots, z_n\} \subset H^\circ$, and where $z_1, z_2, \cdots, z_n$ are distinct singular points of f. Also for each $\rho > 0$, let S_ρ be the cco semicircle

$$S_\rho = \{z : \quad z = \rho e^{i\theta} \quad \text{and} \quad 0 \leq \theta \leq \pi\}.$$

See Figure 7.1. If

$$\lim_{\rho \to \infty} \int_{S_\rho} f(z)\,dz = 0, \tag{7.7}$$

then

$$P \int_{-\infty}^{\infty} f(x)\,dx = 2\pi i \sum_{k=1}^{n} \text{Res}\,(f, z_k). \tag{7.8}$$

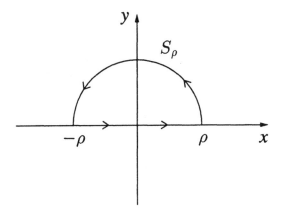

Figure 7.1: S_ρ : $z = \rho e^{i\theta}$ for θ in $[0, \pi]$

Proof. For each $\rho > 0$, let $L_\rho = [-\rho, \rho]$, the closed interval on the x-axis, and let

$$B_\rho = S_\rho \cup L_\rho \quad \text{with the counterclockwise orientation.}$$

By Cauchy's Residue Theorem, 6.3.10, for ρ sufficiently large,

$$
\begin{aligned}
2\pi i \sum_{k=1}^{n} \operatorname{Res}(f, z_k) &= \int_{B_\rho} f(z) \, dz \\
&= \int_{L_\rho} f(z) \, dz + \int_{S_\rho} f(z) \, dz \\
&= \int_{-\rho}^{\rho} f(x) \, dx + \int_{S_\rho} f(z) \, dz.
\end{aligned} \tag{7.9}
$$

Thus by (7.9),

$$
\begin{aligned}
2\pi i \sum_{k=1}^{n} \operatorname{Res}(f, z_k) &= \lim_{\rho \to \infty} \int_{B_\rho} f(z) \, dz \\
&= \lim_{\rho \to \infty} \int_{-\rho}^{\rho} f(x) \, dx \quad \text{by (7.7)} \\
&= P \int_{-\infty}^{\infty} f(x) \, dx
\end{aligned}
$$

which is (7.8).

Example 7.2.4 Show that $\displaystyle P\int_{-\infty}^{\infty} \frac{1}{x^4 + 1}\, dx = \frac{\pi\sqrt{2}}{2}$.

Solution. We show that (7.7) holds when $f(z) = \frac{1}{z^4 + 1}$, that is,

$$\lim_{\rho \to \infty} \int_{S_\rho} \frac{1}{z^4 + 1}\, dz = 0.$$

Now

$$\left|\frac{1}{z^4 + 1}\right| \le \frac{1}{|z^4| - 1} = \frac{1}{\rho^4 - 1} \quad \text{if } \rho = |z| > 1.$$

Thus by 3.4.10(g),

$$\left|\int_{S_\rho} \frac{1}{z^4 + 1}\, dz\right| \le \frac{\pi\rho}{\rho^4 - 1} = \frac{\pi}{\rho^3 - \frac{1}{\rho}} \to 0 \quad \text{as } \rho \to \infty.$$

Hence by (7.8),

$$P\int_{-\infty}^{\infty} \frac{dx}{x^4 + 1} = 2\pi i[\text{Res}\,(f, z_1) + \text{Res}\,(f, z_2)] \tag{7.10}$$

where $z_1 = e^{\frac{\pi i}{4}}$ and $z_2 = e^{\frac{3\pi i}{4}}$ are the only poles of $\frac{1}{z^4 + 1}$ in the upper half-plane $y > 0$. By 6.3.15, for $k = 1, 2$,

$$\text{Res}\,(f, z_k) = \frac{1}{4z_k^3} = \frac{z_k}{4z_k^4} = -\frac{z_k}{4}. \tag{7.11}$$

By (7.10) and (7.11),

$$P\int_{-\infty}^{\infty} \frac{dx}{x^4 + 1} = -\frac{\pi i}{2}\left[e^{\frac{\pi i}{4}} + e^{\frac{3\pi i}{4}}\right] = \frac{\pi\sqrt{2}}{2}.$$

Lemma 7.2.5 Let
$$f(z) = \frac{a_0 + a_1 z + \cdots + a_m z^m}{b_0 + b_1 z + \cdots + b_n z^n}$$
where $m \le n$, $a_m b_n \ne 0$, and the coefficients are complex numbers and where the fraction is in the reduced form and the denominator has no real zeros. Then there are numbers $r > 0$ and $M > 0$ such that

$$|f(z)| < \frac{M}{|z|^s} \quad \text{if } |z| > r \qquad \text{where } s = n - m.$$

Proof. We define g by

$$g(z) = \frac{a_0 z^s + a_1 z^{s+1} + \cdots + a_m z^{s+m}}{b_0 + b_1 z + \cdots + b_n z^n} = z^s f(z). \tag{7.12}$$

We divide both numerator and denominator of (7.12) by z^n and observe that

$$|g(z)| \leq \frac{\left|\frac{a_0}{z^{n-s}}\right| + \left|\frac{a_1}{z^{n-(s+1)}}\right| + \cdots + |a_m|}{\left|\frac{b_0}{z^n} + \frac{b_1}{z^{n-1}} + \cdots + b_n\right|} = H(z) \rightarrow \left|\frac{a_m}{b_n}\right| \quad \text{as } |z| \rightarrow \infty. \tag{7.13}$$

Thus

for each ϵ such that $0 < \epsilon < 1$, there is a number $r > 0$ such that $|z| > r$

implies $|g(z)| \leq H(z) < \left|\frac{a_m}{b_n}\right| + \epsilon.$ \hfill (7.14)

We now let

$$M = \left|\frac{a_m}{b_n}\right| + 1. \tag{7.15}$$

Thus by (7.12), (7.14), and (7.15),

$$|f(z)| = \frac{|g(z)|}{|z|^s} < \frac{M}{|z|^s} \quad \text{if } |z| > r.$$

Lemma 7.2.6 If f is the rational function in 7.2.5 with $s = (n - m) \geq 2$ and if S_ρ is the semicircle defined by

$$z = \rho e^{i\theta} \quad \text{where } \theta \in [0, \pi],$$

then

$$\lim_{\rho \rightarrow \infty} \int_{S_\rho} f(z) \, dz = 0. \tag{7.16}$$

Proof. By 7.2.5 and 3.4.10(g),

$$0 \leq \left|\int_{S_\rho} f(z) \, dz\right| \leq \frac{M}{\rho^2} (\pi\rho) = \frac{\pi M}{\rho} \quad \text{for } \rho \text{ sufficiently large.} \tag{7.17}$$

Thus (7.16) follows by (7.17) and the definition of the limit of a function.

Theorem 7.2.7 Let f be the rational function in 7.2.5 with $s \geq 2$. If $z_1, z_2, \cdots, z_p$ are the poles of f in the upper half-plane $y > 0$, then

$$P\int_{-\infty}^{\infty} f(x) \, dx = 2\pi i \sum_{k=1}^{p} \text{Res}\,(f, z_k). \tag{7.18}$$

Proof. By Cauchy's Residue Theorem, 6.3.10,

$$\int_{-\rho}^{\rho} f(x)\,dx + \int_{S_\rho} f(z)\,dz = 2\pi i \sum_{k=1}^{p} \operatorname{Res}\left(f, z_k\right) \tag{7.19}$$

where

$$\rho > \max\{|z_1|, |z_2|, \cdots, |z_p|\}.$$

Hence (7.18) follows from (7.16) and (7.19).

Exercises 7.2.8

In Ex. $1-7$, use 7.2.7 to verify each result where a and b are distinct positive constants and c and g are any real constants.

1. $\displaystyle\int_0^\infty \frac{dx}{x^2 + a^2} = \frac{\pi}{2a}$ Use the calculus to check your result.

2. $\displaystyle\int_0^\infty \frac{dx}{(x^2 + a^2)^2} = \frac{\pi}{4a^3}$

3. $\displaystyle P\int_{-\infty}^\infty \frac{x^2 + cx + g}{(x^2 + a^2)(x^2 + b^2)}\,dx = \frac{\pi(g + ab)}{ab(a + b)}$

4. $\displaystyle P\int_{-\infty}^\infty \frac{x^2\,dx}{(x^2 + a^2)^2(x^2 + b^2)} = \frac{\pi}{2a(a + b)^2}$

5. $\displaystyle\int_0^\infty \frac{x^2\,dx}{(x^2 + a^2)^2} = \frac{\pi}{4a}$

6. $\displaystyle P\int_{-\infty}^\infty \frac{x^2\,dx}{(x^2 + a^2)(x^2 + b^2)} = \frac{\pi}{a + b}$

7. $\displaystyle P\int_{-\infty}^\infty \frac{x^2 + 7}{(x^2 + 1)(x^2 + 9)}\,dx = \frac{5\pi}{6}$

8. Why is the answer in Ex. 3 independent of c?

9. Let $f(x) \le g(x)$ for each $x > 0$. Prove if $\lim\limits_{x \to \infty} f(x) = a$ and $\lim\limits_{x \to \infty} g(x) = b$, then $a \le b$.

Lemma 7.2.9 For each θ in $\left[0, \frac{\pi}{2}\right]$,

$$e^{-\alpha\rho\sin\theta} \leq e^{-\frac{2\alpha\rho\theta}{\pi}} \quad \text{where } \alpha, \rho > 0. \tag{7.20}$$

Proof.
$$\frac{d}{d\theta}\left(\frac{\sin\theta}{\theta}\right) = \frac{\cos\theta}{\theta^2}(\theta - \tan\theta) < 0 \quad \text{on } \left(0, \frac{\pi}{2}\right) \tag{7.21}$$

since $h(\theta) = \theta - \tan\theta < 0$ on $\left(0, \frac{\pi}{2}\right)$. To see this, note that $h(0) = 0$ and $h'(\theta) = 1 - \sec^2\theta < 0$ on $\left(0, \frac{\pi}{2}\right)$. Hence h is decreasing on $\left[0, \frac{\pi}{2}\right)$. Now by (7.21), we see that $\frac{\sin\theta}{\theta}$ decreases on $\left(0, \frac{\pi}{2}\right]$ and has the value $\frac{2}{\pi}$ at $\theta = \frac{\pi}{2}$. Thus

$$\sin\theta \geq \frac{2\theta}{\pi} \quad \text{on } \left[0, \frac{\pi}{2}\right]. \tag{7.22}$$

Now
$$x_1 \leq x_2 \quad \text{implies} \quad e^{x_1} \leq e^{x_2}.$$

Thus by (7.22),
$$-\alpha\rho\sin\theta \leq -\frac{2\alpha\rho\theta}{\pi}$$

which implies (7.20).

Lemma 7.2.10 If $a \in R$, then

$$\int_0^\pi e^{a\sin\theta}\, d\theta = 2\int_0^{\frac{\pi}{2}} e^{a\sin\theta}\, d\theta. \tag{7.23}$$

Proof. In the second integral, we substitute $\phi = \pi - \theta$. Thus $d\theta = -d\phi$. Also, $\phi = \frac{\pi}{2}$ if $\theta = \frac{\pi}{2}$ and $\phi = 0$ if $\theta = \pi$. Now we have

$$\int_0^\pi e^{a\sin\theta}\, d\theta = \int_0^{\frac{\pi}{2}} e^{a\sin\theta}\, d\theta + \int_{\frac{\pi}{2}}^\pi e^{a\sin\theta}\, d\theta.$$

$$\int_0^\pi e^{a\sin\theta}\, d\theta = \int_0^{\frac{\pi}{2}} e^{a\sin\theta}\, d\theta + \int_{\frac{\pi}{2}}^0 e^{a\sin(\pi-\phi)}(-d\phi)$$

$$= \int_0^{\frac{\pi}{2}} e^{a\sin\theta}\, d\theta + \int_0^{\frac{\pi}{2}} e^{a\sin\phi}\, d\phi = 2\int_0^{\frac{\pi}{2}} e^{a\sin\theta}\, d\theta$$

Theorem 7.2.11 If f is the rational function in 7.2.5, if $s \geq 1$, and if S_ρ is the semicircle in 7.2.6, then

$$\lim_{\rho\to\infty} \int_{S_\rho} e^{i\alpha z} f(z)\, dz = 0 \quad \text{if } \alpha > 0. \tag{7.24}$$

Proof. By 7.2.5, there is a number M such that for ρ sufficiently large,

$$|f(z)| < \frac{M}{\rho} \quad \text{for each } z \text{ on } S_\rho \quad \text{since } s \geq 1. \tag{7.25}$$

On S_ρ,

$$z = \rho e^{i\theta} = \rho(\cos\theta + i\sin\theta) \quad \text{and} \quad dz = \rho i e^{i\theta}\, d\theta.$$

Thus on S_ρ,

$$\begin{aligned}
\left|e^{i\alpha z}\right| &= \left|e^{i\alpha\rho(\cos\theta+i\sin\theta)}\right| = \left|e^{i\alpha\rho\cos\theta}\right|\left|e^{-\alpha\rho\sin\theta}\right| \\
&= e^{-\alpha\rho\sin\theta} \quad \text{since } \alpha\rho\cos\theta \text{ is real.}
\end{aligned} \tag{7.26}$$

By (7.25) and (7.26), for ρ sufficiently large,

$$\left|e^{i\alpha z} f(z)\right| < e^{-\alpha\rho\sin\theta}\frac{M}{\rho} \quad \text{for each } z \text{ on } S_\rho. \tag{7.27}$$

Now by (7.27) and 3.4.10(f),

$$\begin{aligned}
\left|\int_{S_\rho} e^{i\alpha z} f(z)\, dz\right| &\leq \int_0^\pi \left|e^{i\alpha z} f(z)\right| \rho\, d\theta \leq \int_0^\pi e^{-\alpha\rho\sin\theta}\frac{M}{\rho}\rho\, d\theta \\
&= 2M \int_0^{\frac{\pi}{2}} e^{-\alpha\rho\sin\theta}\, d\theta \quad \text{by 7.2.10} \\
&\leq M \int_0^{\frac{\pi}{2}} e^{-\frac{2\alpha\rho\theta}{\pi}} 2\, d\theta \quad \text{by 7.2.9} \\
&= M \left[-\frac{\pi}{\alpha\rho} e^{-\frac{2\alpha\rho\theta}{\pi}}\right]_0^{\frac{\pi}{2}} \\
&= \frac{M\pi}{\alpha\rho}\left[1 - e^{-\alpha\rho}\right] < \frac{\pi M}{\alpha\rho}.
\end{aligned}$$

Hence

$$\lim_{\rho\to\infty} \int_{S_\rho} e^{i\alpha z} f(z)\, dz = 0.$$

Theorem 7.2.12 Let f be the rational function of 7.2.5 with $s \geq 1$. Let $f \in A(R)$ and let $z_1, z_2, \cdots, z_n$ be the distinct poles of f in the upper half-plane $y > 0$. Then for $\alpha > 0$,

$$P\int_{-\infty}^\infty e^{i\alpha x} f(x)\, dx = 2\pi i \sum_{k=1}^n \text{Res}\,(e^{i\alpha z} f, z_k). \tag{7.28}$$

Also, if $f(-x) = f(x)$ on R, then

$$\int_0^\infty f(x)\cos\alpha x\, dx = \pi i \sum_{k=1}^n \text{Res}\,(e^{i\alpha z} f, z_k). \tag{7.29}$$

Finally, if $f(-x) = -F(x)$ on R, then

$$\int_0^\infty f(x) \sin \alpha x \, dx = \pi \sum_{k=1}^n \text{Res} \, (e^{i\alpha z} f, z_k). \tag{7.30}$$

Proof. By Cauchy's Residue Theorem, 6.3.10,

$$\int_{-\rho}^\rho e^{i\alpha x} f(x) \, dx + \int_{S_\rho} e^{i\alpha z} f(z) \, dz = 2\pi i \sum_{k=1}^n \text{Res} \, (e^{i\alpha z} f, z_k) \tag{7.31}$$

where S_ρ is the semicircle $z = \rho e^{i\theta}$ for θ in $[0, \pi]$ with

$$\rho > \max\{|z_1|, |z_2|, \cdots, |z_n|\}.$$

Thus (7.28) follows from (7.24) and (7.31).

To verify (7.29), let $x = -x'$. Then

$$\int_{-\rho}^0 e^{i\alpha x} f(x) \, dx = \int_\rho^0 e^{-i\alpha x'} f(-x')(-dx') = \int_0^\rho e^{-i\alpha x'} f(-x') \, dx'. \tag{7.32}$$

Hence if $f(-x') = f(x')$, then

$$\int_{-\rho}^\rho e^{i\alpha x} f(x) \, dx = 2 \int_0^\rho \frac{(e^{-i\alpha x} + e^{i\alpha x})}{2} f(x) \, dx = 2 \int_0^\rho f(x) \cos \alpha x \, dx. \tag{7.33}$$

But if $f(-x') = -f(x')$, then

$$
\begin{aligned}
\int_{-\rho}^\rho e^{i\alpha x} f(x) \, dx &= \int_{-\rho}^0 e^{i\alpha x} f(x) \, dx + \int_0^\rho e^{i\alpha x} f(x) \, dx \\
&= \int_0^\rho e^{-i\alpha x'} f(-x') \, dx' + \int_0^\rho e^{i\alpha x} f(x) \, dx \quad \text{by (7.32)} \\
&= -\int_0^\rho e^{-i\alpha x'} f(x') \, dx' + \int_0^\rho e^{i\alpha x} f(x) \, dx \\
&= 2i \int_0^\rho \frac{e^{i\alpha x} - e^{-i\alpha x}}{2i} f(x) \, dx \\
&= 2i \int_0^\rho f(x) \sin \alpha x \, dx. \tag{7.34}
\end{aligned}
$$

Hence (7.28) and (7.33) imply (7.29) while (7.28) and (7.34) imply (7.30).

Remarks 7.2.13 Let f and α be given as in 7.2.12.

(a) The poles of $e^{i\alpha z} f$ are the poles of f, but the two functions might not have the same residues at a given pole.

(b) If the coefficients in f are **real**, then

$$\mathcal{R}\left[\,_P\!\!\int_{-\infty}^{\infty} e^{i\alpha x} f(x)\,dx\right] = \,_P\!\!\int_{-\infty}^{\infty} f(x)\cos\alpha x\,dx$$

and

$$\mathcal{I}\left[\,_P\!\!\int_{-\infty}^{\infty} e^{i\alpha x} f(x)\,dx\right] = \,_P\!\!\int_{-\infty}^{\infty} f(x)\sin\alpha x\,dx.$$

Thus by (7.28),

$$_P\!\!\int_{-\infty}^{\infty} f(x)\cos\alpha x\,dx = \mathcal{R}\left[2\pi i\sum_{k=1}^{n}\mathrm{Res}\left(e^{i\alpha z}f, z_k\right)\right]$$

and

$$_P\!\!\int_{-\infty}^{\infty} f(x)\sin\alpha x\,dx = \mathcal{I}\left[2\pi i\sum_{k=1}^{n}\mathrm{Res}\left(e^{i\alpha z}f, z_k\right)\right].$$

(c) Let the coefficients in f be real. If f is even, then

$$_P\!\!\int_{-\infty}^{\infty} f(x)\cos\alpha x\,dx = 2\int_{0}^{\infty} f(x)\cos\alpha x\,dx$$

and

$$_P\!\!\int_{-\infty}^{\infty} f(x)\sin\alpha x\,dx = 0.$$

If f is odd, then

$$_P\!\!\int_{-\infty}^{\infty} f(x)\sin\alpha x\,dx = 2\int_{0}^{\infty} f(x)\sin\alpha x\,dx \quad\text{and}\quad _P\!\!\int_{-\infty}^{\infty} f(x)\cos\alpha x\,dx = 0.$$

Exercises 7.2.14

Evaluate the given integrals in Ex. $1-8$ where $a > 0$ and $b > 0$.

1. $\displaystyle\int_{0}^{\infty} \frac{x\sin ax\,dx}{x^2 + b^2} = \frac{\pi}{2}e^{-ab}$

2. $\displaystyle\,_P\!\!\int_{-\infty}^{\infty} \frac{x\sin ax\,dx}{x^2 + b^2} = \pi e^{-ab}$

3. $\displaystyle\int_{0}^{\infty} \frac{\cos ax\,dx}{x^2 + b^2} = \frac{\pi}{2b}e^{-ab}$

4. $\displaystyle P\int_{-\infty}^{\infty} \frac{\cos ax\,dx}{x^2+b^2} = \frac{\pi}{b}e^{-ab}$

5. $\displaystyle P\int_{-\infty}^{\infty} \frac{\sin ax\,dx}{x^2+b^2} = 0$

6. $\displaystyle P\int_{-\infty}^{\infty} \frac{x\cos ax\,dx}{x^2+b^2} = 0$

7. $\displaystyle \int_{0}^{\infty} \frac{\cos ax\,dx}{(x^2+b^2)^2} = \frac{\pi e^{-ab}(ab+1)}{4b^3}$

8. $\displaystyle P\int_{-\infty}^{\infty} \frac{\cos ax\,dx}{(x^2+b^2)^2} = \frac{\pi e^{-ab}(ab+1)}{2b^3}$

Lemma 7.2.15 Let S_ρ be an arc of the circle given by

$$z = a + \rho e^{i\theta} \quad \text{where } a \in R,\ \theta_1 \le \theta \le \theta_2,\ \text{and } \rho > 0.$$

If f has a simple pole at $z = a$, then

$$\lim_{\rho \to 0} \int_{S_\rho} f(z)\,dz = i(\theta_2 - \theta_1)\operatorname{Res}(f, a).$$

Proof. By (6.67),

$$f(z) = \frac{\operatorname{Res}(f, a)}{z - a} + g(z) \quad \text{for each } z \text{ in } N'(a, r) \tag{7.35}$$

for some $r > 0$ where

$$g(z) = \sum_{n=0}^{\infty} a_n(z - a)^n.$$

Let r_1 be fixed such that $0 < r_1 < r$. Since $g \in A(N(a, r))$, we have $g \in C(\overline{N(a, r_1)})$. Thus by Ex. 2(b) in 2.4.9, there is a real number M such that

$$|g(z)| \le M \quad \text{for each } z \text{ in } \overline{N(a, r_1)}. \tag{7.36}$$

Hence by (7.36) and 3.4.10(g) for $\rho \le r_1$,

$$\left| \int_{S_\rho} g(z)\,dz \right| \le \rho(\theta_2 - \theta_1)M \to 0 \quad \text{as } \rho \to 0. \tag{7.37}$$

By (7.35),

$$\int_{S_\rho} f(z)\, dz = \operatorname{Res}(f,a) \int_{\theta_1}^{\theta_2} \frac{i\rho e^{i\theta}}{\rho e^{i\theta}}\, d\theta + \int_{S_\rho} g(z)\, dz.$$

Thus by (7.37),

$$\lim_{\rho \to 0} \int_{S_\rho} f(z)\, dz = i(\theta_2 - \theta_1) \operatorname{Res}(f,a).$$

Definition 7.2.16 Let f be a real function of a real variable which is Riemann integrable on each closed interval contained in $[a,b] - \{c\}$ where c is some fixed point in the open interval (a,b) and let $|f(x)| \to \infty$ as $x \to c$.

(a) The **improper integral** $\int_a^b f(x)\, dx$ is defined by

$$\int_a^b f(x)\, dx = \lim_{\epsilon \to 0+} \int_a^{c-\epsilon} f(x)\, dx + \lim_{\delta \to 0+} \int_{c+\delta}^b f(x)\, dx$$

if the two limits exist.

(b) The **Cauchy Principal Value Integral** $P\int_a^b f(x)\, dx$ is defined by

$$P\int_a^b f(x)\, dx = \lim_{\epsilon \to 0+} \left[\int_a^{c-\epsilon} f(x)\, dx + \int_{c+\epsilon}^b f(x)\, dx \right]$$

if this limit exists.

(c) Let f be Riemann integrable on each closed interval contained in $R - \{c\}$. If

$$\lim_{\rho \to \infty} P\int_{-\rho}^{\rho} f(x)\, dx$$

exists, then this limit is called the **Cauchy Principal Value Integral** from $-\infty$ to ∞ and is denoted by

$$P\int_{-\infty}^{\infty} f(x)\, dx.$$

(d) Let f be a real function of a real variable such that $|f(x)| \to \infty$ as $x \to c_k$ for each $k = 1, 2, \cdots, n$ where $c_1 < c_2 < \cdots < c_n$. Let f be Riemann integrable on each closed interval contained in $(R - \{c_1, c_2, c_3, \cdots, c_n\})$. Then

$$P\int_{-\infty}^{\infty} f(x)\, dx = \lim_{\rho \to \infty} P\int_{-\rho}^{\rho} f(x)\, dx$$

where

$$P\int_{-\rho}^{\rho} f(x)\,dx \;=\; \lim_{\epsilon_k \to 0^+} \left[\int_{-\rho}^{c_1-\epsilon_1} f(x)\,dx + \int_{c_1+\epsilon_1}^{c_2-\epsilon_2} f(x)\,dx + \cdots \right.$$

$$\left. + \int_{c_n+\epsilon_n}^{\rho} f(x)\,dx \right]$$

if these limits exist.

Theorem 7.2.17 Let f be a rational function with coefficients in R such that the degree of the denominator is greater than the degree of the numerator. Let

$$f \in A(\{z: \; \mathcal{I}(z) \geq 0\} - \{z_1, z_2, \cdots, z_p, x_1, x_2, \cdots, x_q\})$$

where z_j is a pole of f with $\mathcal{I}(z_j) > 0$ for j equal 1 to p and where each x_k is a **simple** pole of f on the real axis with $x_1 < x_2 < \cdots < x_q$. Then

$$P\int_{-\infty}^{\infty} e^{i\alpha x} f(x)\,dx \;=\; 2\pi i \sum_{j=1}^{p} \mathrm{Res}\,(e^{i\alpha z} f, z_j) + \pi i \sum_{k=1}^{q} \mathrm{Res}\,(e^{i\alpha z} f, x_k) \qquad (7.38)$$

where $\alpha > 0$.

Proof. First suppose $q = 1$, that is, there is exactly one simple pole x_1 on the real axis. Let ρ be sufficiently large and r sufficiently small so that $z_j \in I(C)$ for each $j = 1, 2, \cdots, p$ where

$$C = L_1 + (-S_r) + L_2 + S_\rho$$

and where L_1, L_2, S_r, and S_ρ are the line segments and cco semicircular arcs as shown in Figure 7.2.

Then by Cauchy's Residue Theorem, 6.3.10,

$$\int_{-\rho}^{x_1-r} e^{i\alpha x} f(x)\,dx + \int_{-S_r} e^{i\alpha z} f(z)\,dz + \int_{x_1+r}^{\rho} e^{i\alpha x} f(x)\,dx + \int_{S_\rho} e^{i\alpha z} f(z)\,dz \;=\;$$

$$2\pi i \sum_{j=1}^{p} \mathrm{Res}\,(e^{i\alpha z} f, z_j). \qquad (7.39)$$

Taking the limit as $r \to 0^+$, by 7.2.15,

$$P\int_{-\rho}^{\rho} e^{i\alpha x} f(x)\,dx \;=\; \pi i \,\mathrm{Res}\,(e^{i\alpha z} f, x_1)$$

$$- \int_{S_\rho} e^{i\alpha z} f(z)\,dz \;+\; 2\pi i \sum_{j=1}^{p} \mathrm{Res}\,(e^{i\alpha z} f, z_j). \qquad (7.40)$$

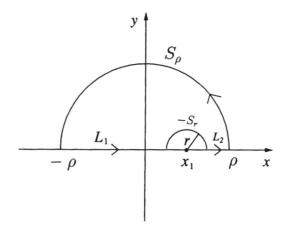

Figure 7.2: $C = L_1 + (-S_r) + L_2 + S_\rho$

[We know that $\displaystyle P\int_{-\rho}^{\rho} e^{i\alpha x} f(x)\, dx$ exists since the right member of (7.40) is a number.]

Now taking the limit in (7.40) as $\rho \to \infty$, we have by 7.2.11,

$$P\int_{-\infty}^{\infty} e^{i\alpha x} f(x)\, dx = 2\pi i \sum_{j=1}^{p} \operatorname{Res}\left(e^{i\alpha z} f, z_j\right) + \pi i \operatorname{Res}\left(e^{i\alpha z} f, x_1\right).$$

Now for the case in which $q > 1$, we insert a semicircle about each x_k and by a similar argument, we obtain (7.38).

Exercises 7.2.18

In Ex. 1–4, verify each result. See 7.2.13(b) and (c) and 7.2.17.

1. $\displaystyle \int_0^\infty \frac{\sin bx}{x}\, dx = \begin{cases} \dfrac{\pi}{2} & \text{if } b > 0 \\[2mm] -\dfrac{\pi}{2} & \text{if } b < 0 \\[2mm] 0 & \text{if } b = 0 \end{cases}$

2. $\displaystyle P\int_{-\infty}^{\infty} \frac{\cos bx}{x}\, dx = 0$

3. $I(a) = \int_0^\infty \dfrac{\sin \alpha x\, dx}{x(x^2 + a^2)} = \dfrac{\pi}{2a^2}\left(1 - e^{-a\alpha}\right)$

4. $\displaystyle\int_0^\infty \dfrac{\sin \alpha x\, dx}{x(x^2 + a^2)^2} = \dfrac{\pi}{2a^4}\left(1 - \dfrac{a\alpha + 2}{2}\, e^{-a\alpha}\right)$

5. Prove if the degree of the denominator of f in 7.2.17 exceeds the degree of the numerator by at least two, then

$$P\int_{-\infty}^\infty f(x)\, dx = 2\pi i \sum_{j=1}^p \operatorname{Res}(f, z_j) + \pi i \sum_{k=1}^q \operatorname{Res}(f, x_k).$$

6. Use Ex. 5 to show that $\displaystyle P\int_{-\infty}^\infty \dfrac{x^2 + x + 1}{x^4 - 1}\, dx = 0.$

7. Show that $\displaystyle P\int_{-\infty}^\infty \dfrac{dx}{x^3 + 1} = \dfrac{\pi}{\sqrt{3}}.$

8. Apply the following theorem to the equation in Ex. 3 to verify the result in Ex. 4.

Theorem. Let $f(a, x), f_a(a, x) \in C(S)$ where $S = \{(a, x) : \quad a_1 \le a \le a_2$ and $b \le x < +\infty\}$, and let $\int_b^\infty f_a(a, x)\, dx$ converge uniformly on $[a_1, a_2]$. If $I(a) = \int_b^\infty f(a, x)\, dx$ for each a in $[a_1, a_2]$, then $I'(a) = \int_b^\infty f_a(a, x)\, dx$ for each a in $[a_1, a_2]$.

Theorem 7.2.19 Let f be a rational function with distinct poles $z_j = x_j + iy_j$ where $j = 1, 2, \cdots, n$ such that none of them is on the positive real axis. Let $a \in (R - J)$ and let

$$g(z) = (-z)^{a-1} f(z) \quad \text{where } -\pi < \operatorname{Arg}(-z) \le \pi. \tag{7.41}$$

If

$$\lim_{|z|\to\infty} |z||g(z)| = 0 = \lim_{z\to 0} |z||g(z)|, \tag{7.42}$$

then

$$\int_0^\infty x^{a-1} f(x)\, dx = \dfrac{\pi}{\sin \pi a} \sum_{j=1}^n \operatorname{Res}(g, z_j). \tag{7.43}$$

Proof. We apply Cauchy's Residue Theorem, (6.3.10), to f on the closed contour C, consisting of the circular arcs C_M and C_m of radii M and m, respectively, and line segments L_1 and L_2 on the rays $\theta = \epsilon$ and $\theta = 2\pi - \epsilon$ as shown in Figure 7.3. We choose m, M, and ϵ so that $z_j \in I(C)$ for each $j = 1, 2, \cdots, n$. Thus by 6.3.10,

$$\begin{aligned}
\int_C g(z)\, dz &= \int_{C_M} g(z)\, dz + \int_{L_2} g(z)\, dz + \int_{C_m} g(z)\, dz + \int_{L_1} g(z)\, dz \\
&= 2\pi i \sum_{j=1}^n \operatorname{Res}(g, z_j). \tag{7.44}
\end{aligned}$$

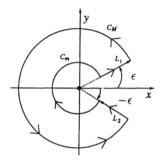

Figure 7.3: $C = L_1 + C_M + L_2 + C_m$

Now

$$\left| \int_{C_M} g(z)\, dz \right| = \left| \int_{C_M} \frac{(-z)^a f(z)\, dz}{(-z)} \right|$$

$$\leq (2\pi M) \frac{K_M}{M} \quad \text{by 3.4.10(g)}$$

$$= 2\pi K_M \quad \text{where } K_M = \max |(-z)^a f(z)| \text{ on } C_M. \quad (7.45)$$

We know that K_M exists by Ex. 2(b) in 2.4.9 and by Ex. 9(a) in 5.4.11.
Similarly,

$$\left| \int_{C_m} g(z)\, dz \right| \leq 2\pi K_m \quad \text{where } K_m = \max |z|^a |f(z)| \text{ on } C_m. \quad (7.46)$$

Also, on L_1, $z = re^{i\epsilon}$, $dz = e^{i\epsilon}\, dr$, and $-z = re^{i(\epsilon-\pi)}$. Thus

$$\int_{L_1} g(z)\, dz = \int_m^M r^{a-1} \left[e^{i(\epsilon-\pi)} \right]^{a-1} f(re^{i\epsilon}) e^{i\epsilon}\, dr$$

$$= -e^{ia(\epsilon-\pi)} \int_m^M r^{a-1} f(re^{i\epsilon})\, dr. \quad (7.47)$$

Likewise, on L_2, we have $z = re^{i(2\pi-\epsilon)} = re^{-i\epsilon}$ and $dz = e^{-i\epsilon}\, dr$. Furthermore, $-z = re^{i(\pi-\epsilon)}$. Hence

$$\int_{L_2} g(z)\, dz = \int_M^m r^{a-1} e^{i(\pi-\epsilon)(a-1)} f(re^{-i\epsilon}) e^{-i\epsilon}\, dr$$

$$= e^{ia(\pi-\epsilon)} \int_m^M r^{a-1} f(re^{-i\epsilon})\, dr. \quad (7.48)$$

By (7.44), (7.47), and (7.48),

$$2\pi i \sum_{j=1}^n \operatorname{Res}(g, z_j) = \int_{C_M} g(z)\, dz + e^{ia(\pi-\epsilon)} \int_m^M r^{a-1} f(re^{-i\epsilon})\, dr$$

$$+ \int_{C_m} g(z)\, dz - e^{ia(\epsilon-\pi)} \int_m^M r^{a-1} f(re^{i\epsilon})\, dr \quad (7.49)$$

In (7.49), we **first** let $\epsilon \to 0$ and then let $m \to 0$ and $M \to \infty$. By (7.42), we see that $K_M \to 0$ as $M \to \infty$, and $K_m \to 0$ as $m \to 0$. Thus by (7.49), we have $e^{\pi a i} \int_0^\infty r^{a-1} f(r) \, dr - e^{-\pi a i} \int_0^\infty r^{a-1} f(r) \, dr = 2\pi i \sum_{j=1}^n \text{Res}\,(g, z_j)$. This implies (7.43) since

$$\frac{2\pi i}{e^{\pi a i} - e^{-\pi a i}} = \pi \frac{2i}{e^{\pi a i} - e^{-\pi a i}} = \frac{\pi}{\sin \pi a} \quad \text{by 5.2.7(g)}.$$

Note. To see that $\int_m^M r^{a-1} f(re^{i\epsilon}) \, dr \to \int_m^M r^{a-1} f(r) \, dr$ as $\epsilon \to 0$, we observe that $r^{a-1} f(re^{iz})$ is continuous on $D \times [m, M]$, where $D = N_\rho(0)$ if ρ is sufficiently small that r in $[m, M]$, re^{iz} is not a pole of f when z is in D and r is in $[m, M]$. Thus by 3.7.8, it follows that $\int_m^M r^{a-1} f(re^{iz}) \, dr \to \int_m^M r^{a-1} f(r) \, dr$ as $z \to 0$ (in any direction). In particular if $z = \epsilon \to 0^+$, the same limit results.

Observe by 5.4.7, **when $x > 0$**,

$$(-z)^{a-1} \to -x^{a-1} e^{-\pi a i} \text{ as } z \to x \text{ from above the } x\text{-axis}$$

and

$$(-z)^{a-1} \to -x^{a-1} e^{\pi a i} \text{ as } z \to x \text{ from below the } x\text{-axis}.$$

Thus g is not continuous and hence not analytic at any point on the positive real axis.

Example 7.2.20 Show that

$$\int_0^\infty \frac{x^{a-1} \, dx}{x^2 + b^2} = \frac{\pi b^{a-2}}{2 \sin \frac{\pi a}{2}} \quad \text{if } 0 < a < 2, \ a \neq 1, \text{ and } b > 0.$$

Solution. In (7.43), we have $f(x) = \frac{1}{x^2 + b^2}$. Observe that (7.42) holds. For $|(-z)^{a-1}| = |-z|^{a-1} = |z|^{a-1}$ since $a \in R$. See the first sentence following 5.4.7. By 6.3.11,

$$\begin{cases} \text{Res}\left(\frac{(-z)^{a-1}}{(z - bi)(z + bi)}, bi\right) &= \lim_{z \to bi} \frac{(-z)^{a-1}}{z + bi} = \frac{(-bi)^{a-1}}{2bi} \\ &= \frac{e^{(a-1)\left[\text{Log}\,|bi| - \frac{\pi}{2}i\right]}}{2bi} \\ \text{Res}\left(\frac{(-z)^{a-1}}{(z - bi)(z + bi)}, -bi\right) &= \lim_{z \to -bi} \frac{(-z)^{a-1}}{z - bi} = \frac{(bi)^{a-1}}{-2bi} \\ &= \frac{e^{(a-1)\left[\text{Log}\,|bi| + \frac{\pi}{2}i\right]}}{-2bi}. \end{cases} \quad (7.50)$$

Thus by (7.43) and (7.50),

$$\int_0^\infty \frac{x^{a-1} \, dx}{x^2 + b^2} = \frac{\pi b^{a-2}}{\sin \pi a} \left[\frac{e^{-(a-1)\frac{\pi}{2}i} - e^{(a-1)\frac{\pi}{2}i}}{2i}\right]$$

$$= \frac{\pi b^{a-2}}{\sin \pi a} \left[\sin \left(\frac{\pi}{2} - \frac{\pi a}{2} \right) \right] \quad \text{by 5.2.7(g)}$$

$$= \frac{\pi b^{a-2} \cos \frac{\pi a}{2}}{2 \sin \frac{\pi a}{2} \cos \frac{\pi a}{2}} = \frac{\pi b^{a-2}}{2 \sin \frac{\pi a}{2}} \left\{ \begin{array}{l} \text{by 5.2.8 and} \\ \text{Ex. 1(d) in 5.2.11.} \end{array} \right.$$

Exercises 7.2.21

In Ex. 1−8, use 7.2.19 to verify each result.

1. $\displaystyle \int_0^\infty \frac{x^{a-1}\, dx}{x+1} = \frac{\pi}{\sin \pi a} \quad$ if $0 < a < 1$

2. $\displaystyle \int_0^\infty \frac{x^a\, dx}{1 + 2x \cos \theta + x^2} = \frac{\pi \sin a\theta}{\sin \pi a \sin \theta} \quad$ if $-1 < a < 1$, $a \neq 0$, and $-\pi < \theta < \pi$

3. $\displaystyle \int_0^\infty \frac{x^{a-1}\, dx}{x^2 + x + 1} = \frac{-2\pi \, \sin(a-1)\frac{\pi}{3}}{\sqrt{3} \sin \pi a} \quad$ if $0 < a < 2$ and $a \neq 1$

4. $\displaystyle \int_0^\infty \frac{x^{a-1}\, dx}{(x+b)(x+c)} = \frac{\pi(b^{a-1} - c^{a-1})}{(c-b)\sin \pi a} \quad$ if $0 < a < 2, a \neq 1, b > 0, c > 0$, and $b \neq c$

5. $\displaystyle \int_0^\infty \frac{x^p\, dx}{x^2 + a^2} = \frac{\pi a^{p-1}}{2 \cos \frac{\pi p}{2}} \quad$ if $-1 < p < 1$, $p \neq 0$, and $a > 0$

6. $\displaystyle \int_0^\infty \frac{x^{a-1}\, dx}{x^3 + b^3} = \frac{\pi b^{a-3}}{3 \sin \pi a} \left(1 + 2 \cos \frac{2\pi a}{3} \right) \quad$ if $0 < a < 3$, $a \neq 1, a \neq 2$, and $b > 0$

7. $\displaystyle \int_0^\infty \frac{x^a\, dx}{(x^2 + b^2)^2} = \frac{\pi b^{a-3}(1-a)}{4 \cos \frac{\pi a}{2}} \quad$ if $-1 < a < 3$, $a \neq 0$, $a \neq 1$, $a \neq 2$, and $b > 0$

8. $\displaystyle \int_0^\infty \frac{x^{a-1}\, dx}{x^4 + 1} = \frac{\pi}{4 \sin \frac{\pi a}{4}} \quad$ if $0 < a < 4$, $a \neq 1$, $a \neq 2$, and $a \neq 3$

Theorem 7.2.22 Let f be a rational function which has only **simple** poles $x_1, x_2, \cdots, x_p$ on the positive real axis (where $x_k \neq x_j$ if $k \neq j$), and let $z_1, z_2, \cdots, z_n$ be the distinct poles of f not on the non-negative real axis. If a and g are defined as in 7.2.19 and if (7.42) holds, then

$$P \int_0^\infty x^{a-1} f(x)\, dx = \pi(\csc \pi a) \sum_{j=1}^n \operatorname{Res}(g, z_j) - \pi(\cot \pi a) \sum_{k=1}^p \operatorname{Res}(h, x_k) \qquad (7.51)$$

where

$$h(z) = z^{a-1} f(z). \qquad (7.52)$$

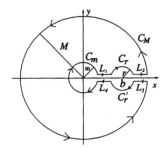

Figure 7.4: $C = C_M + L_3 + C_r' + L_4 + C_m + L_1 + C_r + L_2$

Proof. First suppose that $x_1 = b$ is the only simple pole of f on the positive real axis. We apply the Cauchy Residue Theorem to f on the simple closed contour C consisting of the circular arcs C_M, C_m, C_r, and C_r' and the line segments L_1, L_2, L_3, and L_4 as shown in Figure 7.4.

Now C is chosen so that the poles $z_1, z_2, \cdots, z_n \in I(C)$. Then

$$
\begin{aligned}
\int_C g(z)\, dz &= \int_{C_M} g(z)\, dz + \int_{L_3} g(z)\, dz + \int_{C_r'} g(z)\, dz + \int_{L_4} g(z)\, dz \\
&\quad + \int_{C_m} g(z)\, dz + \int_{L_1} g(z)\, dz + \int_{C_r} g(z)\, dz + \int_{L_2} g(z)\, dz \\
&= 2\pi i \sum_{j=1}^{n} \operatorname{Res}(g, z_j). \tag{7.53}
\end{aligned}
$$

As in (7.45) and (7.46),

$$
\left| \int_{C_M} g(z)\, dz \right| \le 2\pi K_M \quad \text{and} \quad \left| \int_{C_m} g(z)\, dz \right| \le 2\pi K_m \tag{7.54}
$$

where K_M and K_m are given in (7.45) and (7.46).

Since C_r is contained in the first quadrant, we have for each z on C_r,

$$
\operatorname{Arg}(-z) = -\pi + \operatorname{Arg} z.
$$

Thus for each z on C_r

$$
\begin{aligned}
g(z) &= |z|^{a-1} e^{i(a-1)\operatorname{Arg}(-z)} f(z) \quad \text{by 5.4.7} \\
&= |z|^{a-1} e^{i(a-1)[-\pi + \operatorname{Arg} z]} f(z) \\
&= |z|^{a-1} e^{i(a-1)\operatorname{Arg} z} e^{-\pi a i} e^{\pi i} f(z) \\
&= -e^{-\pi a i} z^{a-1} f(z) \\
&= -e^{-\pi a i} h(z) \tag{7.55}
\end{aligned}
$$

where h is defined in 7.52.

Since C'_r is in the fourth quadrant, we have for each z on C'_r,

$$\text{Arg}\,(-z) = \pi + \text{Arg}\,z.$$

Thus for each z on C'_r,

$$
\begin{aligned}
g(z) &= |z|^{a-1} e^{i(a-1)\,\text{Arg}(-z)} f(z) \\
&= |z|^{a-1}\, e^{i(a-1)[\pi+\text{Arg}\,z]} \\
&= |z|^{a-1} e^{i(a-1)\,\text{Arg}(z)}\, e^{\pi i a} e^{-\pi i} f(z) \\
&= -e^{\pi a i} z^{a-1} f(z) \\
&= -e^{\pi a i} h(z).
\end{aligned}
\tag{7.56}
$$

Now by (7.55) and (7.56),

$$
\left\{
\begin{aligned}
\int_{C_r} g(z)\,dz &= -e^{-\pi a i} \int_{C_r} h(z)\,dz \\
&\quad\text{and} \\
\int_{C'_r} g(z)\,dz &= -e^{\pi a i} \int_{C'_r} h(z)\,dz.
\end{aligned}
\right.
\tag{7.57}
$$

Note that b is not a simple pole of g, since g is not analytic at any point on the positive real axis. However, h has a simple pole at b, and hence 7.2.15 may be applied to h but not to g for C_r and C'_r. See Ex. 7(c) and Ex. 9 in 5.4.11.

Now letting L_1, L_2, L_3, and L_4 approach the real axis, we have the following.

$$
\int_{L_1} g(z)\,dz \to e^{-(a-1)\pi i} \int_m^{b-r} x^{a-1} f(x)\,dx = -e^{-\pi a i} \int_m^{b-r} x^{a-1} f(x)\,dx
\tag{7.58}
$$

[since $g(z) = (-z)^{a-1} f(z) = |z|^{a-1} e^{(a-1)\theta i} f(z) \to x^{a-1} e^{-(a-1)\pi i} f(x)$ as $L_1 \to$ the real axis, where $-\pi < \theta = \text{Arg}\,(-z) \le \pi$, and since $z \to x$ and $\text{Arg}\,(-z) \to -\pi$ as $L_1 \to$ the real axis];

$$
\int_{L_2} g(z)\,dz \to e^{-(a-1)\pi i} \int_{b+r}^M x^{a-1} f(x)\,dx = -e^{-\pi a i} \int_{b+r}^M x^{a-1} f(x)\,dx;
\tag{7.59}
$$

$$
\int_{L_3} g(z)\,dz \to e^{(a-1)\pi i} \int_M^{b+r} x^{a-1} f(x)\,dx = -e^{\pi a i} \int_M^{b+r} x^{a-1} f(x)\,dx
\tag{7.60}
$$

[since $\text{Arg}\,(-z) \to \pi$ as $L_3 \to$ the real axis];

$$
\int_{L_4} g(z)\,dz \to e^{(a-1)\pi i} \int_{b-r}^m x^{a-1} f(x)\,dx = -e^{\pi a i} \int_{b-r}^m x^{a-1} f(x)\,dx
\tag{7.61}
$$

See Ex. 10(e) in 7.2.23.

Next, we let $m \to 0$, $r \to 0$, and $M \to \infty$ and we apply 7.2.15 to the integrals in (7.57). We have by (7.53), (7.54), and (7.57)–(7.61),

$$\left[-e^{\pi ai} \int_{\infty}^{b} x^{a-1} f(x)\, dx - e^{\pi ai} [-\pi i \operatorname{Res}(h, b)] \right.$$

$$- e^{\pi ai} \int_{b}^{0} x^{a-1} f(x)\, dx - e^{-\pi ai} \int_{0}^{b} x^{a-1} f(x)\, dx$$

$$\left. - e^{-\pi ai} [-\pi i \operatorname{Res}(h, b)] - e^{-\pi ai} \int_{b}^{\infty} x^{a-1} f(x)\, dx \right]$$

$$= 2\pi i \sum_{j=1}^{n} \operatorname{Res}(g, z_j).$$

Thus

$$\left(e^{\pi ai} - e^{-\pi ai} \right) {}_P\!\int_{0}^{\infty} x^{a-1} f(x)\, dx \;=\; 2\pi i \sum_{j=1}^{n} \operatorname{Res}(g, z_j)$$

$$- \pi i \left[e^{\pi ai} + e^{-\pi ai} \right] \operatorname{Res}(h, b).$$

Hence

$$_P\!\int_{0}^{\infty} x^{a-1} f(x)\, dx \;=\; \frac{\pi}{\frac{1}{2i}\left(e^{\pi ai} - e^{-\pi ai} \right)} \sum_{j=1}^{n} \operatorname{Res}(g, z_j)$$

$$- \pi \, \frac{\frac{1}{2}\left(e^{\pi ai} + e^{-\pi ai} \right)}{\frac{1}{2i}\left(e^{\pi ai} - e^{-\pi ai} \right)} \operatorname{Res}(h, b)$$

which is (7.51) if $x_1 = b$ is the only simple pole of f on the positive real axis.

Finally, if f has a simple pole at x_k for $k = 1, 2, \cdots, p$, we insert circular arcs C_{r_k}, C'_{r_k} centered at x_k and apply the foregoing process to obtain (7.51).

Exercises 7.2.23

In Ex. $1-3$, use (7.51) to verify each result.

1. $\displaystyle {}_P\!\int_{0}^{\infty} \frac{x^{a-1}\, dx}{x-1} = -\pi \cot \pi a \quad \text{if } 0 < a < 1$

2. $\displaystyle {}_P\!\int_{0}^{\infty} \frac{x^{a-1}}{x^2-1}\, dx = -\frac{\pi}{2} \cot \frac{\pi a}{2} \quad \text{if } 0 < a < 2 \text{ and } a \neq 1$

3. $\displaystyle P\int_0^\infty \frac{x^{a-1}\,dx}{(x-1)(x^2+4)} = -\frac{2^a\pi}{40}\left(2\sec\frac{\pi a}{2} + \csc\frac{\pi a}{2}\right) - \frac{\pi}{5}\cot\pi a$

if $0 < a < 3$ and $a \neq 1$ and $a \neq 2$.

In Ex. 4−8, verify each result.

4. $\displaystyle \int_{-\infty}^\infty \frac{e^{ax}\,dx}{1+e^x} = \frac{\pi}{\sin\pi a}$ if $0 < a < 1$
    Hint. Substitute $t = e^x$ and use Ex. 1 in 7.2.21.

5. $\displaystyle \int_0^\infty \frac{\sin ax}{x}\,dx = \frac{\pi}{2}$ if $a > 0$

    Hint. Apply the Cauchy-Goursat Theorem to $\int_C \frac{e^{iaz}}{z}\,dz$ where C is the cco
    contour which consists of the line segment $L_1:$ $y = 0$, $-M \le x \le -m$, the
    semicircle $S_m:$ $z = me^{i\theta}$, $\pi \ge \theta \ge 0$, the line segment $L_2:$ $y = 0$, $m \le$
    $x \le M$, and the semicircle $S_M:$ $z = Me^{i\theta}$, $0 \le \theta \le \pi$; and let $m \to 0$ and
    $M \to \infty$. By 7.2.15, we have $\int_{S_m} \frac{e^{iaz}}{z}\,dz \to -\pi i$ as $m \to 0$. By 7.2.11, we have
    $\int_{S_M} \frac{e^{iaz}}{z}\,dz \to 0$ as $M \to \infty$.

6. $\displaystyle \int_{-\infty}^\infty \frac{\sin x}{x}\,dx = \pi$ **Hint.** $\int_{-\infty}^\infty \frac{\sin x}{x}\,dx = 2\int_0^\infty \frac{\sin x}{x}\,dx$ (Why?) Use Ex. 5.

7. $\displaystyle \int_0^\infty \frac{\sin^2 x}{x^2}\,dx = \frac{\pi}{2}$
    Hints. Apply the Cauchy-Goursat Theorem to $\int_C \frac{1-bie^{2iz}}{z^2}\,dz$ where C is given
    in Ex. 5. By 3.4.10(g) (since $z = M\operatorname{cis}\theta$ on S_M),

$$\left|\int_{S_M} \frac{-e^{2iz}}{z^2}\,dz\right| = \left|\int_{S_M} \frac{e^{-2M\sin\theta}e^{2Mi\cos\theta}}{z^2}\,dz\right| \le \frac{1}{M^2}(\pi M) \to 0$$

as $M \to \infty$.

$$\int_{S_m} \frac{1-e^{2iz}}{z^2}\,dz = \int_{S_m} \frac{\left[1-\sum_{n=0}^\infty \frac{(2iz)^n}{n!}\right]}{z^2}\,dz = -\int_{S_m} \frac{2i}{z}\,dz + \int_{S_m} 2\,dz - \int_{S_m} g(z)\,dz$$

where $g(z) = \sum_{n=3}^\infty \frac{(2i)^n}{n!} z^{n-2}$. Now $\left|\int_{S_m} g(z)\,dz\right| \le Km\pi$ if $m \le 1$ where K is
the maximum of $|g|$ on the closed disk $|z| \le 1$. See Ex. 2(b) in 2.4.9 and 4.3.6.
Thus $\int_{S_m} \frac{1-e^{2iz}}{z^2}\,dz \to -2\pi$ as $m \to 0$.

Finally,

$$\int_{-M}^{-m} \frac{1-e^{2iz}}{x^2}\,dx + \int_m^M \frac{1-e^{2iz}}{x^2}\,dx = \int_m^M \frac{4\sin^2 x}{x^2}\,dx.$$

8. $\displaystyle P\!\int_{-\infty}^{\infty} \frac{dx}{x(x^2+px+q)} = \frac{-\pi p}{q\Delta}$ if $p, q \in R$ such that $\Delta^2 = 4q - p^2 > 0$

Hint. Apply the Cauchy Residue Theorem to $\int_C g(z)\, dz$ where $g(z) = \frac{1}{z(z^2+pz+q)}$ and where C is given in Ex. 5. Then let $m \to 0$ and $M \to \infty$. Use 7.2.15 to show that $\int_{S_m} g(z)\, dz \to -\frac{\pi i}{q}$ as $m \to 0$. See 7.2.6 to show that $\int_{S_M} g(z)\, dz \to 0$ as $M \to \infty$.

Exercises 9 and 10 justify (7.58).

9. Let $f \in A(D)$ where D is a simply connected domain in R^2 and let

$$g(z, w) = \int_z^w f(t)\, dt \quad \text{for } z \text{ and } w \text{ in } D$$

(where the contour of integration is in D).

Prove if $z_0, w_0 \in D$, then

$$\lim_{z, w \to z_0, w_0} g(z, w) = g(z_0, w_0).$$

Hint. Let $b \in D$. Then

$$
\begin{aligned}
g(z, w) &= \int_z^b f(t)\, dt + \int_b^w f(t)\, dt && \text{by 3.4.10(d)} \\
&= \int_b^w f(t)\, dt - \int_b^z f(t)\, dt && \text{by 3.4.10(a).}
\end{aligned}
$$

By 3.6.3, each term on the right is continuous in D.

10. Let

$$G(z) = (-z)^{a-1} = |z|^{a-1} e^{(a-1)\theta i}$$

where $-2\pi < \theta = \arg(-z) \le 0$. Verify each of the following statements.

(a) $G \in A(R^2 - H)$ where H is the nonpositive real axis. **Hint.** See Ex. 9 in 5.4.11.

(b) $G(z)f(z) = g(z)$ for each z in $Q = \{(x, y): \ x > 0 \text{ and } y > 0\}$ where f and g are the same as in 7.2.22.

(c) $\displaystyle \int_{L_1} G(z)f(z)\, dz = \int_{L_1} g(z)\, dz$ where L_1, f, and g are the same as in 7.2.22.

(d) If in the proof of 7.2.22, the segment L_1 approaches L where $L = [m, b-r]$, then

$$
\begin{aligned}
\int_{L_1} G(z)f(z)\, dz \ &\to \ \int_L G(z)f(z)\, dz \quad \text{by Ex. 9} \\
&= -e^{-\pi a i} \int_m^{b-r} x^{a-1} f(x)\, dx.
\end{aligned}
$$

(e) By (c) and (d),

$$\int_{L_1} g(z)\, dz \quad \rightarrow \quad -e^{-\pi a i} \int_m^{b-r} x^{a-1} f(x)\, dx$$

as $L_1 \rightarrow L$ from above.

The Gamma Function and the Beta Function. The gamma function Γ is defined by

$$\begin{aligned}
\Gamma(x) &= \int_0^\infty t^{x-1} e^{-t}\, dt \quad \text{for each } x > 0 \\
&= 2 \int_0^\infty r^{2x-1} e^{-r^2}\, dr \quad \text{where } t = r^2.
\end{aligned} \qquad (7.62)$$

[It is shown in advanced calculus that the improper integral $\int_0^\infty t^{x-1} e^{-t}\, dt$ is defined for each $x > 0$.]

The beta function $\beta(x, y)$ is defined by

$$\begin{aligned}
\beta(x, y) &= \int_0^1 t^{x-1} (1-t)^{y-1}\, dt \quad \text{for each } x, y > 0 \\
&= 2 \int_0^{\frac{\pi}{2}} (\sin\theta)^{2x-1} (\cos\theta)^{2y-1}\, d\theta \quad \text{where } t = \sin^2\theta.
\end{aligned} \qquad (7.63)$$

If $x, y > 0$, then

$$\beta(x, y) = \frac{\Gamma(x)\, \Gamma(y)}{\Gamma(x+y)}. \qquad (7.64)$$

For by (7.62) and (7.63),

$$\begin{aligned}
\Gamma(x+y)\, \beta(x,y) &= 4 \int_0^\infty r^{2x+2y-1} e^{-r^2}\, dr \int_0^{\frac{\pi}{2}} (\sin\theta)^{2x-1} (\cos\theta)^{2y-1}\, d\theta \\
&= 4 \int_0^\infty \int_0^{\frac{\pi}{2}} (r\sin\theta)^{2x-1} (r\cos\theta)^{2y-1} e^{-r^2} r\, d\theta\, dr \\
&= \left(2 \int_0^\infty v^{2x-1} e^{-v^2}\, dv\right) \left(2 \int_0^\infty u^{2y-1} e^{-u^2}\, du\right) \\
&= \Gamma(x)\, \Gamma(y) \quad \text{by (7.62)}.
\end{aligned}$$

Hence (7.64) follows. **Note**. The product of the last two integrals is obtained from the previous integral by substituting $v = r\sin\theta$ and $u = r\cos\theta$. This is the usual change from polar to rectangular coordinates. Thus $r\, d\theta\, dr$ is replaced by $du\, dv$.

We now show that

$$\Gamma(a)\, \Gamma(1-a) = \frac{\pi}{\sin\pi a} \quad \text{for } 0 < a < 1. \qquad (7.65)$$

By (7.62), we see that $\Gamma(1) = 1$. Thus by (7.63) and (7.64),

$$
\begin{aligned}
\Gamma(a)\,\Gamma(1-a) &= \beta(a, 1-a) \\
&= \int_0^1 t^{a-1}(1-t)^{-a}\,dt.
\end{aligned}
$$

Let

$$
t = \frac{x}{x+1} = 1 - \frac{1}{x+1}.
$$

Hence

$$
dt = \frac{dx}{(x+1)^2} \quad \text{and} \quad 1 - t = \frac{1}{x+1}.
$$

Thus

$$
\begin{aligned}
\Gamma(a)\,\Gamma(1-a) &= \int_0^\infty \frac{x^{a-1}}{(x+1)^{a-1}}\,(x+1)^a\,\frac{dx}{(x+1)^2} \\
&= \int_0^\infty \frac{x^{a-1}}{x+1}\,dx = \frac{\pi}{\sin \pi a} \quad \text{by Ex. 1 in 7.2.21.}
\end{aligned}
$$

This proves (7.65).

Exercises 7.2.24

1. Show that $\Gamma\left(\dfrac{1}{2}\right) = 2 \displaystyle\int_0^\infty e^{-x^2}\,dx$. **Hint.** Let $t = x^2$ in $\Gamma\left(\dfrac{1}{2}\right) = \displaystyle\int_0^\infty t^{-\frac{1}{2}} e^{-t}\,dt$.

2. Show that $\displaystyle\int_0^\infty e^{-x^2}\,dx = \dfrac{\sqrt{\pi}}{2}$. **Hint.** Use Ex. 1 and (7.65).

3. Show that $\displaystyle\int_0^\infty e^{-x^2} \cos 2px\,dx = \dfrac{\sqrt{\pi}}{2}\,e^{-p^2}$ if $p > 0$. **Hint.** Apply the Cauchy-Goursat Theorem to $\int_C e^{-z^2}\,dz$ where C is the rectangle with vertices $\pm M$ and $(\pm M + ip)$ and let M approach ∞.

4. Show that $\displaystyle\int_0^\infty \sin x^2\,dx = \int_0^\infty \cos x^2\,dx = \dfrac{\sqrt{2\pi}}{4}$. (Fresnel Integrals)

   Hints. Apply the Cauchy-Goursat Theorem to $\int_C e^{-z^2}\,dz$ where C is the boundary of the circular sector $0 \le \theta \le \frac{\pi}{4}$, $0 \le r \le M$, and let M approach ∞.

$$
\begin{aligned}
\left| \int_0^{\frac{\pi}{4}} e^{-M^2 \cos 2\theta} e^{-iM^2 \sin 2\theta} Mie^{i\theta}\,d\theta \right| &\le M \int_0^{\frac{\pi}{4}} e^{-M^2 \cos 2\theta}\,d\theta \quad \text{by 3.1.7(c)} \\
&\le M \int_0^{\frac{\pi}{4}} e^{-M^2 \sin 2\phi}\,d\phi \quad \text{where } \theta = \frac{\pi}{4} - \phi \\
&\le M \int_0^{\frac{\pi}{4}} e^{-\frac{4M^2\phi}{\pi}}\,d\phi \quad \text{by 7.2.9} \\
&\to 0 \quad \text{as } M \to \infty.
\end{aligned}
$$

Set the real and imaginary parts equal and solve the simultaneous equations.

Theorem 7.2.25 Let P and Q be relatively prime polynomials in 7.2.5 where the degree of Q is at least two greater than the degree of P. Suppose Q has no real zeros and that $\frac{P}{Q}$ is an even function. Let

$$f(z) = \frac{P(z)(\text{Log } z)^n}{Q(z)} \qquad \text{where } n \in N.$$

If $z_1, z_2, z_3, \cdots, z_t$ are the distinct poles of f in the upper half-plane, then

$$\int_0^\infty \frac{P(x)}{Q(x)} (\text{Log } x + \pi i)^n \, dx + \int_0^\infty f(x) \, dx = 2\pi i \sum_{j=1}^t \text{Res}\,(f, z_j). \qquad (7.66)$$

Proof. Let C be the contour given in the Hint for Ex. 5 of 7.2.23. By the Cauchy Residue Theorem,

$$\int_C f(z) \, dz = \int_{-M}^{-m} \frac{P(x)}{Q(x)} (\text{Log } |x| + \pi i)^n \, dx + \int_{S_m} f(z) \, dz + \int_m^M f(x) \, dx$$

$$+ \int_{S_M} f(z) \, dz = 2\pi i \sum_{j=1}^t \text{Res}\,(f, z_j). \qquad (7.67)$$

See Ex. 12 in 7.2.27.

Since $\frac{P}{Q}$ is an even function, on replacing x by $-x$, we have

$$\int_{-M}^{-m} \frac{P(x)}{Q(x)} (\text{Log } |x| + \pi i)^n \, dx = \int_m^M \frac{P(x)}{Q(x)} (\text{Log } x + \pi i)^n \, dx. \qquad (7.68)$$

If a_0 and b_0 are the constant terms in P and Q respectively, then

$$\left| \frac{P(z)}{Q(z)} \right| \to \left| \frac{a_0}{b_0} \right| \quad \text{as } z \to 0.$$

Thus, if

$$G = \left| \frac{a_0}{b_0} \right| + 1,$$

then

$$\left| \frac{P(z)}{Q(z)} \right| < G \quad \text{for } |z| \text{ sufficiently small.}$$

Hence by 3.4.10(g),

$$\left| \int_{S_m} \frac{P(z)}{Q(z)} (\text{Log } z)^n \, dz \right| \leq G |\text{Log } m + \pi i|^n \, \pi m$$

$$\leq \pi G m (|\text{Log } m| + \pi)^n$$

$$\to 0 \quad \text{as } m \to 0 \quad \text{by Ex. 13 in 7.2.27.} \qquad (7.69)$$

By 7.2.5, there is a H such that

$$\left|\frac{P(z)}{Q(z)}\right| < \frac{H}{|z|^s} < \frac{H}{|z|^2} \quad \text{for } |z| \text{ sufficiently large}, \tag{7.70}$$

where s is the degree of Q minus the degree of P. Again by 3.4.10(g) and by (7.70), for M sufficiently large,

$$\begin{aligned}
\left|\int_{S_M} \frac{P(z)}{Q(z)} (\operatorname{Log} z)^n \, dz\right| &\leq \frac{H}{M^2} (\operatorname{Log} M + \pi)^n \pi M \\
&= \frac{\pi H}{M} (\operatorname{Log} M + \pi)^n \\
&\to 0 \text{ as } M \to \infty \quad \text{by Ex. 14 in 7.2.27.} \tag{7.71}
\end{aligned}$$

Hence (7.67), (7.68), (7.69), and (7.71) imply (7.66).

Remarks 7.2.26 If in particular $n = 1$, then (7.66) becomes

$$\int_0^\infty \frac{P(x)}{Q(x)} \operatorname{Log} x \, dx + \frac{\pi i}{2} \int_0^\infty \frac{P(x)}{Q(x)} \, dx = \pi i \sum_{j=1}^t \operatorname{Res}(f, z_j).$$

Exercises 7.2.27

In Ex. $1-11$, verify each result.

1. $\displaystyle\int_0^\infty \frac{\operatorname{Log} x \, dx}{x^2 + 1} = 0$

2. $\displaystyle\int_0^\infty \frac{(\operatorname{Log} x)^2 \, dx}{x^2 + 1} = \frac{\pi^3}{2^3}$

3. $\displaystyle\int_0^\infty \frac{(\operatorname{Log} x)^3 \, dx}{x^2 + 1} = 0$

4. $\displaystyle\int_0^\infty \frac{(\operatorname{Log} x)^4 \, dx}{x^2 + 1} = \frac{5\pi^5}{2^5}$

5. $\displaystyle\int_0^\infty \frac{(\operatorname{Log} x)^5 \, dx}{x^2 + 1} = 0$

6. $\displaystyle\int_0^\infty \frac{(\operatorname{Log} x)^6 \, dx}{x^2 + 1} = \frac{61\pi^7}{2^7}$

7. $\displaystyle\int_0^\infty \frac{(\operatorname{Log} x)^7 \, dx}{x^2 + 1} = 0$

8. $\displaystyle\int_0^\infty \frac{(\operatorname{Log} x)^8 \, dx}{x^2 + 1} = \frac{1385\pi^9}{2^9}$

9. If $n = 2k - 1$ where $k \in N$, then $I_k = \int_0^\infty \dfrac{(\text{Log } x)^n \, dx}{x^2 + 1} = 0$.

Hint. Observe (i) $I_1 = 0$ by Ex. 1, and prove (ii) if $I_k = 0$ for each $k < m$, then $I_m = 0$. To prove (ii), observe that for $n = 2m - 1$, the result in (7.66) becomes

$$2 \int_0^\infty \frac{(\text{Log } x)^n \, dx}{x^2 + 1} + {}_nC_1 \pi i \int_0^\infty \frac{(\text{Log } x)^{n-1} \, dx}{x^2 + 1} - {}_nC_3 \pi^3 i \int_0^\infty \frac{(\text{Log } x)^{n-3} \, dx}{x^2 + 1}$$

$$+ {}_nC_5 \pi^5 i \int_0^\infty \frac{(\text{Log } x)^{n-5} \, dx}{x^2 + 1} + \cdots - (-1)^m \pi^n i \int_0^\infty \frac{dx}{x^2 + 1}$$

$$= -\frac{(-1)^m \pi^{n+1} i}{2^n}$$

where the numbers ${}_nC_k$ are the binomial coefficents. Now equate the real parts in this equation.

10. $\displaystyle\int_0^\infty \frac{\text{Log } x \, dx}{(x^2 + 1)^2} = -\frac{\pi}{4}$

11. $\displaystyle\int_0^\infty \frac{(\text{Log } x)^2 \, dx}{(x^2 + 1)^2} = \frac{\pi^3}{16}$ See Ex. 2 in 7.2.8.

12. Let $g(z) = \text{Log } |z| + i\theta$ where $-\dfrac{\pi}{2} < \theta = \arg z \leq \dfrac{3\pi}{2}$.

   (a) Prove $g \in A(\overline{I(C)})$ where C is the contour given in the Hint for Ex. 5 of 7.2.23.

   (b) Prove $g(z) = \text{Log } z$ for each z in $\overline{I(C)}$.

   (c) Justify the application of the Cauchy Residue Theorem in (7.67) to f, even though f is not analytic on $L_1 \cup L_2$.

13. Show by repeated application of L'Hôpital's rule that

$$\frac{(-\text{Log } m + \pi)^n}{\frac{1}{m}} \to 0 \text{ as } m \to 0^+$$

where $n \in N$.

14. Prove $\dfrac{(\text{Log } M + \pi)^n}{M} \to 0$ as $M \to \infty$ where $n \in N$.

15. Give two examples to show that $\lim\limits_{x \to \infty} [f(x) + g(x)]$ exists but $\lim\limits_{x \to \infty} f(x)$ and $\lim\limits_{x \to \infty} g(x)$ do not exist.

$$\text{Ans.} \quad \begin{cases} f(x) = \sqrt{x + 1} \text{ and } g(x) = -\sqrt{x} \\ f(x) = \cos x \text{ and } g(x) = -\cos x \end{cases}$$

16. In the proof of 7.2.25, we showed that

$$\lim_{m,M\to 0,\infty}\left[\int_m^M \frac{P(x)}{Q(x)}\left(\text{Log}\,x+\pi i\right)^n dx + \int_m^M f(x)\,dx\right]$$

$$= 2\pi i \sum_{j=1}^t \text{Res}\,(f,z_j).$$

From this we concluded that (7.66) holds. In view of Ex. 15, justify this conclusion. **Hint.** We write

$$\int_m^M f(x)\,dx = \int_m^1 f(x)\,dx + \int_1^M f(x)\,dx.$$

Now we let $r > 1$ such that (7.70) holds. Then for each $x > r$,

$$|f(x)| = \left|\frac{P(x)}{Q(x)}\left(\text{Log}\,x\right)^n\right| < \frac{H}{x^2}\left(\text{Log}\,x\right)^n.$$

Observe by successive integration by parts that

$$\begin{aligned}
\int_1^M \frac{(\text{Log}\,x)^n}{x^2}\,dx &= -\frac{(\text{Log}\,M)^n}{M} + n\int_1^M \frac{(\text{Log}\,x)^{n-1}}{x^2}\,dx\\
&= -\frac{(\text{Log}\,M)^n}{M} - \frac{n\,(\text{Log}\,M)^{n-1}}{M} - \frac{n(n-1)(\text{Log}\,M)^{n-2}}{M}\\
&\quad -\cdots - n!\,\frac{\text{Log}\,M}{M} + n!\int_1^M \frac{dx}{x^2}\\
&\to n!\quad \text{as } M\to\infty.\quad \text{See Ex. 14.}
\end{aligned}$$

Now we apply the following theorem from calculus.

Theorem. Let $f \in C([a,\infty))$ and let $|f(x)| \le g(x)$ for each $x \ge b$ for some $b \ge a$. If $\int_b^\infty g(x)\,dx$ converges, then $\int_a^\infty f(x)\,dx$ converges.

7.3 Mittag-Leffler Theorem

If a function is indicated by giving the rule which specifies $g(z)$ for each z in the domain of the function, then we simply use g to denote the function. We recall that a sequence is a function. Thus it is natural simply to use $\boldsymbol{a}$ to stand for the sequence which we often denote by $\{a_n\}$ or by $\{a_n\}_{n=1}^\infty$.

Now, if we say the sequence $\boldsymbol{a}$ is one-to-one, then we understand this means that if $n \ne k$ in N, then $a_n \ne a_k$.

Finally, if $\boldsymbol{a}$ is a sequence, then we understand that $a(N)$ denotes the range of the function $\boldsymbol{a}$. Thus $a(N)$ is the set of all points a_n such that $n \in N$.

Theorem 7.3.1 (Mittag-Leffler). Let a be a one-to-one sequence in R^2 such that $|a_n| \to \infty$ as $n \to \infty$ and for each n in N, let

$$P_n(z) = \frac{b_{n1}}{z - a_n} + \frac{b_{n2}}{(z - a_n)^2} + \cdots + \frac{b_{nk_n}}{(z - a_n)^{k_n}}$$
$$\text{where } b_{nj} \in R^2 \text{ and } b_{nk_n} \neq 0. \tag{7.72}$$

Then there is a function f which has the following properties.

(a) The function f is analytic in $R^2 - a(N)$.

(b) Each point a_n is a pole of f.

(c) For each n in N, the expression P_n is the principal part of f at a_n.

Proof. If $a_n \neq 0$, then by 6.1.4 and 4.4.9, the Maclaurin series for the function P_n converges uniformly to P_n in the disk $N\left(0, \frac{|a_n|}{2}\right)$. Thus for each n such that $a_n \neq 0$, let Q_n be the polynomial which is the sum of the first m_n terms in the Maclaurin series for P_n, where m_n is sufficiently large that

$$|P_n(z) - Q_n(z)| < \frac{1}{2^n} \quad \text{for each } z \text{ in } N\left(0, \frac{|a_n|}{2}\right). \tag{7.73}$$

But if $a_n = 0$, there is no Maclaurin series for P_n. So we take $Q_n \equiv 0$ if $a_n = 0$.

Let $D_r = N(0, r)$ where r is any positive number. Since $|a_n| \to \infty$ as $n \to \infty$, let $M \in N$ such that

$$\begin{cases} |a_n| > 2r & \text{for each } n \geq M, \\ \text{that is, } D_r \subset N\left(0, \frac{|a_n|}{2}\right) & \text{for } n \geq M. \end{cases} \tag{7.74}$$

Now let

$$f(z) = \sum_{n=1}^{\infty} [P_n(z) - Q_n(z)]$$

$$= G(z) + H(z) \tag{7.75}$$

where

$$G(z) = \sum_{n=1}^{M} [P_n(z) - Q_n(z)] \tag{7.76}$$

and

$$H(z) = \sum_{n=M+1}^{\infty} [P_n(z) - Q_n(z)]. \tag{7.77}$$

By (7.73), (7.74), and the Weierstrass M test, the series in (7.77) converges uniformly on D_r. By (7.74), we see that $P_n \in A(D_r)$ for each $n \geq M$. Thus (since $Q_n \in A(R^2)$), we have $H \in A(D_r)$ by 4.4.12. Also, G is analytic at each point of D_r except at those points a_n which are in D_r. Hence by (7.75), we see that f is analytic at each point in D_r except at these points a_n in D_r. Furthermore, if $a_n \in D_r$, then P_n is the principal part of f at a_n.

Now since r is an **arbitrary** positive number, it follows that f is analytic in $R^2 - a(N)$ and P_n is the principal part of f at a_n for each n in N.

To see that f has a pole at a fixed a_k and principal part P_k at a_k, we observe that the removal of the kth term $(P_k - Q_k)$ from the series in (7.75), yields a function g which is analytic at a_k, by the foregoing argument. But if we let $h = P_k - Q_k$, then clearly h has a pole at a_k and has P_k as the principal part at a_k (since Q_k is a polynomial). Now $f = g + h$. Thus f has a pole at a_k and principal part P_k at a_k by Ex. 1 in 7.3.5.

Corollary 7.3.2 If g is any function having the properties stated for f in (a), (b), and (c) of 7.3.1, then there is an entire function h such that $g = f + h$ where f is given in (7.75).

Proof. Since g and f have exactly the same poles and corresponding principal parts and are analytic at each point in R^2 except at these poles, it follows by Ex. 2 in 7.3.5 that a function h may be defined such that $g = f + h$, where $h \in A(R^2)$.

Remarks 7.3.3 A function f is said to be **Meromorphic** in a domain D iff f is analytic at each point of D except possibly for poles in D. The proof of 7.3.1 shows how to construct a function meromorphic in R^2 which has prescribed poles and principal parts at these poles. Corollary 7.3.2 states that any two functions, which are meromorphic in R^2 and which have the same poles and same corresponding principal parts, differ by an entire function.

We recall that, in general, $\{a, b, c, d\}$ is a notation which denotes the set of those things whose names are listed. Thus $\{0\}$ denotes the set containing just the single element indicated. This means that the complex number 0 is the only member of the set $\{0\}$. Hence the set $R^2 - \{0\}$ is the set of all nonzero complex numbers.

We should mention that in this book, we use the notation $\{a_n\}$ as an abbreviation for $\{a_n\}_{n=1}^{\infty}$. However, we explicitly refer to the "sequence $\{a_n\}$" unless the context clearly indicates a sequence. Of course, if n is fixed, then $\{a_n\}$ refers to the set consisting of the single element a_n. In mathematics, we often use the same notation to denote different things in different contexts. For instance, sometimes (a, b) denotes an ordered pair, but sometimes (a, b) denotes an open interval.

Theorem 7.3.4 Let $p \in N$ and let a be a one-to-one sequence in $R^2 - \{0\}$ such that

$$\sum_{n=1}^{\infty} \frac{1}{|a_n|^{p+1}} \tag{7.78}$$

converges. If f is given by

$$f(z) = \sum_{n=1}^{\infty} \left[\frac{1}{z - a_n} + \frac{1}{a_n} + \frac{z}{a_n^2} + \cdots + \frac{z^{p-1}}{a_n^p} \right], \tag{7.79}$$

then f has the following properties.

(a) The function f is analytic in $R^2 - a(N)$.

(b) Each a_n is a simple pole of f.

(c) For each n in N, the fraction $\frac{1}{z - a_n}$ is the principal part of f at a_n.

Note. The convergence of the series in (7.78) implies that $|a_n| \to \infty$ as $n \to \infty$. Thus 7.3.1 implies the existence of some function satisfying (a), (b), and (c). But 7.3.4 gives a specific construction for such a function. Here, $P_n(z) = \frac{1}{z - a_n}$ and $b_{nk_n} = 1 = k_n$. Also, we take Q_n in (7.73) to be the sum of the first p terms in the Maclaurin series for $\frac{1}{z - a_n}$. We have

$$\frac{1}{z - a_n} = -\frac{1}{a_n} \frac{1}{1 - \frac{z}{a_n}} = -\frac{1}{a_n} - \frac{z}{a_n^2} - \cdots \quad \text{if } |z| < |a_n|$$

and

$$Q_n = -\left(\frac{1}{a_n} + \frac{z}{a_n^2} + \cdots + \frac{z^{p-1}}{a_n^p} \right) = -\frac{a_n^p - z^p}{a_n^p(a_n - z)}. \tag{7.80}$$

Thus the nth term in (7.79) is $P_n - Q_n$. The proof of 7.3.4 differs from the proof of 7.3.1 in that the comparative series $\sum_{n=1}^{\infty} \frac{1}{2^n}$ is replaced by the series $\sum_{n=1}^{\infty} \frac{2r^p}{|a_n|^{p+1}}$.

Proof. Let $D_r = N(0, r)$ where r is any positive number. By the note, let $M \in N$ such that

$$|a_n| > 2r \quad \text{for each } n \geq M. \tag{7.81}$$

Now by (7.79) and (7.80),

$$f(z) = \sum_{n=1}^{\infty} \left[\frac{1}{z - a_n} - Q_n \right]$$

$$= G(z) + H(z) \tag{7.82}$$

where

$$G(z) = \sum_{n=1}^{M} \left[\frac{1}{z - a_n} - Q_n \right] \tag{7.83}$$

and

$$H(z) = \sum_{n=M+1}^{\infty} \left[\frac{1}{z - a_n} - Q_n \right]. \tag{7.84}$$

But for each z in D_r and for each $n > M$,

$$\left| \frac{1}{z - a_n} - Q_n \right| = \frac{|z|^p}{|a_n|^p |a_n - z|} \quad \text{by (7.80)}$$

$$\leq \frac{r^p}{|a_n|^p |a_n - z|}$$

$$\leq \frac{r^p}{|a_n|^p \left| |a_n| - \frac{|a_n|}{2} \right|} \quad \left\{ \begin{array}{l} \text{by (7.81), since} \\ |a_n - z| \geq ||a_n| - |z|| \end{array} \right.$$

$$\leq \frac{2r^p}{|a_n|^{p+1}}.$$

Thus by the the Weierstrass M-test, the series in (7.84) converges uniformly in D_r (since r and p are fixed and $\sum_{n=M+1}^{\infty} \frac{1}{|a_n|^{p+1}}$ converges by hypothesis). By (7.81), for each $n \geq M + 1$, we have $a_n \notin D_r$. Thus each term of the series in (7.84) is analytic in D_r. Hence, $H \in A(D_r)$ by 4.4.12. Also, G is analytic at each point of D_r except at those points a_n which are in D_r. Therefore by (7.82), the function f is analytic at each point of D_r except at those points a_n which are in D_r.

Now since r is an arbitrary positive number, it follows that $f \in A(R^2 - a(N))$. Also (b) and (c) follow as in the proof of 7.3.1.

Exercises 7.3.5

1. Let $g \in A(z_0)$ and let h have a pole at z_0 and principal part P at z_0. Prove $g + h$ has a pole at z_0 and P is the principal part of $g + h$ at z_0. **Hint.** See 6.1.2, 6.1.6, and 6.3.4.

2. Let $f, g \in A(R^2 - E)$ such that f and g have poles at each point of E. Prove if at each pole, f and g have the same principal part, then $(g - f) \in A(R^2 - E)$ and $g - f$ has a removable singularity at each point of E. **Hint.** Let p be a fixed point of E and represent f and g by their Laurent series which are valid in some deleted neighborhood of p.

3. Let a be a one-to-one sequence in $R^2 - \{0\}$ such that $\sum_{n=1}^{\infty} \frac{1}{|a_n|}$ converges. Let f be given by $f = \sum_{n=1}^{\infty} \frac{1}{z - a_n}$. Prove that f has properties (a),(b), and (c) stated in 7.3.4. **Hint.** Let $r > 0$. Since $\sum_{n=1}^{\infty} \frac{1}{|a_n|}$ converges, let $M \in N$ such that $|a_n| > 2r$ for each $n \geq M$. Then for each $n \geq M$ and for each z in $N_r(0)$, we have $\frac{1}{|z - a_n|} \leq \frac{1}{|a_n| - |z|} \leq \frac{2}{|a_n|}$. Thus for each z in $N_r(0)$, we have $f(z) = \sum_{n=1}^{M} \frac{1}{z - a_n} + \sum_{n=M+1}^{\infty} \frac{1}{z - a_n}$.

4. Given the hypotheses of 7.3.4, let $b_n \in (R^2 - \{0\})$ such that $|b_n| \leq B$ for each n in N. Prove if g is given by

$$g(z) = \sum_{n=1}^{\infty} \left[\frac{b_n}{z - a_n} + b_n \left(\frac{1}{a_n} + \cdots + \frac{z^{p-1}}{a_n^p} \right) \right],$$

then $g \in A(R^2 - a(N))$, g has a pole at each a_n, and $\frac{b_n}{z-a_n}$ is the principal part of g at a_n. **Hint.** $\sum_{n=1}^{\infty} \frac{2Br^p}{|a_n|^{p+1}}$ is the comparative series for D_r.

5. Given the hypotheses and the function g of Ex. 4, let $h(z) = \frac{1}{z} + g(z)$. Prove $h \in A[R^2 - a(N) - \{0\}]$. **Hint.** Use Ex. 1.

Lemma 7.3.6 Let

$$g_1(z) = \sum_{n=0}^{\infty} \frac{1}{(z-n)^2}.$$

Then

(a) g_1 is analytic at z if z is not a non-negative integer,

(b) g_1 has a pole at each non-negative integer n, and

(c) at each non-negative integer n, the principal part of g_1 is $\frac{1}{(z-n)^2}$.

Proof. Let $D_r = N(0,r)$ where r is **any** positive number. Let $M \in N$ such that $r \le \frac{M}{2}$. Then

$$g_1(z) = G(z) + H(z)$$

where

$$G(z) = \sum_{n=0}^{M} \frac{1}{(z-n)^2} \quad \text{and} \quad H(z) = \sum_{n=M+1}^{\infty} \frac{1}{(z-n)^2}.$$

Now for each z in D_r and for each $n > M$,

$$\frac{1}{|z-n|^2} \le \frac{1}{|n-|z||^2}$$

$$\le \frac{1}{|n-\frac{n}{2}|^2} \quad \text{since } |z| < \frac{M}{2} < \frac{n}{2}$$

$$\le \frac{4}{n^2}.$$

Thus $H \in A(D_r)$ by the Weierstrass M-test and 4.4.12. Now the conclusion of 7.3.6 follows as in the proof of 7.3.4.

Remarks 7.3.7 (a) Just as in 7.3.6, the function

$$g_2(z) = \sum_{n=-1}^{-\infty} \frac{1}{(z-n)^2}$$

is analytic at each point of $R^2 - \{n : -n \in N\}$. Also, g_2 has a pole at $z = -1, -2, -3, \cdots$, and $\frac{1}{(z-n)^2}$ is the principal part of g_2 at $z = n$ for $n = -1, -2, -3, \cdots$.

(b) Let

$$g(z) = g_1(z) + g_2(z) = \sum_{n=-\infty}^{\infty} \frac{1}{(z-n)^2}. \tag{7.85}$$

Then by Part (a), 7.3.6, and Ex. 1 in 7.3.5, we have $g \in A(R^2 - J)$, and g has a pole at n with principal part $\frac{1}{(z-n)^2}$ at each n in J.

(c) Let

$$f(z) = \frac{\pi^2}{\sin^2 \pi z} \quad \text{for each } z \text{ in } (R^2 - J). \tag{7.86}$$

Then $f \in A(R^2 - J)$ and f has a pole at $z = n$ for each n in J. Since

$$\frac{1}{\sin^2 \pi z} = \frac{1}{\pi^2 z^2 - \frac{\pi^4 z^4}{3} + \cdots} = \frac{1}{\pi^2 z^2} + \frac{1}{3} + \cdots,$$

it follows that $\frac{1}{z^2}$ is the principal part of f at $z = 0$. Also, since $\sin^2 \pi(z-n) = \sin^2 \pi z$ for each n in J, it follows that $\frac{1}{(z-n)^2}$ is the principal part of $\frac{\pi^2}{\sin^2 \pi z}$ at $z = n$.

(d) By (b), (c), and Ex. 2 in 7.3.5, there is an entire function h such that

$$f = g + h. \tag{7.87}$$

(e) We shall show $h \equiv 0$ on R^2. When this is done, it will follow from $(7.85)-(7.87)$ that

$$\frac{\pi^2}{\sin^2 \pi z} = \sum_{n=-\infty}^{\infty} \frac{1}{(z-n)^2} \quad \text{for each } z \text{ in } R^2 - J. \tag{7.88}$$

Proof of (7.88). By (7.87), we know that $h = f - g$ on $R^2 - J$. Hence by (7.85) and (7.86),

$$\frac{1}{4}\left[h\left(\frac{z}{2}\right) + h\left(\frac{z-1}{2}\right) \right]$$

$$= \frac{1}{4}\left[\left(\frac{\pi^2}{\sin^2 \frac{\pi z}{2}} - \sum_{n=-\infty}^{\infty} \frac{4}{(z-2n)^2} \right) + \left(\frac{\pi^2}{\sin^2 \frac{\pi}{2}(z-1)} - \sum_{n=-\infty}^{\infty} \frac{4}{(z-1-2n)^2} \right) \right]$$

$$= \frac{1}{4}\left[\left(\frac{\pi^2}{\sin^2 \frac{\pi z}{2}} + \frac{\pi^2}{\cos^2 \frac{\pi z}{2}} \right) - \sum_{n=-\infty}^{\infty} \frac{4}{(z-n)^2} \right] \quad \text{(sum of odd and even terms)}$$

$$= \frac{\pi^2}{\sin^2 \pi z} - g(z) \quad \text{since } \sin \pi z = 2\sin \frac{\pi z}{2} \cos \frac{\pi z}{2}$$

$$= h(z). \tag{7.89}$$

Since $h \in A(R^2)$, let $M = \max |h(z)|$ on the closed disk $D = \{\overline{N_1(0)}\}$. By (7.89) for each z in D,

$$
\begin{aligned}
|h(z)| &= \frac{1}{4}\left|h\left(\frac{z}{2}\right) + h\left(\frac{z-1}{2}\right)\right| \\
&\leq \frac{1}{4}[M + M] = \frac{M}{2} \quad \left(|z| \leq 1 \text{ implies } \left|\frac{z-1}{2}\right| \leq 1\right).
\end{aligned}
$$

Thus $M \leq \frac{M}{2}$, which implies that $M = 0$. Hence $h \equiv 0$ on D. Finally by 6.2.8, it follows that $h \equiv 0$ on R^2, and (7.88) follows from (7.87).

Example 7.3.8 We now use (7.88) to obtain

$$
\pi \cot \pi z = \frac{1}{z} + \sum_{n \neq 0} \left(\frac{1}{z - n} + \frac{1}{n}\right) \quad \text{for each } z \text{ in } (R^2 - J) \tag{7.90}
$$

where $\sum_{n \neq 0}$ means $\sum_{n=-1}^{-\infty} + \sum_{n=1}^{\infty}$. Let

$$
F(z) = \sum_{n \neq 0} \frac{1}{(z - n)^2} \quad \text{for each } z \text{ in } D \quad \text{where } D = \{0\} \cup (R^2 - J). \tag{7.91}
$$

As in the proof of 7.3.7(b), we conclude that $F \in A(D)$. Now let z be a fixed point of D and let D_0 be a bounded simply connected domain contained in D such that D_0 contains the origin and the point z. [Let $r > 0$ such that $D_0 \subset N(0, r)$. As in the proof of 7.3.4, if $M \in N$ such that $M > 2r$, then

$$
\sum_{n=M+1}^{\infty} \frac{1}{(z - n)^2} \quad \text{and} \quad \sum_{n=-M-1}^{-\infty} \frac{1}{(z - n)^2} \tag{7.92}
$$

converge uniformly in $N(0, r)$ and hence in D_0. Hence the two series in (7.92) may be integrated termwise along any contour in D_0 from 0 to z. (See 4.4.3.) Since the finite sum

$$
\sum_{n=-M}^{M} \frac{1}{(z - n)^2} \quad (n \neq 0)
$$

can be integrated termwise, it follows that the series in (7.91) can be integrated termwise in D_0.] Now by (7.91) for the fixed z in D,

$$
\begin{aligned}
\int_0^z F(w)\, dw &= \int_0^z \sum_{n \neq 0} \frac{1}{(w - n)^2}\, dw \tag{7.93} \\
&= -\sum_{n \neq 0} \left(\frac{1}{z - n} + \frac{1}{n}\right) = G(z)
\end{aligned}
$$

where the integral is along any contour from 0 to z which is contained in D_0.

By (7.93),

$$\begin{aligned}
G(z) &= -\sum_{n=1}^{\infty}\left[\frac{1}{z-n}+\frac{1}{n}\right] - \sum_{n=-1}^{-\infty}\left[\frac{1}{z-n}+\frac{1}{n}\right] \\
&= -\sum_{n=1}^{\infty}\left[\frac{1}{z-n}+\frac{1}{n}\right] - \sum_{n=1}^{\infty}\left[\frac{1}{z+n}-\frac{1}{n}\right] \\
&= -\sum_{n=1}^{\infty}\frac{2z}{z^2-n^2}.
\end{aligned} \tag{7.94}$$

Now by (7.88) and (7.91),

$$\frac{\pi^2}{\sin^2 \pi z} - \frac{1}{z^2} = F(z) \quad \text{for each } z \text{ in } R^2 - J. \tag{7.95}$$

If

$$H(z) = -\pi \cot \pi z + \frac{1}{z}, \tag{7.96}$$

then

$$\begin{aligned}
\frac{d}{dz} H(z) &= \frac{\pi^2}{\sin^2 \pi z} - \frac{1}{z^2} \\
&= F(z) \quad \text{for each } z \text{ in } R^2 - J \text{ by (7.95).}
\end{aligned} \tag{7.97}$$

By (7.93) and (7.97),

$$H(z) \quad \text{and} \quad G(z)$$

have the same derivative on $R^2 - J$. Hence

$$G(z) - H(z) = C \quad \text{on } R^2 - J \quad \text{by Ex. 20 in 2.3.16} \tag{7.98}$$

where C is a constant. By (7.94) and (7.96), we see that $G - H$ is an odd function. Thus $C = 0$ by (7.98). [If $\phi(z) = C$ is odd, then $C = 0$. For $C = \phi(-z) = -\phi(z) = -C$. Hence $C = 0$.] Thus (7.90) follows from (7.93), (7.96), and (7.98).

Exercises 7.3.9

In Ex. 1–4, prove each result.

1. $\pi \cot \pi z = \dfrac{1}{z} + \displaystyle\sum_{n=1}^{\infty} \frac{2z}{z^2-n^2}$ on $R^2 - J$
   Hint. Use (7.90) and the ideas in (7.94).

2. $\dfrac{\pi^2}{\cos^2 \pi z} = \displaystyle\sum_{n=-\infty}^{\infty} \frac{1}{\left(z-n-\frac{1}{2}\right)^2}$ for each z such that $\left(z - \frac{1}{2}\right) \in (R^2 - J)$
   Hint. Use (7.88) and the identity $\sin\left(\theta - \frac{\pi}{2}\right) = -\cos\theta$.

3. $\pi \tan \pi z = \sum_{n=0}^{\infty} \dfrac{2z}{(n + \frac{1}{2})^2 - z^2}$

Hint.

$$\begin{aligned}
\pi \tan \pi z &= -\pi \cot \pi \left(z + \frac{1}{2} \right) \\
&= -\left\{ \frac{1}{z + \frac{1}{2}} + \sum_{n=1}^{\infty} \left(\frac{1}{z + n + \frac{1}{2}} - \frac{1}{n} \right) + \sum_{n=1}^{\infty} \left(\frac{1}{z - n + \frac{1}{2}} + \frac{1}{n} \right) \right\} \\
&= -\left\{ \left(\frac{1}{z + \frac{1}{2}} + \sum_{n=1}^{\infty} \frac{1}{z + n + \frac{1}{2}} \right) + \sum_{n=1}^{\infty} \frac{1}{z - n + \frac{1}{2}} \right\} \\
&= -\left\{ \sum_{n=0}^{\infty} \frac{1}{z + n + \frac{1}{2}} + \sum_{n=0}^{\infty} \frac{1}{z - n - \frac{1}{2}} \right\} \quad (n \to n + 1 \text{ in 2nd series})
\end{aligned}$$

4. $\dfrac{\pi}{\sin \pi z} = -\dfrac{1}{z} + \sum_{n=0}^{\infty} \dfrac{(-1)^n \, 2z}{z^2 - n^2}$

Hint. $\dfrac{\pi}{\sin \pi z} = \dfrac{1}{2} \left(\pi \cot \dfrac{\pi z}{2} + \pi \tan \dfrac{\pi z}{2} \right)$ Use Ex. 1 and 3.

Solution to Ex. 3.

$$\begin{aligned}
\pi \tan \pi z &= -\pi \cot \pi \left(z + \frac{1}{2} \right) \\
&= -\left[\frac{1}{z + \frac{1}{2}} + \sum_{n=1}^{\infty} \left(\frac{1}{z + n + \frac{1}{2}} - \frac{1}{n} \right) + \sum_{n=1}^{\infty} \left(\frac{1}{z - n + \frac{1}{2}} + \frac{1}{n} \right) \right]
\end{aligned}$$

$$\text{by (7.90) and the ideas in (7.94)}$$

$$= -\frac{1}{z + \frac{1}{2}} - \sum_{n=1}^{\infty} \left(\frac{1}{z + n + \frac{1}{2}} - \frac{1}{n} \right) - \sum_{n=0}^{\infty} \left(\frac{1}{z - n - \frac{1}{2}} + \frac{1}{n + 1} \right)$$

$$\text{replacing } n \text{ by } n + 1 \text{ in the second sum}$$

$$= -\sum_{n=0}^{\infty} \left(\frac{1}{z + n + \frac{1}{2}} - b_n \right) - \sum_{n=0}^{\infty} \left(\frac{1}{z - \left(n + \frac{1}{2} \right)} + \frac{1}{n + 1} \right)$$

$$b_0 = 0 \text{ and } b_n = \frac{1}{n} \text{ if } n > 0$$

$$= -\sum_{n=0}^{\infty} \left(\frac{2z}{z^2 - \left(n + \frac{1}{2} \right)^2} - b_n + \frac{1}{n + 1} \right)$$

$$= \sum_{n=0}^{\infty} \frac{2z}{\left(n + \frac{1}{2} \right)^2 - z^2} + \sum_{n=0}^{\infty} \left(b_n - \frac{1}{n + 1} \right)$$

To show that the $\sum_{n=0}^{\infty} \left(b_n - \dfrac{1}{n + 1} \right) = 0$, we write

$$\sum_{n=0}^{\infty} \left(b_n - \frac{1}{n + 1} \right) = -1 + \sum_{n=1}^{\infty} \left(\frac{1}{n} - \frac{1}{n + 1} \right).$$

Next we show that the sum of the telescoping series $\sum_{n=1}^{\infty} \left(\frac{1}{n} - \frac{1}{n+1} \right)$ is 1. The kth partial sum S_k for this telescoping series is

$$S_k = \left(1 - \frac{1}{2} \right) + \left(1 - \frac{1}{3} \right) + \cdots + \left(\frac{1}{k} - \frac{1}{k+1} \right) = 1 - \frac{1}{k+1}.$$

Clearly $S_k \to 1$ as $k \to \infty$.

Solution to Ex. 4.

$$
\begin{aligned}
\frac{\pi}{\sin \pi z} &= \frac{\pi}{2 \sin \frac{\pi z}{2} \cos \frac{\pi z}{2}} = \frac{\pi \left(\cos^2 \frac{\pi z}{2} + \sin^2 \frac{\pi z}{2} \right)}{2 \sin \frac{\pi z}{2} \cos \frac{\pi z}{2}} \\
&= \frac{1}{2} \left(\pi \cot \frac{\pi z}{2} + \pi \tan \frac{\pi z}{2} \right) \\
&= \frac{1}{2} \frac{2}{z} + \sum_{n=1}^{\infty} \frac{2z}{z^2 - (2n)^2} + \sum_{n=0}^{\infty} \frac{-2z}{z^2 - (2n+1)^2} \quad \text{by Ex. 1 and 3} \\
&= \frac{1}{z} + \sum_{n=0}^{\infty} \frac{2z}{z^2 - (2n)^2} + \sum_{n=0}^{\infty} \frac{-2z}{z^2 - (2n+1)^2} \\
&= \frac{1}{z} + \sum_{n=0}^{\infty} \frac{(-1)^n 2z}{z^2 - n^2}.
\end{aligned}
$$

Chapter 8

Rouche's Theorem and Open Mapping Theorem

8.1 The Principle of Reflection

Let D be a domain in R^2. Then f satisfies the **reflection principle** in D iff

$$f(\bar{z}) = \overline{f(z)} \quad \text{for each } z \text{ in } D. \tag{8.1}$$

Remarks 8.1.1 (a) If f satisfies the reflection principle in a domain D which contains an interval (a, b) on the x-axis, then $f(x)$ is real for each x in (a, b). For by hypothesis, $f(x) = f(\bar{x}) = \overline{f(x)}$. Thus $f(x)$ must be real.

(b) The functions

$$2z, \quad 5z^2 + 7, \quad \cos z, \quad \text{and} \quad e^z$$

satisfy the reflection principle in R^2, whereas the functions

$$2z + 5i, \quad (3 + 2i)\sin z, \quad \text{and} \quad e^{iz}$$

do not. For example, by (5.12), if $f(z) = \cos z$, then

$$
\begin{aligned}
f(\bar{z}) &= \cos \bar{z} = \cos x \cosh(-y) - i \sin x \sinh(-y) \\
&= \cos x \cosh y + i \sin x \sinh y = \overline{\cos z} \\
&= \overline{f(z)}.
\end{aligned}
$$

But if $f(z) = e^{iz}$, then for $y = \mathcal{I}(z) \neq 0$,

$$f(\bar{z}) = e^{i\bar{z}} = e^{i(x-iy)} = e^{y+ix} \neq e^{-y-ix} = e^{\overline{ix-y}} = \overline{e^{iz}} = \overline{f(z)}.$$

Theorem 8.1.2 Let f be analytic in a domain D which contains an interval (a, b) on the x-axis and which is symmetric to the x-axis. Then

$$f \text{ satisfies the reflection principle in } D \tag{8.2}$$

219

if and only if

$$f(x) \text{ is real} \qquad \text{for each } x \text{ in } (a,b). \tag{8.3}$$

Proof. Remark 8.1.1(a) shows that (8.2) implies (8.3).

To prove the other half, we suppose that (8.3) holds. We next let

$$f(z) = u(x,y) + iv(x,y) \qquad \text{for each } z \text{ in } D,$$

and we let

$$G(z) = \overline{f(\overline{z})} = U(x,y) + iV(x,y) \qquad \text{for each } z \text{ in } D. \tag{8.4}$$

Thus, since $G(z) = \overline{f(\overline{z})}$, we have

$$U(x,y) = u(x,t) \quad \text{and} \quad V(x,y) = -v(x,t) \qquad \text{where } t = -y. \tag{8.5}$$

Now, f is analytic in D. Thus $u(x,t)$, $v(x,t)$ and their partial derivatives are continuous on D and satisfy the Cauchy-Riemann equations

$$u_x = v_t \quad \text{and} \quad u_t = -v_x \qquad \text{for each } z = (x,t) \text{ in } D. \tag{8.6}$$

By (8.5),

$$U_x = u_x \quad \text{and} \quad V_y = -v_t \frac{dt}{dy} = v_t$$

which implies

$$U_x = V_y \quad \text{by (8.6)}. \tag{8.7}$$

Also by (8.5),

$$U_y = u_t \frac{dt}{dy} = -u_t \quad \text{and} \quad V_x = -v_x$$

which implies

$$U_y = -V_x \quad \text{by (8.6)}. \tag{8.8}$$

Thus G is analytic in D.

By (8.3) and (8.5),

$$V(x,0) = 0 \qquad \text{for each } x \text{ in } (a,b). \tag{8.9}$$

Hence

$$\begin{aligned} G(x) &= \overline{f(\overline{x})} &= U(x,0) + iV(x,0) \\ &&= u(x,-0) + i\,0 \qquad \text{by (8.5) and (8.9)}. \end{aligned}$$

Thus

$$G(x) = u(x,0) = f(x) \qquad \text{for each } x \text{ in } (a,b).$$

Now since f and G are analytic in D and since $f(x) = G(x)$ on (a,b), it follows from the Identity Theorem, 6.2.8, that $G(z) \equiv f(z)$ on D. This means, by (8.4), that

$$\overline{f(\overline{z})} = f(z) \qquad \text{on } D.$$

Hence

$$f(\overline{z}) = \overline{f(z)} \qquad \text{on } D,$$

and (8.2) is proved.

8.2 Rouche's Theorem

Theorem 8.2.1 Let C be a contour given by

$$z = z(t) \qquad \text{for each } t \text{ in } [a, b]$$

and let f be analytic on C where f is never zero on C. Let

$$g(t) = \int_a^t \frac{f'(z(s))\, z'(s)\, ds}{f(z(s))} + \text{Log}\, f(z(a)) \quad \text{for each } t \text{ in } [a, b]. \tag{8.10}$$

(a) The function g is continuous on the interval $[a, b]$.

(b) Furthermore, $e^{g(t)} = f(z(t))$ for each t in $[a, b]$.

Proof. By 2.2.15 and 3.4.1(b), we know that $\frac{f'}{f} z'$ is continuous at all points of $[a, b]$ except for at most a finite number of points of discontinuity of $z'(t)$. Hence Part (a) follows from 3.1.7(d). Also, for each t in $[a, b]$ at which $z'(t)$ is continuous,

$$g'(t) = \frac{f'(z(t))\, z'(t)}{f(z(t))}. \tag{8.11}$$

See 3.4.2.

To prove Part (b), we define the function ϕ by

$$\phi(t) = f(z(t))\, e^{-g(t)} \quad \text{for each } t \text{ in } [a, b], \tag{8.12}$$

and we prove that

$$\phi(t) \equiv 1 \qquad \text{on } [a, b]. \tag{8.13}$$

Now $\phi'(t)$ exists at each t in $[a, b]$ except for those points t at which $z'(t)$ does not exist. Except for these points,

$$\phi'(t) = e^{-g(t)}[f'(z(t))z'(t) - f(z(t))g'(t)] = 0 \quad \text{by (8.11)}.$$

Hence

$$\phi(t) = K \quad \text{for some constant } K \text{ in } R^2. \tag{8.14}$$

By (8.10),

$$g(a) = \text{Log}\, f(z(a)), \text{ that is, } e^{g(a)} = f(z(a)). \tag{8.15}$$

Hence by (8.12) and (8.15), $\phi(a) = 1$. Thus (8.14) implies (8.13). Now (8.12) (with $\phi(t) \equiv 1$) implies Part (b).

Definition 8.2.2 Let C, f, and g be given as in 8.2.1. Then the **variation of the logarithm of $f(z)$ along C**, denoted by $\Delta_C \log f(z)$, is the value of

$$\begin{aligned} g(b) - g(a) &= \int_a^b \frac{f'(z(t))\, z'(t)\, dt}{f(z(t))} \\ &= \Delta_C \log f(z) = \int_C \frac{f'(z)}{f(z)}\, dz. \end{aligned} \tag{8.16}$$

Corollary 8.2.3 Let C, f, and g be given as in 8.2.1. If C is closed then

$$\frac{1}{2\pi i} \Delta_C \log f(z) = k \qquad \text{for some integer } k. \tag{8.17}$$

Proof. By 8.2.1(b) (since $z(b) = z(a)$),

$$e^{g(a)} = f(z(a)) = f(z(b)) = e^{g(b)}.$$

Thus by 5.2.7(f),

$$g(b) - g(a) = 2\pi i k \qquad \text{for some integer } k,$$

and hence (8.17) holds.

Definition 8.2.4 (Winding Number or Index). Let C be a contour and let $z_0 \in R^2 - C$. The **winding number** $W(C, z_0)$ of C with respect to z_0 is given by

$$W(C, z_0) = \frac{1}{2\pi i} \int_C \frac{dz}{z - z_0}. \tag{8.18}$$

Remarks 8.2.5 Let C be a closed contour. By (8.16) and (8.17) (with $f(z) = z - z_0$), we see that $W(C, z_0)$ is an integer given by

$$W(C, z_0) = \frac{1}{2\pi i} \Delta_C \log (z - z_0).$$

In case C is a cco **simple** closed contour by (8.18),

$$W(C, z_0) = \begin{cases} 1 & \text{if } z_0 \in I(C) \quad \text{by Cauchy's Integral Formula} \\ 0 & \text{if } z_0 \in E(C) \quad \text{by the Cauchy-Goursat Theorem.} \end{cases} \tag{8.19}$$

If C is a clockwise oriented simple closed contour, then $W(C, z_0) = -1$ when $z_0 \in I(C)$. Now when C is a closed contour (not necessarily **simple**), the integer $W(C, z_0)$

> "is the number of times C winds around the point z_0 in the counter-clockwise direction minus the number of times C winds around z_0 in the clockwise direction." Briefly, "$W(C, z_0)$ is the number of times C winds around z_0."

Theorem 8.2.6 Let C be a closed contour given by $z = z(t)$ on $[a, b]$. If h is analytic on C and $|h(z)| < 1$ on C, then

$$\Delta_C \log [1 + h(z)] = 0. \tag{8.20}$$

Proof. If $h = h_1 + ih_2$, then $|h_1(z)| \leq |h(z)| < 1$. Thus $-1 < h_1(z)$ implies $1 + h_1(z) > 0$. Hence $1 + h(z)$ is in the half-plane $\mathcal{R}(z) > 0$. Thus

$$-\frac{\pi}{2} < \operatorname{Arg}[1 + h(z)] < \frac{\pi}{2}. \tag{8.21}$$

Let g be given by (8.10) with $f = 1 + h$. Then by 8.2.1(b),

$$e^{g(t)} = 1 + h(z(t)) \quad \text{for each } t \text{ in } [a, b].$$

Hence by (8.21), (8.15), and 8.2.1(a),

$$g(t) = \operatorname{Log}[1 + h(z(t))] = \operatorname{Log}|1 + h(z(t))| + i\operatorname{Arg}[1 + h(z(t))].$$

But C is closed so that $z(b) = z(a)$, and thus

$$g(b) - g(a) = \operatorname{Log}[1 + h(z(b))] - \operatorname{Log}[1 + h(z(a))] = 0.$$

Hence (8.20) follows by (8.16).

Theorem 8.2.7 Let C be a cco simple closed contour, and let S be the set of all poles in $I(C)$ of a function f where f is analytic in $\overline{I(C)} - S$ and where f is never zero on C. If S is finite, then

$$\frac{1}{2\pi i} \int_C \frac{f'(z)}{f(z)}\, dz = n - p \tag{8.22}$$

where n and p are, respectively, the sum of the orders of the zeros of f in $I(C)$ and the sum of the orders of the poles of f in $I(C)$.

Proof. Let z_0 be a **zero** of f of order j in $I(C)$. By 6.2.2 and 6.2.3, there is some $r > 0$ such that

$$f(z) = (z - z_0)^j g(z) \quad \text{on } N_r(z_0)$$

where $g \in A(N_r(z_0))$ and where g is never zero in $N_r(z_0)$. Hence for each z in $N_r(z_0)$,

$$f'(z) = j(z - z_0)^{j-1}g(z) + (z - z_0)^j g'(z).$$

Thus

$$\frac{f'(z)}{f(z)} = \frac{j}{z - z_0} + \frac{g'(z)}{g(z)} \quad \text{on } N_r'(z_0).$$

Now since $\frac{g'}{g}$ is analytic at z_0 it follows from the last equation that $\frac{f'}{f}$ has a simple pole at z_0 with residue j.

Let z_1 be a **pole** of f of order k in $I(C)$. By 6.3.5, we let $s > 0$ such that

$$f(z) = (z - z_1)^{-k}q(z) \quad \text{on } N_s'(z_1) \tag{8.23}$$

where q is analytic in $N_s(z_1)$ and q is never 0 in $N_s(z_1)$. Thus by (8.23),

$$\frac{f'(z)}{f(z)} = \frac{-k}{z - z_1} + \frac{q'(z)}{q(z)} \quad \text{on } N_s'(z_1).$$

Hence by 6.3.4 (Case II) and 6.3.8, it follows that $\frac{f'}{f}$ has a simple pole at z_1 with residue $-k$. Now if $j_1, j_2, \cdots, j_{n_1}$ are the orders of the zeros of f in $I(C)$ and $k_1, k_2, \cdots, k_{n_2}$ are the orders of the poles of f in $I(C)$, then by the Cauchy Residue Theorem

$$\begin{aligned}
\int_C \frac{f'(z)\,dz}{f(z)} &= 2\pi i(j_1 + j_2 + \cdots + j_{n_1} - k_1 - k_2 - \cdots - k_{n_2}) \\
&= 2\pi i(n - p).
\end{aligned}$$

Hence (8.22) is proved.

In a particular setting, the phrase *number of zeros* is often used to refer to **the sum of the orders of the specified zeros.** We shall enclose the phrase in quotation marks when we want the expression to denote the sum of the orders of the specified zeros.

Theorem 8.2.8 (Rouche's Theorem). Let f and g be analytic in the closure of the interior of C where C is a cco simple closed contour given by

$$C: \quad z = z(t) \quad \text{on } [a, b].$$

If $|f(z)| > |g(z)|$ on C, then f and $f + g$ have the same "number of zeros" in $I(C)$.

Proof. We note that

$$|f| > |g| \geq 0 \text{ on } C \text{ implies } f \text{ is never } 0 \text{ on } C$$

and

$$|f + g| \geq |f| - |g| > 0 \text{ on } C \text{ implies } f + g \text{ is never } 0 \text{ on } C.$$

If n and n' denote, respectively, the "numbers of zeros" of f and $f + g$ in $I(C)$, then by (8.16) and (8.22) with $p = 0$,

$$2\pi i n = \Delta_C \log f(z) \tag{8.24}$$

and

$$2\pi i n' = \Delta_C \log [f(z) + g(z)] = \Delta_C \log \left[f(z) \left(1 + \frac{g(z)}{f(z)} \right) \right]. \tag{8.25}$$

By 8.2.1, there are functions α and β continuous on $[a, b]$ such that

$$e^{\alpha(t)} = f(z(t)) \quad \text{and} \quad e^{\beta(t)} = 1 + \frac{g(z(t))}{f(z(t))}. \tag{8.26}$$

Let
$$\gamma(t) = \alpha(t) + \beta(t). \qquad (8.27)$$

Then
$$e^{\gamma(t)} = e^{\alpha(t)} e^{\beta(t)} = f(z(t)) \left(1 + \frac{g(z(t))}{f(z(t))} \right).$$

Hence
$$
\begin{aligned}
\Delta_C \log \left[f(z) \left(1 + \frac{g(z)}{f(z)} \right) \right] &= \gamma(b) - \gamma(a) \quad \text{by (8.16)} \\
&= [\alpha(b) - \alpha(a)] + [\beta(b) - \beta(a)] \quad \text{by (8.27)} \\
&= \Delta_C \log f(z) + \\
&\qquad \Delta_C \log \left(1 + \frac{g(z)}{f(z)} \right) \quad \text{by (8.16) and (8.26)} \\
&= \Delta_C \log f(z) \quad \text{by 8.2.6.} \qquad (8.28)
\end{aligned}
$$

Thus $n' = n$ by (8.24), (8.25), and (8.28).

Theorem 8.2.9 An nth degree polynomial ($n > 0$) with coefficients in R^2 has "n zeros" in R^2.

    Proof. Let
$$f(z) = a_n z^n \qquad \text{where } a_n \neq 0,$$

and let
$$g(z) = a_{n-1} z^{n-1} + \cdots + a_0.$$

There is some $r_0 > 0$ such that for each $r \geq r_0$,
$$|a_n| > \frac{|a_{n-1}|}{r} + \frac{|a_{n-2}|}{r^2} + \cdots + \frac{|a_0|}{r^n}. \qquad (8.29)$$

Let γ be an arbitrary circle given by
$$\gamma = \{ z : \ |z| = r \} \qquad \text{where } r \geq r_0.$$

Then for each z on γ,
$$|f(z)| = |a_n| r^n > |a_{n-1}| r^{n-1} + \cdots + |a_0| \geq |g(z)| \quad \text{by (8.29)}.$$

Now $a_n z^n$ has "n zeros" in $I(\gamma)$ (namely, $z = 0$ of order n). Hence by 8.2.8,
$$f(z) + g(z) = \sum_{k=0}^{n} a_k z^k$$

has "n zeros" in $I(\gamma)$. But since γ was **any** circle with center at $z = 0$ and with an arbitrarily large radius $r \geq r_0$, it follows that
$$\sum_{k=0}^{n} a_k z^k$$

has "n zeros" in R^2.

Corollary 8.2.10 Let $P(z) = a_n z^n + a_{n-1} z^{n-1} + \cdots + a_1 z + a_0$ be a polynomial with coefficients in R^2 such that $a_n a_0 \neq 0$. Let r be the largest integer such that

$$|a_n|r^n + |a_{n-1}|r^{n-1} + \cdots + |a_1|r < |a_0|, \tag{8.30}$$

and let s be the least positive integer such that

$$|a_n|s^n > |a_{n-1}|s^{n-1} + \cdots + |a_1|s + |a_0|. \tag{8.31}$$

Then all of the zeros of P are in the annular region $A = \{z : \ r < |z| < s\}$.

Proof. First, suppose $r = 0$. Since $a_0 \neq 0$, each zero of $P(z)$ is in $\{z : \ |z| > r = 0\}$.

Now, suppose $r > 0$. Let $f(z) = a_0$, let $g(z) = a_n z^n + \cdots + a_1 z$, and let $\gamma_1 = \{z : \ |z| = r\}$. From the definition of r, for each z on γ_1,

$$|g(z)| \leq |a_n|r^n + \cdots + |a_1|r < |a_0| = |f(z)|.$$

By Rouche's Theorem, $P(z) = f(z) + g(z)$ has the same "number of zeros" in $I(\gamma_1)$ as $f(z)$. But $f(z)$, and hence $P(z)$, has no zeros in $I(\gamma_1)$. Now let

$$F(z) = a_n z^n, \quad G(z) = a_{n-1} z^{n-1} + \cdots + a_0, \quad \text{and} \quad \gamma_2 = \{z : \ |z| = s\}. \tag{8.32}$$

Then by (8.32), for each z on γ_2,

$$|F(z)| = |a_n|s^n > |a_{n-1}|s^{n-1} + \cdots + |a_1|s + |a_0| \geq |G(z)|.$$

Thus by 8.2.8, we see that $P = F + G$ has "n zeros" in $I(\gamma_2)$ since $F(z) = a_0 z^n$ has "n zeros" at $z = 0$ in $I(\gamma_2)$. Hence the "n zeros" of P are in the annular region A.

Theorem 8.2.11 Let z_0 be a zero of $f(z) - w_0$ of order m where f is analytic at z_0. Then there is some $\delta > 0$ such that the following holds.

If $0 < r < \delta$, then there is some $s > 0$ such that for each fixed w_1 in $N_s'(w_0)$,

there are exactly m **distinct** values of z in $N_r'(z_0)$ for which $f(z) = w_1$. (8.33)

Proof. Since $f^{(m)}(z_0) \neq 0$, we know that f is not constant in a neighborhood of z_0. Thus z_0 is an isolated zero of $f - w_0$ by 6.2.5. Also by 6.2.5, the zeros of f' are isolated. Thus we choose $\delta > 0$ such that f is analytic in $N_\delta(z_0)$ and such that

$$f - w_0 \text{ is never zero in } N_\delta'(z_0) \tag{8.34}$$

and

$$f' \text{ is never zero in } N_\delta'(z_0). \tag{8.35}$$

Now we let r be such that $0 < r < \delta$, and we let C be the cco circle given by

$$C : \quad |z - z_0| = r. \tag{8.36}$$

Also, we let

$$s = \min\{|f(z) - w_0| : \ z \in C\}$$
$$> 0 \quad \text{by (8.34) and by Ex. 2(c) in 2.4.9.} \tag{8.37}$$

Now let

$$w_1 \in N'_s(w_0) \quad \text{(where } w_1 \text{ is fixed but arbitrary).} \tag{8.38}$$

By (8.37) and (8.38),

$$\text{for each } z \text{ on } C, \ |w_0 - w_1| < s \leq |f(z) - w_0|.$$

Now we apply Rouche's Theorem to $f - w_0$ and the constant function $g(z)$ given by

$$g(z) = w_0 - w_1.$$

Now

$$f - w_1 = (f - w_0) + (w_0 - w_1).$$

Thus by Rouche's Theorem, $f(z) - w_1$ has the same "number of zeros" in $I(C)$ as $f(z) - w_0$. By (8.34), it follows that z_0 is the only zero of $f(z) - w_0$ in $I(C) = N_r(z_0)$, since $r < \delta$. But this zero is of order m. Thus $f(z) - w_1$ has exactly "m zeros" in $I(C) = N_r(z_0)$. Since $w_1 \neq w_0$, these "m zeros" are in $N'_r(z_0)$.

Now $(f - w_1)' = f'$ is never zero in $N'_r(z_0)$ by (8.35), since $r < \delta$. Thus each zero of $f - w_1$ in $N'_r(z_0)$ is of order 1. This means there are exactly m distinct values of z in $N'_r(z_0)$ such that $f(z) = w_1$.

Corollary 8.2.12 If $f \in A(z_0)$ and $f'(z_0) \neq 0$, then there is some $\rho > 0$ such that f is one-to-one on $N_\rho(z_0)$.

Proof. Let $w_0 = f(z_0)$ and let $\delta > 0$ such that (8.33) and (8.34) hold. Let r be such that $0 < r < \delta$ and let $s > 0$ as given in (8.33). Since $f \in C(z_0)$, let ρ be such that

$$f(N_\rho(z_0)) \subset N_s(w_0) \quad \text{where } 0 < \rho < r.$$

Then by (8.33) and (8.34), it follows that f is one-to-one on $N_\rho(z_0)$, since $f'(z_0) \neq 0$ implies $m = 1$. Note that (8.34) implies there is no z in $N'_\rho(z_0)$ such that $f(z) = w_0$.

Corollary 8.2.13 Let $f \in A(G)$ where G is an open subset of R^2. If f is one-to-one on G, then f' is never zero in G.

Proof. Suppose $f'(z_0) = 0$ for some z_0 in G. Then z_0 is a zero of $f - f(z_0)$ of order $m \geq 2$. Let $\delta > 0$ such that (8.33) holds. Let r be such that

$$0 < r < \delta \quad \text{and such that } N_r(z_0) \subset G.$$

Then by 8.2.11,

$$f \text{ is not one-to-one on } N_r(z_0) \quad \text{(since } m \geq 2\text{)}.$$

This means that f is not one-to-one on G, contrary to the hypothesis.

Theorem 8.2.14 (Open Mapping Theorem). Let $f \in A(D)$ where D is a domain in R^2, and let f be nonconstant on D. If G is an open subset of D, then $f(G)$ is open.

Proof. Suppose $w_0 \in f(G)$, and let $z_0 \in G$ such that $f(z_0) = w_0$. By 8.2.11, let $\delta > 0$ such that (8.33) holds. Let r be such that $0 < r < \delta$ and such that $N_r(z_0) \subset G$. By 8.2.11, let $s > 0$ such that

if $w \in N_s(w_0)$, then $w = f(z)$ for some z in $N_r(z_0) \subset G$.

This means that each w in $N_s(w_0)$ is in $f(G)$, that is,

$$N_s(w_0) \subset f(G).$$

Thus $f(G)$ is open.

Definition 8.2.15 Let f be a one-to-one function having any set G as its domain. The **inverse function** f^{-1} is the function with domain $f(G)$ determined by the following criterion.

If $w \in f(G)$, then $f^{-1}(w)$ is the unique z in G such that $f(z) = w$.

Theorem 8.2.16 (Inverse Function Theorem). If f is one-to-one and analytic on an open set G in R^2, then each of the following holds.

(a) The inverse function f^{-1} is continuous on $f(G)$.

(b) Also, f^{-1} is analytic in $f(G)$.

(c) If $w \in f(G)$, then $(f^{-1})'(w) = \frac{1}{f'(z)}$ where $z = f^{-1}(w)$.

(Note that $f'(z) \neq 0$ by 8.2.13.)

Proof of (a). Suppose $w_0 \in f(G)$. To prove f^{-1} is continuous at w_0, we let $\epsilon > 0$ such that
$$N_\epsilon(z_0) \subset G \quad \text{where } z_0 = f^{-1}(w_0).$$
By 8.2.14, it follows that

$$U = f[N_\epsilon(z_0)] \text{ is an open subset of } f(G). \tag{8.39}$$

Since $w_0 \in U$ and U is open, let $\delta > 0$ such that

$$N_\delta(w_0) \subset U. \tag{8.40}$$

By (8.39) and (8.40), we have

$$f^{-1}(N_\delta(w_0)) \subset N_\epsilon(z_0) = N_\epsilon(f^{-1}(w_0)).$$

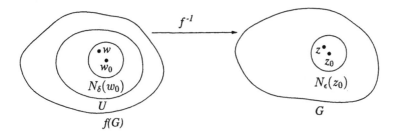

Figure 8.1: $f^{-1}(N_\delta(w_0)) \subset N_\epsilon(z_0) = N_\epsilon(f^{-1}(w_0))$

Thus

$$|w - w_0| < \delta \quad \text{implies} \quad |f^{-1}(w) - f^{-1}(w_0)| < \epsilon.$$

Hence f^{-1} is continuous at w_0. Since w_0 is an arbitary point in $f(G)$, it follows that f^{-1} is continuous on $f(G)$. See Figure 8.1.

Proof of (b) and (c). Suppose $w_0 \in f(G)$. Then for each w in $f(G)$ such that $w \neq w_0$,

$$\begin{aligned}
\frac{f^{-1}(w) - f^{-1}(w_0)}{w - w_0} &= \frac{z - z_0}{f(z) - f(z_0)} \\
&= \frac{1}{\frac{f(z)-f(z_0)}{z-z_0}} \quad \text{where } z \neq z_0 \text{ since } w \neq w_0 \\
&\to \frac{1}{f'(z_0)} \quad \text{as } w \to w_0.
\end{aligned}$$

The last limit statement follows since $z \to z_0$ as $w \to w_0$ by Part (a).

Since w_0 was an arbitrary point of $f(G)$, Parts (b) and (c) are proved.

Theorem 8.2.17 Let $E \subset R^2$. Then $z \in \overline{E}$ iff

$$E \cap N_r(z) \neq \emptyset \qquad \text{for each } r > 0. \tag{8.41}$$

Proof. First suppose $z \in \overline{E} = E \cup E^*$. Then $z \in E$ or $z \in E^*$. In either case, each neighborhood of z contains a point of E.

Next suppose that (8.41) holds. Then either (i) $z \in E$, or (ii) $E \cap N'_r(z) \neq \emptyset$ for each $r > 0$. Clearly, condition (i) implies that $z \in E \cup E^* = \overline{E}$. But condition (ii) means that $z \in E^*$, and hence $z \in \overline{E}$. Thus in either case $z \in \overline{E}$.

Theorem 8.2.18 Let $E \subset D \subset R^2$ and let f be continuous on D. If $z \in D \cap \overline{E}$, then $f(z) \in \overline{f(E)}$.

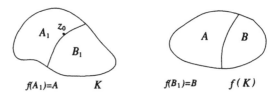

Figure 8.2: $A_1 = K \cap f^{-1}(A)$ and $B_1 = K \cap f^{-1}(B)$

Proof. Suppose z is an element of $D \cap \overline{E}$ and let $\epsilon > 0$. Since $f \in C(z)$, let $\delta > 0$ such that

$$f(D \cap N_\delta(z)) \subset N_\epsilon(f(z)). \qquad (8.42)$$

By 8.2.17, let $p \in E \cap N_\delta(z) \subset D \cap N_\delta(z)$. Thus by (8.42),

$$f(p) \in f(E) \cap N_\epsilon(f(z)) \quad \text{which implies} \quad f(E) \cap N_\epsilon(f(z)) \neq \emptyset.$$

Hence $f(z) \in \overline{f(E)}$ by 8.2.17.

Theorem 8.2.19 If $f \in C(K)$ where K is a connected subset of R^2, then $f(K)$ is connected.

Proof. Let A and B be **any** nonempty sets such that

$$f(K) = A \cup B \quad \text{and} \quad A \cap B = \emptyset.$$

We shall prove

$$A \cap \overline{B} \neq \emptyset \quad \text{or} \quad \overline{A} \cap B \neq \emptyset. \qquad (8.43)$$

This will show there exists no separation of $f(K)$, that is, $f(K)$ is connected. Let

$$A_1 = \{z : z \in K \text{ and } f(z) \in A\} \quad \text{and} \quad B_1 = \{z : z \in K \text{ and } f(z) \in B\}.$$

Then

$$K = A_1 \cup B_1 \quad \text{and} \quad A_1 \neq \emptyset \neq B_1.$$

Since K is connected, $A_1 \cap \overline{B_1} \neq \emptyset$ or $\overline{A_1} \cap B_1 \neq \emptyset$. Suppose the sets are named so that $A_1 \cap \overline{B_1} \neq \emptyset$ and let

$$z_0 \in A_1 \cap \overline{B_1}.$$

Then $f(z_0) \in A$ and by 8.2.18, we have $f(z_0) \in \overline{f(B_1)} = \overline{B}$ [since $f(B_1) = B$]. Thus $f(z_0) \in A \cap \overline{B}$ and (8.43) is proved. See Figure 8.2.

Combining 8.2.14 and 8.2.19, we have the following theorem.

Theorem 8.2.20 Let $f \in A(D)$ where D is a domain in R^2. If f is nonconstant on D, then $f(D)$ is a domain.

Exercises 8.2.21

1. Show that $az^n + be^z$ has "n zeros" in the interior of the unit circle if $|a| > |b|e$.

2. Show that the "seven zeros" of $3z^7 + 5z - 1$ lie in the interior of the circle $|z| = 2$.

3. Show that the polynomial $P(z)$ in Ex. 2 has "one zero" in $N_1(0)$ and that zero is a real number between 0 and 1. **Hint.** Apply Rouche's Theorem with $f(z) = 5z$ and $g(z) = 3z^7 - 1$. Also note that $P(0) < 0$ and $P(1) > 0$.

4. Show that the polynomial $P(z)$ in Ex. 2 has the six remaining zeros in the annular region $1 < |z| < 2$. **Hint.** $|P(z)| = |(3z^7 + 5z) - 1| \geq ||3z^7 + 5z| - 1| \geq |||3z^7| - |5z|| - 1| \geq 2 - 1 = 1$ if $|z| = 1$.

5. Let $h(z) = \frac{1}{2}\left(z + \frac{1}{z}\right)$. Prove that if w is any complex number not in the closed interval $[-1, 1]$, then there is exactly one z in $N_1(0)$ such that $h(z) = w$.

Solution. For a fixed w not in $[-1, 1]$, let

$$f(z) = h(z) - w. \tag{8.44}$$

Let C be the cco circle given by

$$C: \qquad z(\theta) = \cos\theta + i\sin\theta \quad \text{on } [0, 2\pi]. \tag{8.45}$$

Then for each $z = e^{i\theta}$ on C,

$$h(e^{i\theta}) = \frac{1}{2}(e^{i\theta} + e^{-i\theta}) = \cos\theta \in [-1, 1] \quad \text{by 5.2.7(g).} \tag{8.46}$$

Thus $f(z)$ is never zero on C since w is not in $[-1, 1]$. Since f has only the simple pole at $z = 0$ in $I(C)$, it follows by 8.2.7 that if n is the "number of zeros" of f in $I(C)$, then

$$
\begin{aligned}
\frac{1}{2\pi i}\int_C \frac{f'(z)\,dz}{f(z)} &= n - 1 \\
&= \frac{1}{2\pi i}\Delta_C \log f(z) \qquad \text{by 8.2.2} \\
&= \frac{1}{2\pi i}[g(2\pi) - g(0)] \quad \text{by 8.2.2} \tag{8.47}
\end{aligned}
$$

where g is given in 8.2.1. By 8.2.1(b) and (8.10),

$$g(\theta) = \log f(z(\theta)) \quad \text{and} \quad g(0) = \operatorname{Log} f(z(0)). \tag{8.48}$$

Now for each θ in $[0, 2\pi]$, we have

$$
\begin{aligned}
f(z(\theta)) &= h(z(\theta)) - w \quad \text{by (8.44)} \\
&= -w + \cos\theta \quad \text{by (8.46).}
\end{aligned}
$$

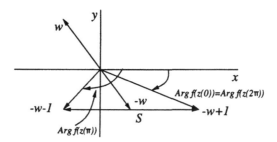

Figure 8.3: $f(z(\theta)) = -w + \cos\theta$ which varies along the segment S from $-w + 1$ to $-w - 1$ and back to $-w + 1$.

Thus for a fixed w not in $[-1, 1]$ (see Figure 8.3),

> as θ varies from 0 to π, $f(z(\theta))$ varies along the horizontal segment S from $-w + 1$ to $-w - 1$, $\hspace{2cm}$ (8.49)

and

> as θ varies from π to 2π, $f(z(\theta))$ varies along S from $-w - 1$ back to $-w + 1$. $\hspace{2cm}$ (8.50)

By 8.2.1(a), it follows that $g \in C\left([0, 2\pi]\right)$. Thus by (8.48)–(8.50),

$$g(\theta) = \text{Log}\, f(z(\theta)) \quad \text{for each } \theta \text{ in } [0, 2\pi]. \hspace{1cm} (8.51)$$

By (8.51) and (8.45), we have

$$\begin{aligned}
g(2\pi) - g(0) &= \text{Log}\, f(z(2\pi)) - \text{Log}\, f(z(0)) \\
&= \text{Log}\, f(1) - \text{Log}\, f(1) = 0. \hspace{1cm} (8.52)
\end{aligned}$$

By (8.47) and (8.52), we have $n = 1$ where n is the "number of zeros" of f in $I(C)$. This means there is exactly one z in $I(C) = N_1(0)$ such that $h(z) = w$.

Chapter 9

Harmonic Functions

9.1 Harmonic Functions

Definition 9.1.1 Let D be a domain in R^2. A real-valued function u of (x, y) is harmonic on D iff the partial derivatives u_{xx}, u_{xy}, u_{yx} and u_{yy} are continuous in D and satisfy Laplace's equation

$$u_{xx} + u_{yy} = 0 \quad \text{on } D.$$

We sometimes indicate that u is harmonic on D by writing $u \in H(D)$. We also write Laplace's equation as

$$\nabla^2 u = 0$$

where ∇^2 denotes the Laplacian operator $\frac{\partial^2}{\partial x^2} + \frac{\partial^2}{\partial y^2}$.

Theorem 9.1.2 If $f = u + iv$ is analytic in a domain D and if u and v are real-valued, then u and v are harmonic on D.

Proof. By 2.3.7, we have $u_x = v_y$ and $u_y = -v_x$ for each (x, y) in D. Thus for each (x, y) in D,

$$u_{xx} = v_{yx} = v_{xy} = -u_{yy}. \quad \text{See Ex. 1 in 3.6.18.}$$

Thus $u_{xx} + u_{yy} = 0$ on D. Hence u is harmonic on D. Similarly v is harmonic on D.

Definition 9.1.3 Let $u \in H(D)$. Then a real-valued function v is a **harmonic conjugate of u on D** iff

$$f = u + iv \text{ is analytic in } D.$$

Thus v is a harmonic conjugate of u on D iff v is the imaginary part of some analytic function in D where u is the real part.

Remarks 9.1.4 (a) If v is a harmonic conjugate of u on D, then $-u$ is a harmonic conjugate of v on D. For if $u + iv$ is analytic in D, then so is $-i(u + iv) = v + i(-u)$.

(b) If v is a harmonic conjugate of u on D, then clearly so is $v + k$ where k is any real constant.

(c) If v and w are harmonic conjugates of u on a domain D, then $v - w$ is constant on D. For if $f = u + iv$ and $g = u + iw$ are analytic in D, then $f - g = i(v - w)$ is analytic in D. Thus $v - w$ is constant on D by Ex. 16(b) in 2.3.16.

Theorem 9.1.5 Let $u \in H(D)$ where D is a simply connected domain in R^2. Then there is a function f which is analytic in D such that $u = \mathcal{R}(f)$ on D. One such f is given by

$$f = u + iv \tag{9.1}$$

where $v(z) = \mathcal{I} \left(\int_{z_0}^{z} [u_x(w) - iu_y(w)]\, dw \right)$ and where $z_0 \in D$.

Proof. Applying the Cauchy-Riemann conditions to

$$h = u_x - iu_y, \tag{9.2}$$

we have

$$\frac{\partial}{\partial x} \mathcal{R}(h) \;=\; u_{xx} \;=\; -u_{yy} \quad \text{since } u \in H(D)$$

$$= \frac{\partial}{\partial y} \mathcal{I}(h)$$

and

$$\frac{\partial}{\partial y} \mathcal{R}(h) = u_{xy} = u_{yx} = -\frac{\partial}{\partial x} \mathcal{I}(h).$$

Thus $h \in A(D)$ and hence by 3.6.3, we have $g \in A(D)$, where

$$\begin{aligned} g(z) &= \int_{z_0}^{z} [u_x(w) - iu_y(w)]\, dw \\ &= \phi(z) + iv(z) \end{aligned} \tag{9.3}$$

where we let $\phi = \mathcal{R}(g)$ and $v = \mathcal{I}(g)$. Also

$$\begin{aligned} g' &= u_x - iu_y \quad \text{by 3.6.3} \\ &= \phi_x + iv_x \quad \text{by 2.3.7.} \end{aligned} \tag{9.4}$$

Now

$$u_y = -v_x \quad \text{and} \quad u_x \;=\; \phi_x \quad \text{by (9.4)}$$
$$= v_y \quad \text{by (9.3)} \tag{9.5}$$

since $\frac{\partial}{\partial x} \mathcal{R}(g) = \frac{\partial}{\partial y} \mathcal{I}(g)$. Thus (9.5) implies that $f \in A(D)$ where f is given in (9.1).

In view of 9.1.4(b) and (c), it follows that $v + k$ (where $k \in R$) represents all harmonic conjugates of u on D where v is given in (9.1).

Remarks 9.1.6 (a) Let D be a simply connected domain with the following property.

> There is a point $z_0 = (x_0, y_0)$ in D such that for each $z = (x, y)$ in D,
> $L_1 \cup L_2 \subset D$ where L_1 is the line segment from z_0 to (x, y_0)
> and L_2 is the line segment from (x, y_0) to z. $\qquad$ (9.6)

Then $L_1 \cup L_2$ may be used as the contour of integration in (9.1). We have

$$\begin{array}{llll} L_1: & w = s + iy_0 & \text{and} & dw = ds \\ L_2: & w = x + ti & \text{and} & dw = idt \end{array} \qquad (9.7)$$

where s varies from x_0 to x and t varies from y_0 to y. Then (9.1) becomes

$$\begin{aligned} v &= \mathcal{I}\left(\int_{L_1} (u_x - iu_y) \, dw + \int_{L_2} (u_x - iu_y) \, dw \right) \\ &= -\int_{x_0}^{x} u_y(s, y_0) \, ds + \int_{y_0}^{y} u_x(x, t) \, dt \quad \text{by (9.7).} \end{aligned} \qquad (9.8)$$

(b) Let $u \in H(D)$ and let $z_0 \in D$ where D is any domain in R^2, not necessarily simply connected. Since D is open, let $r > 0$ such that $N_r(z_0) \subset D$. Now $N_r(z_0)$ is simply connected and has property (9.6). Thus a harmonic conjugate v of u in $N_r(z_0)$ is given by (9.8).

(c) Let

$$u(z) = \text{Log} |z| \quad \text{for each } z \text{ in } D$$

where

$$D = \{z : |z| > 0\}.$$

It is easy to verify $\nabla^2 u = 0$ on D. Thus $u \in H(D)$. But Ex. 1 in 9.1.7 shows that there is no harmonic conjugate of u in D. Thus the hypothesis that D is simply connected in 9.1.5 is essential.

Exercises 9.1.7

1. Show that there is no harmonic conjugate of u in D where u and D are given in 9.1.6(c).

2. Prove if A and B are nonempty open sets in R^2 such that $A \cap B = \emptyset$, then $A \cup B$ is not connected.

Solution of Ex. 1. Let $E = D - L$ where $L = \{z : z = x < 0\}$. By 9.1.3, the function $v = \mathcal{I}(\text{Log } z) = \text{Arg } z$ is a harmonic conjugate of u in E. By 9.1.4(c), each harmonic conjugate of u in E is of the form $v + k$ (where $k \in R$). But

$$v(z) \to \pi \quad \text{as} \quad z \to x \text{ from above } L$$

and
$$v(z) \to -\pi \quad \text{as} \quad z \to x \text{ from below } L.$$

Thus the domain of v cannot be extended to include L in such a way that $v \in C(D)$. Furthermore $v + k$ cannot be continuous on D. Thus there exists no harmonic conjugate of u in D. [For the "restriction" to E of a harmonic conjugate of u in D must be a harmonic conjugate of u in E.]

Solution of Ex. 2. Let $z \in A$ and let $r > 0$ such that $N_r(z) \subset A$. Then $B \cap N_r(z) = \emptyset$. Thus z is not a limit point of B. Hence no point of A is a limit point of B. Similarly no point of B is a limit point of A. Thus A and B form a separation of $A \cup B$.

Theorem 9.1.8 Let $u \in H(D)$ where D is a domain in R^2. If u is constant on some disk $N_r(z_0)$ which is contained in D, then u is constant on D.

Proof. Let

$$E = \{z : \ z \in D \text{ and } u \text{ is constant on some neighborhood of } z\}. \qquad (9.9)$$

Clearly
$$E \text{ is open and } E \neq \emptyset \text{ (since } z_0 \in E). \qquad (9.10)$$

Now assume
$$D - E \neq \emptyset. \qquad (9.11)$$

Let
$$z_1 \in (D - E) \qquad (9.12)$$

and let $r > 0$ such that $N_r(z_1) \subset D$. We shall see that

$$N_r(z_1) \subset D - E \qquad (9.13)$$

which implies
$$D - E \text{ is open.} \qquad (9.14)$$

[To verify (9.13), let $f \in A(N_r(z_1))$ such that $\mathcal{R}(f) = u$ on $N_r(z_1)$ by 9.1.5, and suppose

$$E \cap N_r(z_1) \neq \emptyset. \qquad (9.15)$$

Then $N_r(z_1)$ contains a disk D_0 on which u is constant. By 2.3.15(b), it follows that f is constant on D_0. Therefore, f is constant on $N_r(z_1)$ by the Identity Theorem 6.2.8. Hence $u = \mathcal{R}(f)$ is constant on $N_r(z_1)$. Thus $z_1 \in E$ by (9.9). But this contradicts (9.12). Hence we must reject (9.15) and thus (9.13) is proved.]

Now the assumption (9.11) implies that the set

$$D = E \cup (D - E)$$

is not connected, in view of (9.10), (9.14), and Ex. 2 in 9.1.7. This contradicts the hypothesis that D is a domain. Therefore, our assumption (9.11) is false and hence $E = D$. Thus u is constant on D by (9.9).

Theorem 9.1.9 (Maximum-Minimum Principle for Harmonic Functions). Let $u \in H(D)$ where D is a domain in R^2. If u attains a maximum or a minimum in D, then u is constant on D.

Proof. (An alternative proof is indicated just before 9.1.15.) Suppose there is some point z_0 in D such that

$$u(z) \le u(z_0) \quad \text{for each } z \text{ in } D. \tag{9.16}$$

Let $r > 0$ such that $N_r(z_0) \subset D$. We shall prove that u is constant on $N_r(z_0)$ and hence, by 9.1.8, u is constant on D. For assume

$$u \text{ is not constant on } N_r(z_0). \tag{9.17}$$

By 9.1.5, let $f \in A(N_r(z_0))$ such that

$$u = \mathcal{R}(f) \quad \text{on } N_r(z_0).$$

Then f is not constant on $N_r(z_0)$ and by the Open Mapping Theorem, 8.2.14, the set $f(N_r(z_0))$ is open. Thus there is some $\delta > 0$ such that

$$N_\delta(f(z_0)) \subset f(N_r(z_0)).$$

Hence there is a horizontal line segment S with left endpoint at $f(z_0) = u(z_0) + iv_0$ such that

$$S \subset N_\delta(f(z_0)).$$

Now for each w in $(S - \{f(z_0)\})$, there is some z in $N_r(z_0)$ such that

$$w = f(z) = u(z) + i\mathcal{I}(f(z)).$$

Clearly for each such z,

$$u(z) > u(z_0).$$

This contradicts (9.16). Hence (9.17) is false. Thus if u attains a maximum in D, then u is constant on D.

Next suppose u attains a minimum in D. Then the harmonic function $-u$ attains a maximum in D. By the previous result, $-u$ is constant on D, which implies u is constant in D.

Corollary 9.1.10 Let $u \in H(D)$ and let $u \in C(\overline{D})$ where D is a bounded domain in R^2. Then u attains its maximum and its minimum on $B(D)$, the boundary of D.

Proof. By 2.4.4(b), the continuous function u attains a maximum and a minimum in the compact set $\overline{D}$. If u is not constant in D, then by 9.1.9, the maximum and the minimum must occur on $B(D)$. But if u is constant on D, then (by continuity) u is constant on $\overline{D}$ and the conclusion is clear.

Corollary 9.1.11 Let $u, v \in H(D)$ and let $u, v \in C(\overline{D})$ where D is a bounded domain in R^2. If $u = v$ on $B(D)$, then $u = v$ on $\overline{D}$.

Proof. By the present hypotheses, $u - v$ satisfies the hypotheses on u in 9.1.10. Thus by 9.1.10, the function $u - v$ attains its maximum and its minimum on $B(D)$. But since $u = v$ on $B(D)$, we have $u - v \equiv 0$ on $B(D)$. Hence

$$\max_{\text{on } \overline{D}} (u - v) \; = \; 0 \; = \; \min_{\text{on } \overline{D}} (u - v)$$

Thus $u - v \equiv 0$ on $\overline{D}$ and hence $u = v$ on $\overline{D}$.

Remark 9.1.12 Let $u(z) = \text{Log}\,|z|$ on $\overline{D}$ and let $v(z) = 0$ on $\overline{D}$ where $D = \{z : |z| > 1\}$. Now $u, v \in H(D)$. [See 9.1.6(c).] Also u and v are continuous on $\overline{D}$ and $u = v$ on $B(D)$. But $u \not\equiv v$ on D. Thus boundedness is essential in 9.1.11.

Theorem 9.1.13 (Mean-Value Property for Analytic Functions). Let $f \in A(\overline{N_r(a)})$ where $r > 0$ and $a \in R^2$. Then

$$f(a) = \frac{1}{2\pi} \int_0^{2\pi} f(a + re^{i\theta}) \, d\theta.$$

Proof. Let C be the cco circle $|w - a| = r$. By Cauchy's Integral Formula,

$$\begin{aligned}
f(a) &= \frac{1}{2\pi i} \int_C \frac{f(w)\,dw}{w - a} = \frac{1}{2\pi i} \int_0^{2\pi} \frac{f(a + re^{i\theta}) r i e^{i\theta} \, d\theta}{re^{i\theta}} \\
&= \frac{1}{2\pi} \int_0^{2\pi} f(a + re^{i\theta}) \, d\theta.
\end{aligned}$$

Theorem 9.1.14 (Mean-Value Property for Harmonic Functions). Let $u \in H(D)$ and let $a \in D$ where D is a domain in R^2. There is some $\rho > 0$ such that if $0 < r < \rho$, then

$$u(a) = \frac{1}{2\pi} \int_0^{2\pi} u(a + re^{i\theta}) \, d\theta.$$

Proof. Let $\rho > 0$ such that $N_\rho(a) \subset D$ and let $0 < r < \rho$. By 9.1.6(b), let $f \in A(N_\rho(a))$ such that $u = \mathcal{R}(f)$ on $N_\rho(a)$. Then

$$\begin{aligned}
u(a) &= \mathcal{R}(f(a)) = \mathcal{R}\left[\frac{1}{2\pi} \int_0^{2\pi} f(a + re^{i\theta})\,d\theta\right] \quad \text{by 9.1.13} \\
&= \frac{1}{2\pi} \int_0^{2\pi} u(a + re^{i\theta}) \, d\theta \quad \text{by 3.2.2.}
\end{aligned}$$

The maximum-minimum principle in 9.1.9 follows immediately from 9.1.14 and 9.1.15 (without using 9.1.8).

Theorem 9.1.15 Let u be real-valued and continuous on a domain D and let u satisfy the mean-value property in D, that is, for each a in D there is some $\rho > 0$ such that

$$u(a) = \frac{1}{2\pi} \int_0^{2\pi} u(a + re^{i\theta})\, d\theta \quad \text{for } 0 < r < \rho.$$

(a) If u attains a maximum or a minimum in D, then u is constant on D.

(b) If D is bounded and $u \in C(\overline{D})$, then u attains its maximum and its minimum on $B(D)$.

Proof of (a). Suppose u assumes a maximum M in D. Let

$$G = \{z : \ z \in D \text{ and } u(z) = M\}. \tag{9.18}$$

Then by our assumption,

$$G \neq \emptyset. \tag{9.19}$$

To see that G is open, let $a \in G$. By hypothesis, let $\rho > 0$ such that if $0 < r < \rho$, then

$$M = u(a) = \frac{1}{2\pi} \int_0^{2\pi} u(a + re^{i\theta})\, d\theta. \tag{9.20}$$

By 3.6.16 and by (9.20), for a fixed $r < \rho$,

$$u(z) = M \quad \text{for each } z \text{ such that } |z - a| = r. \tag{9.21}$$

[For if there is some θ such that $u(a + re^{i\theta}) < M$, then by 3.6.16, we have $u(a) < M$, which contradicts (9.20).] Since (9.21) holds for each positive $r < \rho$, it follows that $u(z) = M$ for each z in $N_\rho(a)$. Hence $N_\rho(a) \subset G$. Thus

$$G \text{ is open.} \tag{9.22}$$

But by the continuity of u,

$$D - G \text{ is open.} \tag{9.23}$$

For suppose $z_0 \in (D - G)$. Let

$$\epsilon = \frac{1}{2}\,[M - u(z_0)]. \tag{9.24}$$

Then $\epsilon > 0$. Since $u \in C(z_0)$, let $\delta > 0$ such that

$$N_\delta(z_0) \subset D \quad \text{and} \quad u(N_\delta(z_0)) \subset N_\epsilon(u(z_0)). \tag{9.25}$$

Now $M \notin N_\epsilon(u(z_0))$ by (9.24). Thus by (9.18) and (9.25), it follows that $N_\delta(z_0) \subset (D - G)$. Hence $D - G$ is open. See Figure 9.1. Now by Ex. 2 in 9.1.7 and by (9.19), (9.22), and (9.23), it follows that $G = D$. This means that $u(z) = M$ on D by (9.18). [If $G \neq D$, then $D - G \neq \emptyset$ and $G \cup (D - G) = D$ is not connected.]

Since $-u$ satisfies the mean-value property, the proof for the case in which u attains a minimum in D follows as in 9.1.9.

The proof of (b) follows as in the proof of 9.1.10.

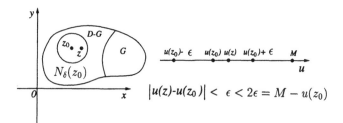

Figure 9.1: $u[N_\delta(z_0)] \subset N_\epsilon[u(z_0)]$

Theorem 9.1.16 Let f be analytic in a domain D and suppose f' is never zero in D. Let $E = f(D)$. Let g be a real-valued function defined on E with continuous second partial derivatives on E and let

$$h(z) = g(f(z)) \quad \text{for each } z \text{ in } D.$$

Then

$$\nabla^2 h = |f'(z)|^2 \, \nabla^2 g \quad \text{on } D. \tag{9.26}$$

[The set E is a domain by 8.2.20.]

Proof. Let $f = u + iv$. Then

$$h(x, y) = g(u, v) \quad \text{where } u = u(x, y) \quad \text{and} \quad v = v(x, y).$$

By the chain rule for functions of two variables, we have

$$h_x = g_u u_x + g_v v_x. \tag{9.27}$$

Also

$$
\begin{aligned}
h_{xx} &= g_u u_{xx} + u_x[g_{uu} u_x + g_{uv} v_x] \\
&\quad + g_v v_{xx} + v_x[g_{vu} u_x + g_{vv} v_x] \\
&= g_u u_{xx} + g_{uu}(u_x)^2 + g_{uv} u_x v_x + g_v v_{xx} \\
&\quad + g_{uv} u_x v_x + g_{vv}(v_x)^2.
\end{aligned}
\tag{9.28}
$$

Similarly,

$$h_{yy} = g_u u_{yy} + g_{uu}(u_y)^2 + g_{uv} u_y v_y + g_v v_{yy} + g_{vu} u_y v_y + g_{vv}(v_y)^2. \tag{9.29}$$

Combining (9.28) and (9.29), we have

$$
\begin{aligned}
h_{xx} + h_{yy} &= g_u(u_{xx} + u_{yy}) + g_{uu}[(u_x)^2 + (u_y)^2] + g_{uv}(u_x v_x + u_y v_y) \\
&\quad + g_v(v_{xx} + v_{yy}) + g_{vu}(u_x v_x + u_y v_y) + g_{vv}[(v_x)^2 + (v_y)^2].
\end{aligned}
\tag{9.30}
$$

Now $f \in A(D)$. Thus u and v are harmonic on D by 9.1.2. Hence

$$u_{xx} + u_{yy} = 0 = v_{xx} + v_{yy}. \tag{9.31}$$

Since $u_x = v_y$ and $u_y = -v_x$, we have

$$(u_x)^2 + (u_y)^2 = (u_x)^2 + (v_x)^2 = (v_x)^2 + (v_y)^2 \tag{9.32}$$

and

$$u_x v_x + u_y v_y = 0. \tag{9.33}$$

Thus by $(9.30)-(9.33)$,

$$\begin{aligned} h_{xx} + h_{yy} &= g_{uu}[(u_x)^2 + (v_x)^2] + g_{vv}[(u_x)^2 + (v_x)^2] \\ &= (g_{uu} + g_{vv})|f'(z)|^2 \\ &= |f'(z)|^2 \, \nabla^2 g. \end{aligned}$$

Corollary 9.1.17 Let f be analytic in a domain D and suppose f' is never zero in D. Let g be a real-valued function defined on $E = f(D)$ and let h be defined on D by

$$h(z) = g(f(z)) \quad \text{for each } z \text{ in } D. \tag{9.34}$$

Then $h \in H(D)$ iff $g \in H(E)$.

Proof. Let $g \in H(E)$. Then $\nabla^2 g = 0$ on E. Thus by (9.26), it follows that $\nabla^2 h = 0$ on D so that $h \in H(D)$.

Conversely, suppose $h \in H(D)$. Let $w_0 \in E$ and let $z_0 \in D$ such that $f(z_0) = w_0$. By 8.2.12, let $\rho > 0$ such that f is one-to-one on $N_\rho(z_0)$. Let $G = f(N_\rho(z_0))$. By 8.2.20, we know that G is a domain and clearly $w_0 \in G$. Now

$$\begin{aligned} &\text{for each } w \text{ in } G, \text{ let } F(w) = z \text{ where } z \text{ is the unique} \\ &\text{point in } N_\rho(z_0) \text{ such that } f(z) = w. \end{aligned} \tag{9.35}$$

Then by (9.34) and (9.35),

$$g(w) = h(F(w)) \quad \text{for each } w \text{ in } G. \tag{9.36}$$

See Figure 9.2.

[Note that $F = f_\rho^{-1}$ where the domain of f_ρ is $N_\rho(z)$ and $f_\rho(z) = f(z)$ for each z in $N_\rho(z_0)$. The function f_ρ is called the **restriction** of f to $N_\rho(z_0)$.] By 8.2.16(b), we see that $F \in A(G)$. Also, $F' \neq 0$ in G by 8.2.16(c). Thus by (9.36) and (9.26) (with g and h interchanged),

$$\nabla^2 g = |F'(w)|^2 \, \nabla^2 h \quad \text{on } G.$$

Thus $h \in H(D)$ implies $\nabla^2 g = 0$ on G and hence $g \in H(G)$. Since w_0 is an arbitrary point of E, it follows that $g \in H(E)$.

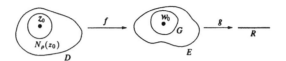

Figure 9.2: $g(w) = h(F(w))$ for each w in G

Exercises 9.1.18

1. Prove if $u \in H(D)$ where D is a simply connected domain, then $u \in C(D)$.
   Hint. See 9.1.5.

2. Let u be harmonic in a domain D. Does it follow that u is continuous in
   D? **Hint.** Let z_0 be any point in D. Since D is open, let $r > 0$ such that
   $N_r(z_0) \subset D$. Apply the result in Ex. 1 to the simply connected domain $N_r(z_0)$.

3. Let $r_0 > 0$ and let C be the circle given by

$$t = r_0\, e^{i\phi} \quad \text{for each } \phi \text{ in } [0, 2\pi].$$

   Let $z = re^{i\theta} \in I(C)$ such that $r > 0$. Let $z_1 = \frac{r_0^2}{r}\, e^{i\theta}$. The point z_1 is called the
   inverse point of z with respect to the circle C.

   (a) Describe the location of z_1 relative to z and C.

   (b) In the defining equation for z_1, multiply both sides by $re^{-i\theta}$ (which is $\bar{z}$)
       to show that $\bar{z} z_1 = r_0^2 = t\bar{t}$ for each point t on C.

   (c) Conclude from Part (b) that

$$\frac{\bar{z} z_1}{t} = \bar{t} \quad \text{for each } t \text{ on } C.$$

Answer to Part (a). We see that $\arg z_1 = \theta = \arg z$ and $|z_1| = \frac{r_0^2}{r}$. These
properties completely determine z_1 as the point in $E(C)$ such that (i) z_1 is on the ray
through z with endpoint at the origin, and (ii) $|z||z_1| = r_0^2$. See Figure 9.3.

9.2 Poisson's Integral Formula and Dirichlet's Problem

Remarks 9.2.1 Cauchy's Integral Formula

$$f(z) = \frac{1}{2\pi i} \int_C \frac{f(t)}{t - z}\, dt \quad \text{for each } z \text{ in } I(C) \tag{9.37}$$

shows that if $f \in A(\overline{I(C)})$, then the values of f along C completely determine f
throughout $I(C)$. From Cauchy's Integral Formula when C is a circle, we can obtain
a corresponding formula for harmonic functions, known as Poisson's Integral Formula.

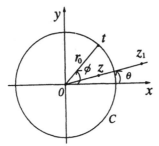

Figure 9.3: $z_1 = \frac{r_0^2}{r} e^{i\theta}$ is the inverse point of $z = re^{i\theta}$ with respect to C.

Theorem 9.2.2 Let u be harmonic in a domain containing the closed disk $\overline{I(C)}$ where C is the cco circle

$$C: \qquad t = r_0 \, e^{i\phi} \qquad \text{for each } \phi \text{ in } [0, 2\pi]. \qquad (9.38)$$

Then for each $z = re^{i\theta}$ in $I(C)$, $u(z)$ is given by

$$u(z) = u(r, \theta) = \frac{1}{2\pi} \int_0^{2\pi} \frac{(r_0^2 - r^2)\, u(r_0, \phi)}{r_0^2 - 2rr_0 \cos(\phi - \theta) + r^2}\, d\phi. \qquad (9.39)$$

Proof. By 9.1.5, let $f \in A(\overline{I(C)})$ such that $\mathcal{R}(f) = u$. Let

$$z = re^{i\theta} \quad \text{be a fixed point in } I(C) \text{ where } r \geq 0. \qquad (9.40)$$

Case 1. Suppose $r > 0$. By the Cauchy Integral Formula,

$$\begin{aligned}
f(z) &= \frac{1}{2\pi i} \int_C \frac{f(t)\, dt}{t - z} \qquad \text{where } t = r_0 \, e^{i\phi} \\
&= \frac{1}{2\pi i} \int_0^{2\pi} \frac{f(t) t i\, d\phi}{t - z} \qquad \text{since } dt = r_0\, e^{i\phi}\, id\phi = ti\, d\phi \\
&= \frac{1}{2\pi} \int_0^{2\pi} \frac{t}{t - z}\, f(t)\, d\phi. \qquad (9.41)
\end{aligned}$$

Now let z_1 be the inverse point of z with respect to C given by

$$z_1 = \frac{r_0^2}{r} e^{i\theta}, \qquad (9.42)$$

which was discussed in Ex. 3 of 9.1.18. (See Figure 9.3.) In particular, by Part (c) of that exercise,

$$\frac{\overline{z} z_1}{t} = \overline{t} \quad \text{for each point } t \text{ on } C. \qquad (9.43)$$

Since $z_1 \in E(C)$, the function $g(t)$ given by

$$g(t) = \frac{f(t)}{t - z_1} \tag{9.44}$$

is analytic in $\overline{I(C)}$. Thus by the Cauchy-Goursat Theorem,

$$
\begin{aligned}
0 &= \frac{1}{2\pi i} \int_C \frac{f(t)\,dt}{t - z_1} \\
&= \frac{1}{2\pi i} \int_0^{2\pi} \frac{f(t)ti\,d\phi}{t - z_1} \\
&= \frac{1}{2\pi} \int_0^{2\pi} \frac{t}{t - z_1} f(t)\,d\phi \\
&= \frac{1}{2\pi} \int_0^{2\pi} \frac{\overline{z}}{\overline{z} - \frac{\overline{z}z_1}{t}} f(t)\,d\phi
\end{aligned}
$$

multiplying previous numerator and denominator by $\dfrac{\overline{z}}{t}$

$$
\begin{aligned}
&= \frac{1}{2\pi} \int_0^{2\pi} \frac{\overline{z}}{\overline{z} - \overline{t}} f(t)\,d\phi \quad \text{by (9.43)} \\
&= -\frac{1}{2\pi} \int_0^{2\pi} \frac{\overline{z}}{\overline{t} - \overline{z}} f(t)\,d\phi. \tag{9.45}
\end{aligned}
$$

Combining (9.41) and (9.45), we have

$$
\begin{aligned}
f(z) &= \frac{1}{2\pi} \int_0^{2\pi} \left(\frac{t}{t - z} + \frac{\overline{z}}{\overline{t} - \overline{z}} \right) f(t)\,d\phi \\
&= \frac{1}{2\pi} \int_0^{2\pi} \frac{(r_0^2 - r^2)\, f(r_0\, e^{i\phi})\,d\phi}{r_0^2 - (t\overline{z} + \overline{t}z) + r^2} \\
&= \frac{1}{2\pi} \int_0^{2\pi} \frac{(r_0^2 - r^2)\, f(r_0\, e^{i\phi})\,d\phi}{r_0^2 - 2rr_0 \cos(\phi - \theta) + r^2} \tag{9.46}
\end{aligned}
$$

since

$$t\overline{z} + \overline{t}\overline{z} = 2\mathcal{R}(t\overline{z}) = 2\mathcal{R}[rr_0\, e^{i(\phi - \theta)}] = 2rr_0 \cos(\phi - \theta).$$

Thus if $z \neq 0$, then (9.39) follows by setting the real parts of the first and last members in (9.46) equal to each other.

Case 2. Suppose $r = 0$. Then $z = 0$. Again by the Cauchy Integral Formula,

$$f(z) = f(0) = \frac{1}{2\pi i} \int_C \frac{f(t)\,dt}{t - 0} = \frac{1}{2\pi i} \int_0^{2\pi} \frac{f(t)ti\,d\phi}{t} = \frac{1}{2\pi} \int_0^{2\pi} f(r_0\, e^{i\phi})\,d\phi.$$

Taking the real part u of f in the first and last members of this equation, we obtain (9.39) for the case in which $z = 0$.

Remarks 9.2.3 (a) The integral in (9.39) is known as **Poisson's integral** and equation (9.39) is known as **Poisson's Integral Formula**. Note that (9.39) expresses $u(z)$ for z in $I(C)$ in terms of the values of $u(z)$ for z on C.

(b) The factor

$$\frac{r_0^2 - r^2}{r_0^2 - 2rr_0\cos(\theta - \phi) + r^2}$$

is called the **Poisson kernel** and is denoted by $P(r_0, r, \theta - \phi)$ or by $K(z, t)$.

(c) By the second member in equation (9.46),

$$P(r_0, r, \theta - \phi) = \frac{t}{t - z} + \frac{\bar{z}}{\bar{t} - \bar{z}} = \frac{|t|^2 - |z|^2}{|t - z|^2} = K(z, t).$$

(d) $P(r_0, r, \theta - \phi) = 1 + 2\sum_{n=1}^{\infty} \left(\frac{r}{r_0}\right)^n \cos n(\theta - \phi)$ for $0 \le r < r_0$.

Proof of (d). From the second member in equation (9.46), we see that

$$
\begin{aligned}
P(r_0, r, \theta - \phi) &= \frac{t}{t - z} + \frac{\bar{z}}{\bar{t} - \bar{z}} \\[2mm]
&= \frac{1}{1 - \frac{z}{t}} + \frac{\left(\frac{\bar{z}}{\bar{t}}\right)}{1 - \left(\frac{\bar{z}}{\bar{t}}\right)} \\[2mm]
&= 1 + \sum_{n=1}^{\infty} \left(\frac{z}{t}\right)^n + \sum_{n=1}^{\infty} \overline{\left(\frac{z}{t}\right)^n} \qquad \text{(geometric series)} \\[2mm]
&= 1 + \sum_{n=1}^{\infty} \left[\left(\frac{z}{t}\right)^n + \overline{\left(\frac{z}{t}\right)^n}\right] \\[2mm]
&= 1 + \sum_{n=1}^{\infty} 2\mathcal{R}\left[\left(\frac{z}{t}\right)^n\right] \\[2mm]
&= 1 + 2\sum_{n=1}^{\infty} \left(\frac{r}{r_0}\right)^n \cos n(\theta - \phi).
\end{aligned}
$$

(e) $P(r_0, r, \theta - \phi) = \sum_{n=-\infty}^{\infty} \left(\frac{r}{r_0}\right)^{|n|} e^{in(\theta - \phi)}$ for $0 \le r < r_0$.

Proof of (e). As in the proof of Part (d),

$$P(r_0, r, \theta - \phi) = 1 + \sum_{n=1}^{\infty} \left(\frac{z}{t}\right)^n + \sum_{n=1}^{\infty} \overline{\left(\frac{z}{t}\right)^n}$$

$$= 1 + \sum_{n=1}^{\infty} \left(\frac{r}{r_0}\right)^n e^{in(\theta-\phi)} + \sum_{n=1}^{\infty} \left(\frac{r}{r_0}\right)^n e^{-in(\theta-\phi)}$$

$$= \sum_{n=0}^{\infty} \left(\frac{r}{r_0}\right)^n e^{in(\theta-\phi)} + \sum_{n=-1}^{-\infty} \left(\frac{r}{r_0}\right)^{-n} e^{in(\theta-\phi)}$$

$$= \sum_{n=-\infty}^{\infty} \left(\frac{r}{r_0}\right)^{|n|} e^{in(\theta-\phi)}.$$

(f) By Part (c), we see that

$$0 < P(r_0, r, \theta - \phi) = P(r_0, r, \phi - \theta) \quad \text{by (9.39)}.$$

Thus the function P is positive, even, and periodic in $(\theta - \phi)$.

(g) It follows from (9.39), by taking $u(r, \theta) \equiv 1$, that

$$\frac{1}{2\pi} \int_0^{2\pi} P(r_0, r, \theta - \phi) \, d\phi = 1.$$

(h) For each fixed t on C, the Poisson kernel $K(z, t)$ is harmonic in $I(C)$ where C is given in (9.38).

Proof of (h). By Part (c),

$$
\begin{aligned}
K(z, t) &= \frac{r_0^2 - r^2}{|t - z|^2} = \mathcal{R}\left(\frac{r_0^2 - r^2}{|t - z|^2}\right) \\
&= \mathcal{R}\left(\frac{t}{t - z} + \frac{\overline{z}}{\overline{t} - \overline{z}}\right) \quad \text{by (9.46)} \\
&= \mathcal{R}\left(\frac{t}{t - z}\right) + \mathcal{R}\left(\frac{\overline{z}}{\overline{t} - \overline{z}}\right) \\
&= \mathcal{R}\left(\frac{t}{t - z}\right) + \mathcal{R}\left(\frac{z}{t - z}\right) \quad \text{since } \mathcal{R}(\overline{w}) = \mathcal{R}(w) \\
&= \mathcal{R}\left(\frac{t + z}{t - z}\right). \quad\quad\quad\quad\quad\quad\quad\quad\quad (9.47)
\end{aligned}
$$

Now for a fixed t on C,

$$\text{the function } \frac{t + z}{t - z} \text{ is analytic in } I(C).$$

Thus (9.47) and 9.1.2 imply that $K(z, t)$ is harmonic in $I(C)$.

Remarks 9.2.4 (Dirichlet's Problem). Let f be a real-valued function defined on $B(D)$ where $B(D)$ is the boundary of a domain D in R^2. **Dirichlet's problem** is the problem of finding u such that

(i) u is harmonic in D,

(ii) $u = f$ on $B(D)$, and

(iii) u is continuous at those points on $B(D)$ at which f is continuous.

This is an extremely important boundary value problem in applied mathematics. It is important because many types of flow are harmonic in their domains. By 9.1.11, for a given f on $B(D)$, the solution u of Dirichlet's problem is uniquely determined throughout D. Thus if the behavior of a harmonic flow is known on $B(D)$, then its behavior is known throughout D.

The solution to Dirichlet's problem for the disk is given in 9.2.5 and 9.2.6.

Theorem 9.2.5 (Dirichlet Problem for the Disk $N_{r_0}(0)$). Let f be real-valued and piecewise continuous on the circle C in (9.38) and for each z in $I(C)$, let

$$u(z) = \frac{1}{2\pi} \int_0^{2\pi} K(z,t)\, f(t)\, d\phi \quad \text{where } t = r_0\, e^{i\phi}. \tag{9.48}$$

Then

$$u \text{ is harmonic in } I(C), \tag{9.49}$$

and for each fixed t on C at which f is continous,

$$\lim_{z \to t} u(z) = f(t). \tag{9.50}$$

Proof. To prove (9.49), let H be an arc of C such that $f \in C(H)$ and let H be given by

$$H: \quad t = r_0\, e^{i\phi} \quad \text{for } \phi \text{ in } [a,b]$$

where $[a,b] \subset [0,2\pi]$. Let

$$g(z) = \frac{1}{2\pi} \int_H K(z,t)\, \frac{f(t)}{ti}\, dt \quad \text{for each } z \text{ in } I(C).$$

Then

$$\begin{aligned}
g(z) &= \frac{1}{2\pi} \int_a^b \mathcal{R}\left(\frac{t+z}{t-z}\right) f(t)\, d\phi \quad \text{by (9.47)} \\
&= \frac{1}{2\pi} \mathcal{R}\left(\int_a^b \frac{t+z}{t-z}\, f(t)\, d\phi\right) \quad \text{by 3.2.2} \\
&= \frac{1}{2\pi} \mathcal{R}\left(\int_H \frac{t+z}{t-z}\, \frac{f(t)}{ti}\, dt\right).
\end{aligned}$$

By 3.7.7, the last integral is analytic in $I(C)$. Thus by 9.1.2, we see that $g \in H(I(C))$. Since f is piecewise continuous on C, it follows that u is the sum of a finite number of harmonic functions (of the form g) and hence u is harmonic in $I(C)$.

To prove (9.50), let $t_0 = r_0\, e^{i\alpha} \in C$ such that f is continuous at t_0 and let $\epsilon > 0$. Let $0 < \delta < \pi$ such that

$$|f(t) - f(t_0)| < \frac{\epsilon}{2} \tag{9.51}$$

when $t = r_0\, e^{i\phi} \in C$ such that $|\phi - \alpha| < \delta$. Since f is piecewise continuous on C, let $M > 0$ such that

$$|f(t) - f(t_0)| < M \quad \text{for each } t \text{ on } C. \tag{9.52}$$

Now by 9.2.3(g),

$$
\begin{aligned}
|u(z) - f(t_0)| &= \left| \frac{1}{2\pi} \int_0^{2\pi} K(z,t)\, f(t)\, d\phi - \frac{f(t_0)}{2\pi} \int_0^{2\pi} K(z,t)\, d\phi \right| \\
&= \frac{1}{2\pi} \left| \int_0^{2\pi} K(z,t)[f(t) - f(t_0)]\, d\phi \right| \\
&= \frac{1}{2\pi} \left| \int_{\alpha-\delta}^{\alpha-\delta+2\pi} K(z,t)[f(t) - f(t_0)]\, d\phi \right| \quad \text{by 9.2.3(f)} \\
&= \frac{1}{2\pi} \left| \int_{\alpha-\delta}^{\alpha+\delta} K(z,t)[f(t) - f(t_0)]\, d\phi \right. \\
&\qquad \left. + \int_{\alpha+\delta}^{\alpha-\delta+2\pi} K(z,t)[f(t) - f(t_0)]\, d\phi \right| \\
&\leq \frac{1}{2\pi} \left| \int_{\alpha-\delta}^{\alpha+\delta} K(z,t)[f(t) - f(t_0)]\, d\phi \right| \\
&\qquad + \frac{1}{2\pi} \left| \int_{\alpha+\delta}^{\alpha-\delta+2\pi} K(z,t)[f(t) - f(t_0)]\, d\phi \right|. \tag{9.53}
\end{aligned}
$$

By (9.51), 9.2.3(f), and 3.1.7(c), for each z in $I(C)$,

$$
\begin{aligned}
\frac{1}{2\pi} \left| \int_{\alpha-\delta}^{\alpha+\delta} K(z,t)[f(t) - f(t_0)]\, d\phi \right| &< \frac{\epsilon}{2} \cdot \frac{1}{2\pi} \int_0^{2\pi} K(z,t)\, d\phi \\
&= \frac{\epsilon}{2} \quad \text{by 9.2.3(g).} \tag{9.54}
\end{aligned}
$$

To see that the last integral in (9.53) can be made arbitrarily small, we observe by 9.2.3(c) that

$$K(z,t) = \frac{r_0^2 - |z|^2}{|t - z|^2}. \tag{9.55}$$

(See Figure 9.4.)

Now there is some $\rho_1 > 0$ and some $\eta > 0$ such that

$$|t - z| > \eta \tag{9.56}$$

for each z in $N_{\rho_1}(t_0) \cap I(C)$ and for each $t = r_0\, e^{i\phi}$ with $\alpha + \delta \leq \phi \leq \alpha - \delta + 2\pi$. Let $0 < \rho \leq \rho_1$ such that

$$r_0^2 - |z|^2 < \frac{\eta^2 \epsilon}{2M} \quad \text{for each } z \text{ in } N_\rho(t_0) \cap I(C). \tag{9.57}$$

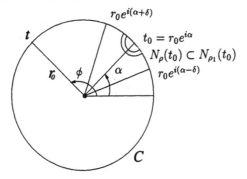

Figure 9.4: $K(z,t) = \frac{r_0^2 - |z|^2}{|t-z|^2}$

Then by (9.52) and (9.55)–(9.57),

$$\frac{1}{2\pi}\left|\int_{\alpha+\delta}^{\alpha-\delta+2\pi} K(z,t)[f(t) - f(t_0)]\,d\phi\right|$$

$$< \frac{1}{2\pi} \cdot \frac{\eta^2\epsilon}{2M\eta^2} \cdot M(2\pi) = \frac{\epsilon}{2} \quad \text{for each } z \text{ in } N_\rho(t_0) \cap I(C). \quad (9.58)$$

Thus by (9.53), (9.54), and (9.58),

$$|u(z) - f(t_0)| < \frac{\epsilon}{2} + \frac{\epsilon}{2} = \epsilon \quad \text{for each } z \text{ in } N_\rho(t_0) \cap I(C).$$

Hence (9.50) is proved.

Theorem 9.2.6 (Dirichlet Problem for the Disk $N_{r_0}(a)$). Let f be real-valued and piecewise continuous on C_a where C_a is given by

$$C_a: \quad \tau = a + r_0\,e^{i\phi} \quad \text{for each } \phi \text{ in } [0, 2\pi]. \quad (9.59)$$

Then the solution u of Dirichlet's problem in $I(C_a)$ is given by

$$u(z) = \begin{cases} \dfrac{1}{2\pi}\displaystyle\int_0^{2\pi} K(z-a, \tau-a)\,f(\tau)\,d\phi & \text{for each } z \text{ in } I(C_a) \\ \\ f(z) & \text{for each } z \text{ on } C_a \end{cases} \quad (9.60)$$

where

$$K(z-a, \tau-a) = \frac{r_0^2 - r^2}{|\tau - z|^2}, \quad \tau = a + r_0\,e^{i\phi} \quad \text{and} \quad r = |z - a|. \quad (9.61)$$

Proof. Let $g(t) = f(a + t)$ for each t on C where C is the circle given by $|t| = r_0$. Then g is piecewise continuous on C since f is piecewise continuous on C_a. Thus by 9.2.5, the solution of Dirichlet's problem in $I(C)$ for g is given by

$$
\begin{aligned}
v(z) &= \frac{1}{2\pi} \int_0^{2\pi} K(z,t)\, g(t)\, d\phi \quad \text{for each } z \text{ in } I(C) \text{ where } t = r_0\, e^{i\phi} \\
&= \frac{1}{2\pi} \int_0^{2\pi} K(z,t)\, f(\tau)\, d\phi \quad \text{where } \tau = a + t.
\end{aligned}
$$

Now for each z in $I(C_a)$, let

$$
\begin{aligned}
u(z) = v(z - a) &= \frac{1}{2\pi} \int_0^{2\pi} K(z - a, t)\, f(\tau)\, d\phi \\
&= \frac{1}{2\pi} \int_0^{2\pi} K(z - a, \tau - a)\, f(\tau)\, d\phi.
\end{aligned}
$$

Now (9.61) follows from 9.2.3(c). [By 9.1.1, we know that $u \in H(I(C_a))$ since $v \in H(I(C))$. For if $v(x, y)$ satisfies Laplace's equation in $I(C)$, then $u(x, y) = v(x - \mathcal{R}(a), y - \mathcal{I}(a))$ satisfies Laplace's equation in $I(C_a)$. Or we may use 9.1.17 with $f(z) = z - a$.]

By 9.2.6 and 9.1.11, we have the following generalization of 9.2.2.

Theorem 9.2.7 Let $u \in H(D)$ where D is a domain. Let $a \in D$ and let C_a be the circle given by

$$
C_a: \quad \tau = a + r_0\, e^{i\phi} \quad \text{for each } \phi \text{ in } [0, 2\pi]
$$

where $\overline{I(C_a)} \subset D$. Then for each z in $I(C_a)$,

$$
u(z) = \frac{1}{2\pi} \int_0^{2\pi} K(z - a, \tau - a)\, u(\tau)\, d\phi \quad \text{where } \tau = a + r_0\, e^{i\phi}. \tag{9.62}
$$

Proof. Let

$$
v(z) = \begin{cases}
\dfrac{1}{2\pi} \displaystyle\int_0^{2\pi} K(z - a, \tau - a)\, u(\tau)\, d\phi & \text{for each } z \text{ in } I(C_a) \\[2ex]
u(z) & \text{for each } z \text{ on } C_a.
\end{cases} \tag{9.63}
$$

Then by 9.2.6 and 9.2.4,

$$
v \in H(I(C_a)) \quad \text{and} \quad v = u \quad \text{on } C_a.
$$

Thus by 9.1.11 and Ex. 1 in 9.1.18,

$$
u = v \quad \text{in } \overline{I(C_a)}.
$$

Thus (9.62) follows by (9.63).

Theorem 9.2.8 Let u be real-valued and continuous in a domain D. Then $u \in H(D)$ iff u satisfies the mean value property in D, that is, iff for each a in D, there is some $\rho > 0$ such that

$$u(a) = \frac{1}{2\pi} \int_0^{2\pi} u(a + re^{i\theta})\, d\theta \quad \text{for each fixed } r \text{ such that } 0 < r < \rho. \qquad (9.64)$$

Proof. If $u \in H(D)$, then (9.64) follows by 9.1.14 or by Ex. 1 in 9.2.9.

Now suppose (9.64) holds for an arbitrary a in D. Let $0 < r_0 < \rho$ such that $\overline{I(C_a)} \subset D$ where
$$C_a = \{z : \; |z - a| = r_0\}.$$
By 9.2.6, let v be the solution of Dirichlet's problem in $I(C_a)$ such that $v = u$ on C_a. Then

$$v \in H(I(C_a)), \quad v = u \text{ on } C_a, \quad \text{and} \quad v \in C(\overline{I(C_a)}). \qquad (9.65)$$

By 9.1.14, we know that v satisfies the mean-value property in $I(C_a)$. By hypothesis, u satisfies the mean-value property in $I(C_a)$. Thus $u - v$ has the mean-value property in $I(C_a)$. Hence $u - v$ takes on its maximum M and its minimum m on C_a by 9.1.15(b). By (9.65),

$$0 = m \le u - v \le M = 0 \quad \text{in } I(C_a).$$

This implies that
$$u = v \quad \text{in } I(C_a) = N_{r_0}(a).$$

Thus by (9.65), the function u is harmonic in $N_{r_0}(a)$. Since a was an arbitrary point in D, it now follows that $u \in H(D)$.

Exercises 9.2.9

1. Show that
$$u(a) = \frac{1}{2\pi} \int_0^{2\pi} u(\tau)\, d\phi \quad \text{where } \tau = a + r_0\, e^{i\phi}$$
   and where u is given in 9.2.7. **Hint.** Observe that in 9.2.7,

$$K(z - a, \tau - a) \equiv K(0, \tau - a) \equiv 1 \quad \text{when } z = a.$$

2. Let u, C_a, and τ be as given in 9.2.7.

   (a) Show that for each z in $I(C_a)$,

$$0 < \frac{r_0 - r}{r_0 + r} \le K(z - a, \tau - a) \le \frac{r_0 + r}{r_0 - r} \quad \text{where } r = |z - a|.$$

   Hint. By (9.61), we have $K(z - a, \tau - a) = \dfrac{r_0^2 - r^2}{|(\tau - a) - (z - a)|^2}$. Now
$$r_0 - r \le |(\tau - a) - (z - a)| \le r_0 + r.$$

(b) Show that if $u(\tau) \geq 0$ on C_a, then for each z in $I(C_a)$,

$$\frac{r_0 - r}{r_0 + r} u(a) \leq u(z) \leq \frac{r_0 + r}{r_0 - r} u(a).$$

[In Part (a), multiply through by $\frac{1}{2\pi} u(\tau)$ and integrate with respect to ϕ from 0 to 2π. Then use Ex. 1 and 9.2.7.]

The inequalities in Ex. 2(a) and 2(b) are called **Harnach's inequalities**.

3. (a) Show that $\dfrac{t + z}{t - z} = K(z,t) + iL(z,t)$ where $t = r_0 e^{i\phi}$, $z = r e^{i\theta}$ and

$$L(z,t) = \frac{2rr_0 \sin(\theta - \phi)}{r_0^2 + r^2 - 2rr_0 \cos(\theta - \phi)}.$$

(L is called the conjugate kernel of u.) **Hint**. Find the real and imaginary parts $\frac{t+z}{t-z} \cdot \frac{\bar{t}-\bar{z}}{\bar{t}-\bar{z}}$.

(b) Show that $\displaystyle\int_0^{2\pi} L(z,t)\, d\phi = 0$. **Hint**. $\displaystyle\int \frac{du}{u} = \mathrm{Log}\,|u| + C.$

9.3   Examples of Dirichlet's Problem

Example 9.3.1 Find the solution of the following Dirichlet problem. For each $t = e^{i\phi}$ on the unit circle C,

$$f(e^{i\phi}) = \begin{cases} 0 & \text{if } 0 < \phi < \pi \\ 1 & \text{if } \pi < \phi < 2\pi. \end{cases}$$

Solution. We shall use the following integration formula

$$\int \frac{dx}{a + b\cos x} = \frac{2}{\sqrt{a^2 - b^2}} \tan^{-1}\left(\frac{\sqrt{a^2 - b^2}\,\tan\frac{x}{2}}{a + b}\right) \quad \text{for } a^2 > b^2. \tag{9.66}$$

By 9.2.5, the solution u in $I(C)$ is given by

$$u(z) = \frac{1}{2\pi} \int_\pi^{2\pi} K(z,t)\, d\phi = \frac{1}{2\pi} \int_\pi^{2\pi} \frac{(1 - r^2)\, d\phi}{(1 + r^2) - 2r\cos(\theta - \phi)}$$

for each $z = r e^{i\theta}$ in $I(C)$. By (9.66) with the appropriate choice of the inverse tangent,

$$\begin{aligned}
\pi\, u(z) &= -\left[\tan^{-1}\left(A\tan\frac{\theta - \phi}{2}\right)\right]_\pi^{2\pi} \quad \text{where } A = \frac{1 + r}{1 - r} \\
&= \tan^{-1}\left[A\tan\left(\pi - \frac{\theta}{2}\right)\right] - \tan^{-1}\left[A\tan\left(\frac{\pi}{2} - \frac{\theta}{2}\right)\right] \\
&= -\tan^{-1}\left(A\tan\frac{\theta}{2}\right) - \tan^{-1}\left(A\cot\frac{\theta}{2}\right).
\end{aligned}$$

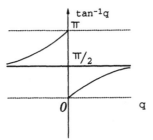

Figure 9.5: $\tan^{-1} q$

Thus

$$
\begin{aligned}
-\tan \pi u &= \frac{A\left(\tan\frac{\theta}{2} + \cot\frac{\theta}{2}\right)}{1 - \left(A\tan\frac{\theta}{2}\right)\left(A\cot\frac{\theta}{2}\right)} = \frac{A}{1 - A^2}\frac{\sec^2\frac{\theta}{2}}{\tan\frac{\theta}{2}} \\
&= \frac{2A}{(1 - A^2)\sin\theta} = -\frac{1 - r^2}{2r\sin\theta}.
\end{aligned}
$$

Hence

$$
u(z) = \frac{1}{\pi}\tan^{-1}\left(\frac{1 - r^2}{2r\sin\theta}\right) \quad \text{for each } z = re^{i\theta} \text{ in } I(C), \tag{9.67}
$$

where the expression on the right is interpreted so that $u \in H(I(C))$ and so that (9.50) holds. To insure that (9.50) is satisfied, $\tan^{-1} q$ should be chosen as shown in Figure 9.5.

Furthermore, to insure that $u \in C[R \cap I(C)]$ (which is necessary by Ex. 1 in 9.1.18), we must interpret $\tan^{-1}\left(\frac{1-r^2}{2y}\right)$ to be $\frac{\pi}{2}$ when $y = r\sin\theta = 0$. Thus we have for $r < 1$,

$$
u(z) = \begin{cases}
\dfrac{1}{\pi}\tan^{-1}\left(\dfrac{1 - r^2}{2y}\right) & \text{if } y = r\sin\theta \neq 0 \\[2mm]
\dfrac{1}{2} & \text{if } y = 0.
\end{cases}
$$

Remarks 9.3.2 If we use the simple forms for $K(z, t)$ given in 9.2.3(b) and (c), the integral in (9.48) may be very difficult to determine. But to overcome this difficulty we may use the series form of $K(z, t)$ given in 9.2.3(e). We first express (9.48) as the line integral

$$
u(z) = \frac{1}{2\pi}\int_S K(z, t)\, f(t)\, d\phi,
$$

where $t = r_0\, e^{i\phi}$ and S is the line segment from 0 to 2π on the real axis in R^2. Thus assuming the hypothesis of 9.2.5, it follows by 9.2.5 and 9.2.3(e) that for each z in

$I(C)$,

$$
\begin{aligned}
u(z) &= \frac{1}{2\pi} \int_S \left[\sum_{n=-1}^{-\infty} \left(\frac{r}{r_0} \right)^{-n} e^{in(\theta-\phi)} \right] f(t)\, d\phi \;+\; \frac{1}{2\pi} \int_S \left[\sum_{n=0}^{\infty} \left(\frac{r}{r_0} \right)^{n} e^{in(\theta-\phi)} \right] f(t)\, d\phi \\
&= \frac{1}{2\pi} \int_S \left[\sum_{n=1}^{\infty} \left(\frac{r}{r_0} \right)^{n} e^{-in(\theta-\phi)} f(t) \right]\, d\phi \;+\; \frac{1}{2\pi} \int_S \left[\sum_{n=0}^{\infty} \left(\frac{r}{r_0} \right)^{n} e^{in(\theta-\phi)} f(t) \right]\, d\phi \\
&= \sum_{n=1}^{\infty} \left[\frac{1}{2\pi} \int_S f(t) e^{in\phi}\, d\phi \right] \left(\frac{r}{r_0} \right)^{n} e^{-in\theta} \\
&\quad + \sum_{n=0}^{\infty} \left[\frac{1}{2\pi} \int_S f(t) e^{-in\phi}\, d\phi \right] \left(\frac{r}{r_0} \right)^{n} e^{-in\theta}
\end{aligned}
$$

(by 4.4.3 and 4.4.4 since $|f(t)| \le B$ and $\left| \frac{r}{r_0} \right| < 1$)

$$
\begin{aligned}
&= \sum_{n=1}^{\infty} C_{-n} \left(\frac{r}{r_0} \right)^{n} e^{-in\theta} \;+\; \sum_{n=0}^{\infty} C_n \left(\frac{r}{r_0} \right)^{n} e^{in\theta} \\
&= \sum_{n=-\infty}^{\infty} C_n \left(\frac{r}{r_0} \right)^{|n|} e^{in\theta}
\end{aligned}
$$

where

$$
C_n = \frac{1}{2\pi} \int_0^{2\pi} f(r_0\, e^{i\phi}) e^{-in\phi}\, d\phi.
$$

The coefficient C_n is called the nth **Fourier coefficent** for $f(r_0\, e^{i\phi})$.

Example 9.3.3 Let C be the circle $|t| = r_0$ and let f be given by

$$
f(r_0\, e^{i\phi}) = \begin{cases} 0 & \text{on the interval } (0, \pi) \\ 1 & \text{on the interval } (\pi, 2\pi). \end{cases}
$$

Give the series solution in $I(C)$ of the Dirichlet problem for f.

Solution. For each n in $J - \{0\}$, it follows by 9.3.2 that

$$
\begin{aligned}
C_n &= \frac{1}{2\pi} \int_0^{2\pi} f(r_0\, e^{i\phi}) e^{-in\phi}\, d\phi \\
&= \frac{1}{2\pi} \int_\pi^{2\pi} e^{-in\phi}\, d\phi \;=\; -\frac{1}{2\pi in} \left[e^{-in\phi} \right]_\pi^{2\pi} \\
&= -\frac{1}{2\pi in} [1 - (-1)^n] \;=\; \begin{cases} -\dfrac{1}{\pi in} & \text{if } n \text{ is odd} \\[2mm] 0 & \text{if } n \text{ is even and } n \ne 0. \end{cases}
\end{aligned}
$$

Also,

$$
C_0 = \frac{1}{2\pi} \int_\pi^{2\pi} 1\, d\phi = \frac{1}{2}.
$$

By 9.3.2, for each z such that $|z| < r_0$,

$$
\begin{aligned}
u(z) &= \frac{1}{2} - \frac{1}{\pi i} \sum_{k=-\infty}^{\infty} \left(\frac{r}{r_0}\right)^{|2k-1|} \frac{e^{i(2k-1)\theta}}{2k-1} \\
&= \frac{1}{2} - \frac{1}{\pi i} \sum_{k=-\infty}^{\infty} \left(\frac{r}{r_0}\right)^{|2k-1|} \left(\frac{\cos(2k-1)\theta + i\sin(2k-1)\theta}{2k-1}\right) \\
&= \frac{1}{2} - \frac{2}{\pi} \sum_{k=1}^{\infty} \left(\frac{r}{r_0}\right)^{2k-1} \frac{\sin(2k-1)\theta}{2k-1}.
\end{aligned}
$$

Exercises 9.3.4

In Ex. 1–5, find the series solution to the given Dirichlet problem for $N_{r_0}(0)$.

1. $f(r_0\, e^{i\phi}) = \phi$ on $[0, 2\pi)$

$$u(z) = 1 - \frac{1}{\pi} \sum_{n=1}^{\infty} \left(\frac{r}{r_0}\right)^n \frac{\sin n\theta}{n} \text{ on } I(C)$$

2. $f(r_0\, e^{i\phi}) = e^\phi$ on $[0, 2\pi)$

$$u(z) = \frac{e^{2\pi}-1}{\pi}\left[\frac{1}{2} + \sum_{n=1}^{\infty} \left(\frac{r}{r_0}\right)^n \frac{\cos n\theta - n\sin n\theta}{n^2+1}\right] \text{ on } I(C)$$

3. $f(r_0\, e^{i\phi}) = \phi^2$ on $[0, 2\pi)$

$$u(z) = \frac{4}{3}\pi^2 + 4\sum_{n=1}^{\infty} \left(\frac{r}{r_0}\right)^n \left[\frac{1}{n^2}\cos n\theta - \frac{\pi}{n}\sin n\theta\right] \text{ on } I(C)$$

4. $f(r_0\, e^{i\phi}) = \sin\phi$

$$u(z) = \frac{r}{r_0}\sin\theta \text{ on } I(C)$$

5. $f(r_0\, e^{i\phi}) = \cos\phi$

$$u(z) = \frac{r}{r_0}\cos\theta \text{ on } I(C)$$

9.4 Dirichlet's Problem for a Half-Plane

We shall need the following definition and theorems.

Definition 9.4.1 (Uniform Convergence of an Improper Integral). Let $f(x,y)$ be a real-valued function defined on the semi-infinite strip

$$S = \{(x,y): \ a \le x \le b \quad \text{and} \quad c \le y < +\infty\}. \tag{9.68}$$

Let

$$\int_c^\infty f(x,y)\, dy = g(x) \quad \text{for each } x \text{ in } [a,b].$$

Then we say that $\int_c^\infty f(x,y)\, dy$ **converges uniformly** to g on $[a, b]$ iff for each $\epsilon > 0$, there is a number r (independent of x) such that if $d \ge r$, then

$$\left|g(x) - \int_c^d f(x,y)\, dy\right| = \left|\int_d^\infty f(x,y)\, dy\right| < \epsilon \quad \text{on } [a,b]. \tag{9.69}$$

We indicate this uniform convergence on $[a, b]$ by writing

$$\int_c^\infty f(x, y)\, dy \xrightarrow{\;U\ [a,b]\;} g(x). \tag{9.70}$$

Theorem 9.4.2 Let f be real-valued such that $f \in C(S)$ and such that (9.70) holds where S is given in (9.68). Then

(i) g is continuous on $[a, b]$, and

(ii) $\displaystyle\int_a^b \left[\int_c^\infty f(x, y)\, dy \right] dx = \int_c^\infty \left[\int_a^b f(x, y)\, dx \right] dy.$

Proof of (i). Let $x_0 \in [a, b]$ and let $\epsilon > 0$. By 9.4.1, let $d \geq r > c$ such that

$$\left| \int_d^\infty f(x, y)\, dy \right| < \frac{\epsilon}{3} \quad \text{on } [a, b]. \tag{9.71}$$

By 2.4.4, we know that f is uniformly continuous on the closed and bounded set

$$T = \{(x, y) : \ a \leq x \leq b \ \text{and} \ c \leq y \leq d\}.$$

Thus we let $\delta > 0$ such that if (x, y) and (t, y) are in T and if $|x - t| < \delta$, then

$$|f(x, y) - f(t, y)| < \frac{\epsilon}{3(d - c)}. \tag{9.72}$$

Now for each x in $[a, b]$ such that $|x - x_0| < \delta$,

$$
\begin{aligned}
|g(x) - g(x_0)| &= \left| \left(g(x) - \int_c^d f(x, y)\, dy \right) + \left(\int_c^d f(x, y)\, dy - \int_c^d f(x_0, y)\, dy \right) \right. \\
&\qquad \left. + \left(\int_c^d f(x_0, y)\, dy - g(x_0) \right) \right| \\
&\leq \left| \int_d^\infty f(x, y)\, dy \right| + \left| \int_c^d [f(x, y) - f(x_0, y)]\, dy \right| \\
&\qquad + \left| \int_d^\infty f(x_0, y)\, dy \right| \\
&< \frac{\epsilon}{3} + \frac{\epsilon}{3} + \frac{\epsilon}{3} = \epsilon \quad \text{by (9.71) and (9.72).}
\end{aligned}
$$

Thus $g \in C([a, b])$.

Proof of (ii). Let $\epsilon > 0$. By 9.4.1, let $r \in R$ such that if $d \geq r$, then

$$\left| g(x) - \int_c^d f(x, y)\, dy \right| < \frac{\epsilon}{b - a} \quad \text{on } [a, b]. \tag{9.73}$$

Now

$$\left| \int_a^b g(x)\,dx - \int_c^d \left[\int_a^b f(x,y)\,dx \right]\,dy \right| = \left| \int_a^b g(x)\,dx - \int_a^b \left[\int_c^d f(x,y)\,dy \right]\,dx \right|$$

$$= \left| \int_a^b \left[g(x) - \int_c^d f(x,y)\,dy \right]\,dx \right|$$

$$\leq \int_a^b \left| g(x) - \int_c^d f(x,y)\,dy \right|\,dx$$

$$< \frac{\epsilon}{b-a}(b-a) = \epsilon \quad \text{for each } d > r$$

$$\text{by (9.73).} \quad (9.74)$$

But (9.74) means that

$$\int_a^b g(x)\,dx = \int_c^\infty \left[\int_a^b f(x,y)\,dx \right]\,dy.$$

Theorem 9.4.3 If

(i) $f \in C(S)$ and $f_x \in C(S)$ where S is given in (9.68),

(ii) $\int_c^\infty f(x,y)\,dy = g(x)$ on $[a,b]$, and

(iii) $\int_c^\infty f_x(x,y)\,dy \xrightarrow{U\ [a,b]} h(x)$,

then

$$g'(x) = h(x) \quad \text{on } [a,b].$$

Proof. By (i), (iii), and 9.4.2(i),

$$\int_a^x h(t)\,dt = \int_a^x \left[\int_c^\infty f_t(t,y)\,dy \right]\,dt$$

$$= \int_c^\infty \left[\int_a^x f_t(t,y)\,dt \right]\,dy \quad \text{by } 9.4.2(\text{ii})$$

$$= \int_c^\infty [f(x,y) - f(a,y)]\,dy$$

$$= g(x) - g(a) \quad \text{on } [a,b] \quad \text{by (ii).}$$

Thus by 3.1.7(b) and 9.4.2(i),

$$g'(x) = h(x) \quad \text{on } [a,b].$$

Theorem 9.4.4 (a) (Cauchy Condition for Improper Integrals). Let $\int_c^d f(t)\,dt$ exist for each $d > c$. Then $\int_c^\infty f(t)\,dt$ exists iff

(i) for each $\epsilon > 0$, there is a number $B > c$ such that $\left| \int_a^b f(t)\,dt \right| < \epsilon$

if $b > a > B$.

(b) (Weierstrass M-Test for Improper Integrals). Let $\int_c^r f(x,t)\,dt$ exist for each $r \geq c$ and for each x in $[a,b]$. If $|f(x,t)| \leq M(t)$ for each x in $[a,b]$ and for each t in $[c,+\infty)$ and if $\int_c^\infty M(t)\,dt$ exists, then

$$\int_c^\infty f(x,t)\,dt \xrightarrow{\ U\ [a,b]\ } g(x) \quad \left(\text{where } g(x) = \int_c^\infty f(x,t)\,dt \right).$$

Proof of (a). Assume $\int_c^\infty f(t)\,dt = L$ where L is a real number. Let $\epsilon > 0$. By the definition of L, let $B > c$ such that if $b > a > B$, then

$$\left| \int_c^a f(t)\,dt - L \right| < \frac{\epsilon}{2} \quad \text{and} \quad \left| \int_c^b f(t)\,dt - L \right| < \frac{\epsilon}{2}.$$

Now if $b > a > B$, then

$$\begin{aligned} \left| \int_a^b f(t)\,dt \right| &= \left| \int_c^b f(t)\,dt - \int_c^a f(t)\,dt \right| \\ &= \left| \left(\int_c^b f(t)\,dt - L \right) + \left(L - \int_c^a f(t)\,dt \right) \right| < \frac{\epsilon}{2} + \frac{\epsilon}{2} = \epsilon. \end{aligned}$$

Thus (i) holds.

Conversely, suppose (i) holds and let

$$a_n = \int_c^n f(t)\,dt \quad \text{for each } n \text{ in } N \text{ such that } n > c.$$

Then by (i) and 4.5.1, it follows that $\{a_n\}$ is a Cauchy sequence since

$$|a_m - a_n| = \left| \int_c^m f(t)\,dt - \int_c^n f(t)\,dt \right| = \left| \int_n^m f(t)\,dt \right|.$$

Hence by 4.5.2, let

(ii) $A = \lim\limits_{n \to \infty} a_n.$

Let $\epsilon > 0$. By (ii) and (i), let $B > c$ such that if $n > B$ and $b > a > B$ then,

$$|a_n - A| < \frac{\epsilon}{2} \quad \text{and also} \quad \left| \int_a^b f(t)\,dt \right| < \frac{\epsilon}{2}.$$

Now if $n > B$ and $b > B$, then

$$
\begin{aligned}
\left| \int_c^b f(t)\, dt - A \right| &= \left| \int_c^n f(t)\, dt - A + \int_n^b f(t)\, dt \right| \\
&\leq |a_n - A| + \left| \int_n^b f(t)\, dt \right| \\
&< \frac{\epsilon}{2} + \frac{\epsilon}{2} = \epsilon.
\end{aligned}
$$

Proof of (b). Let $\epsilon > 0$. By hypothesis, $\int_c^\infty M(t)\, dt$ exists. Thus by (a), let $B > c$ such that

$$
\left| \int_r^d M(t)\, dt \right| < \frac{\epsilon}{2} \quad \text{if } d > r > B.
$$

Then for each fixed x in $[a, b]$ and for $d > r > B$,

$$
\text{(iii)} \quad \left| \int_r^d f(x, t)\, dt \right| \leq \int_r^d |f(x, t)|\, dt \leq \int_r^d M(t)\, dt < \frac{\epsilon}{2} \quad \text{by 3.1.7(c).}
$$

Thus by (a) for each x in $[a, b]$,

$$
\int_c^\infty f(x, t)\, dt \text{ converges to some number } g(x).
$$

Hence $\int_r^\infty f(x, t)\, dt$ converges for each $r > c$. Since (iii) holds for each x in $[a, b]$, we have, letting $d \to \infty$ in (iii),

$$
\left| \int_r^\infty f(x, t)\, dt \right| \leq \frac{\epsilon}{2} < \epsilon \quad \text{for each } x \text{ in } [a, b] \text{ and for each } r > B.
$$

Thus by 9.4.1,

$$
\int_c^\infty f(x, t)\, dt \xrightarrow{\;U\;[a,b]\;} g(x).
$$

We shall use H to denote the upper half-plane $\{z : \ \mathcal{I}(z) \geq 0\}$.

Theorem 9.4.5 (The Cauchy Integral Formula for the Half-Plane). Let $f \in A(H)$ such that

$$
|z|^k\, |f(z)| < M \quad \text{on } H \quad \text{for some } k > 0. \tag{9.75}
$$

If $z \in H^\circ$, then

$$
f(z) = \frac{1}{2\pi i} \int_{-\infty}^\infty \frac{f(t)}{t - z}\, dt. \tag{9.76}
$$

Proof. For a fixed $z = r_0 e^{i\theta}$ (where $r_0 > 0$), let C_r be the cco semicircle with radius $r > r_0$ which is given by

$$C_r : \quad t = re^{i\phi} \quad \text{for } \phi \text{ in } [0, \pi]. \tag{9.77}$$

By the Cauchy Integral Formula,

$$f(z) = \frac{1}{2\pi i} \left[\int_{-r}^{r} \frac{f(t)\,dt}{t - z} + \int_{C_r} \frac{f(t)\,dt}{t - z} \right]. \tag{9.78}$$

Now

$$\left| \int_{C_r} \frac{f(t)\,dt}{t - z} \right| \leq \frac{\pi M}{r^k \left(1 - \frac{|z|}{r} \right)} \to 0 \quad \text{as} \quad r \to +\infty. \tag{9.79}$$

Since $f(z)$ is a fixed number, it follows by (9.78) and (9.79) that the Cauchy principle value

$$P \int_{-\infty}^{\infty} \frac{f(t)\,dt}{t - z} \tag{9.80}$$

exists. Thus

$$f(z) = \frac{1}{2\pi i} \left[P \int_{-\infty}^{\infty} \frac{f(t)\,dt}{t - z} \right]. \tag{9.81}$$

Now let $a > r_0$. Then for $b > a$,

$$\left| \int_{a}^{b} \frac{f(t)\,dt}{t - z} \right| \leq M \int_{a}^{b} \frac{dt}{t^k (t - r_0)} < M \int_{a}^{b} \frac{dt}{(t - r_0)^{k+1}}$$
$$\to \frac{M}{k} \cdot \frac{1}{(a - r_0)^k} \quad \text{as} \quad b \to \infty.$$

Thus $\int_{a}^{\infty} \frac{f(t)\,dt}{t-z}$ exists. Similarly, $\int_{-\infty}^{-a} \frac{f(t)\,dt}{t-z}$ exists. Finally, $\int_{-a}^{a} \frac{f(t)\,dt}{t-z}$ exists, since $\frac{f}{t-z}$ is continuous on the interval $[-a, a]$. Hence the improper integral exists and is equal to its Cauchy principle value. Thus (9.76) follows from (9.81).

Theorem 9.4.6 Let $f = u + iv$ be analytic in H such that (9.75) holds. Then for each z in $H°$,

$$u(x, y) = \frac{y}{\pi} \int_{-\infty}^{\infty} \frac{u(t, 0)\,dt}{|t - z|^2} = \frac{y}{\pi} \int_{-\infty}^{\infty} \frac{u(t, 0)\,dt}{(t - x)^2 + y^2} \tag{9.82}$$

and

$$v(x, y) = \frac{1}{\pi} \int_{-\infty}^{\infty} \frac{(x - t)\,u(t, 0)\,dt}{|t - z|^2}. \tag{9.83}$$

Formula (9.82) is known as Poisson's Integral Formula for the half-plane or the Schwarz Integral Formula.

Proof. Let $z \in H^\circ$. Then $\bar{z}$ is in the lower half-plane and hence $\bar{z} \in E(S_r)$ where

$$S_r = C_r \cup [-r, r]$$

and where C_r is given by (9.77). Hence by the Cauchy-Goursat Theorem,

$$\int_{-r}^{r} \frac{f(t)\,dt}{t - \bar{z}} + \int_{C_r} \frac{f(t)\,dt}{t - \bar{z}} = 0. \tag{9.84}$$

As in the proof of 9.4.5, we use (9.75) to conclude that the second integral in (9.84) approaches zero as $r \to \infty$ and hence that

$$\int_{-\infty}^{\infty} \frac{f(t)\,dt}{t - \bar{z}} = 0. \tag{9.85}$$

Thus by (9.76) and (9.85),

$$\begin{aligned}
f(z) &= \frac{1}{2\pi i} \int_{-\infty}^{\infty} \left[\frac{1}{t - z} - \frac{1}{t - \bar{z}} \right] f(t)\,dt \\
&= \frac{1}{\pi} \int_{-\infty}^{\infty} \frac{y\, f(t)\,dt}{|t - z|^2}. \tag{9.86}
\end{aligned}$$

Hence (9.82) follows by equating real parts in (9.86). Also by (9.76) and (9.85),

$$\begin{aligned}
f(z) &= \frac{1}{2\pi i} \int_{-\infty}^{\infty} \left[\frac{1}{t - z} + \frac{1}{t - \bar{z}} \right] f(t)\,dt. \\
&= \frac{1}{\pi i} \int_{-\infty}^{\infty} \frac{(t - x)f(t)\,dt}{|t - z|^2}. \tag{9.87}
\end{aligned}$$

Thus (9.83) follows by equating imaginary parts in (9.87).

Now let f be real-valued, continuous, and bounded on the real axis (the boundary of H°). We shall use (9.82) as a guide in writing a "trial" solution to Dirichlet's problem for H°. Poisson's Formula (9.82) suggests that

$$u(x, y) = \frac{1}{\pi} \int_{-\infty}^{\infty} \frac{y\, f(t)\,dt}{|t - z|^2} \quad \text{for each } (x, y) = z \text{ in } H^\circ \tag{9.88}$$

may be a solution to Dirichlet's problem for the half-plane. Indeed, our next theorem shows that this is the case.

Theorem 9.4.7 (A Dirichlet Problem for the Half-Plane). Let f be real-valued, continuous, and bounded on the real axis. If $u(x, y)$ is given by (9.88), then

$$u \in H(H^\circ)$$

and

$$\lim_{y \to 0} u(x, y) = f(x) \quad \text{for each } x \text{ in } R. \tag{9.89}$$

Proof. Let $r > 1$ and let

$$Q_r = \{(x,y): \ |x| < r \ \text{and} \ \frac{1}{r} < y < r\}.$$

We shall show that $u \in H(Q_r)$ for each $r > 1$, and hence it will follow that $u \in H(H^\circ)$.
Now by (9.88) for each z in Q_r,

$$
\begin{aligned}
\nabla^2 u &= \frac{1}{\pi} \int_{-\infty}^{\infty} \nabla^2 \left(\frac{y f(t)}{(t-x)^2 + y^2} \right) dt \quad \text{by Ex. 9 in 9.4.8} \\
&= 0 \quad \text{by 9.1.2,}
\end{aligned}
$$

since $\frac{y}{(t-x)^2+y^2} = \mathcal{I}\left(\frac{1}{t-z}\right)$ by Ex. 10 in 9.4.8 and since for each fixed t in R, the
quotient $\frac{1}{t-z}$ is an analytic function of z in Q_r. Thus $u \in H(Q_r)$ for each $r > 1$ and
hence $u \in H(H^\circ)$.

To prove (9.89), let $\epsilon > 0$ and let x be fixed. Since f is continuous at x, let $\delta > 0$
such that

$$|f(t) - f(x)| < \frac{\epsilon}{3} \quad \text{when} \ |t - x| < \delta. \tag{9.90}$$

Since

$$\int_{-\infty}^{\infty} \frac{y \, dt}{(t-x)^2 + y^2} = \left[\text{Tan}^{-1}\left(\frac{t-x}{y} \right) \right]_{-\infty}^{\infty} = \pi,$$

we have

$$
\begin{aligned}
u(x,y) - f(x) &= \frac{1}{\pi} \int_{-\infty}^{\infty} \frac{y[f(t) - f(x)] \, dt}{(t-x)^2 + y^2} \\
&= \frac{1}{\pi} (I_1 + I_2 + I_3) \tag{9.91}
\end{aligned}
$$

where

$$I_1 = \int_{-\infty}^{x-\delta} G \, dt, \quad I_2 = \int_{x-\delta}^{x+\delta} G \, dt \quad \text{and} \quad I_3 = \int_{x+\delta}^{\infty} G \, dt$$

and where G is the integrand in (9.91). By hypothesis, let M be an upper bound for
$|f|$ on R. Then

$$|f(t) - f(x)| \leq 2M \quad \text{for each } t \text{ in } R.$$

Thus

$$|I_1| \leq 2M y \int_{-\infty}^{x-\delta} \frac{dt}{(t-x)^2} = \frac{2M y}{\delta} < \frac{\epsilon}{3} \quad \text{if } 0 < y < \frac{\delta\epsilon}{6M}. \tag{9.92}$$

Similarly,

$$|I_3| \leq 2M y \int_{x+\delta}^{\infty} \frac{dt}{(t-x)^2} = \frac{2M y}{\delta} < \frac{\epsilon}{3} \quad \text{if } 0 < y < \frac{\delta\epsilon}{6M}. \tag{9.93}$$

Also,

$$|I_2| \leq \frac{\epsilon}{3} \int_{x-\delta}^{x+\delta} \frac{y\, dt}{(t-x)^2 + y^2} \quad \text{by (9.90)}$$

$$= \frac{\epsilon}{3} \left[\mathrm{Tan}^{-1}\left(\frac{t-x}{y}\right) \right]_{x-\delta}^{x+\delta}$$

$$= \frac{2\epsilon}{3} \mathrm{Tan}^{-1} \frac{\delta}{y} < \frac{2\epsilon}{3}\left(\frac{\pi}{2}\right) = \frac{\pi\epsilon}{3} \quad \text{if } y > 0. \qquad (9.94)$$

Finally by (9.91)–(9.94),

$$|u(x,y) - f(x)| < \frac{\epsilon}{3} + \frac{\epsilon}{3} + \frac{\epsilon}{3} = \epsilon \quad \text{if } 0 < y < \frac{\delta\epsilon}{6M}.$$

Thus (9.89) is proved.

Exercises 9.4.8

In Ex. 1–9, let $I(x,y,t) = \frac{y f(t)}{(t-x)^2 + y^2}$ which is the integrand in (9.88), let f satisfy the hypotheses in 9.4.7, and let M be an upper bound for $|f(t)|$. Let $(x,y) \in Q_r$ where Q_r is given in the proof of 9.4.7. Finally, let $d > r > 1$.

1. Verify each of the following.

   (a) $I_x = \dfrac{2y\,(t-x)\,f(t)}{[(t-x)^2 + y^2]^2}$

   (b) $|I_x| \leq \dfrac{2M\,y}{(t-r)^3} \quad$ for each $t \geq d$.

   (c) For a fixed y, the integral $\displaystyle\int_d^\infty I_x\, dt$ converges uniformly on $[-r,r]$ with respect to x. **Hint.** Use (b) and 9.4.4.

   (d) If $w(x,y) = \dfrac{1}{\pi} \displaystyle\int_d^\infty I\, dt$, then $w_x = \dfrac{1}{\pi} \displaystyle\int_d^\infty I_x\, dt$ on Q_r.   **Hint.** Use 9.4.3.

2. Verify each of the following.

   (a) $I_{xx} = \dfrac{8y\,(t-x)^2\, f(t)}{[(t-x)^2 + y^2]^3} - \dfrac{2y\, f(t)}{[(t-x)^2 + y^2]^2}$

   (b) $|I_{xx}| \leq \dfrac{8M\,y}{(t-r)^4} + \dfrac{2M\,y}{(t-r)^4} = \dfrac{10M\,y}{(t-r)^4} \quad$ for each $t \geq d$

   (c) For a fixed y, the integral $\displaystyle\int_d^\infty I_{xx}\, dt$ converges uniformly on $[-r,r]$ with respect to x.

   (d) $w_{xx} = \dfrac{1}{\pi} \displaystyle\int_d^\infty I_{xx}\, dt \quad$ on Q_r

3. Verify each of the following.

(a) $|I_x| \leq \dfrac{2M\,y}{(x-t)^3} \leq \dfrac{2M\,y}{(-r-t)^3}$ for each $t \leq -d$

Hint. Since $x > -r$, $x - t > -r - t$. Also, $t \leq -d$ implies $-t \geq d > r$. Hence $-r - t > 0$.

(b) For a fixed y, the integral $\displaystyle\int_{-\infty}^{-d} I_x\, dt$ "converges uniformly on $[-r, r]$."

Hint. Use an appropriate modification of 9.4.1.

(c) If $s = \dfrac{1}{\pi}\displaystyle\int_{-\infty}^{-d} I\, dt$, then $s_x = \dfrac{1}{\pi}\displaystyle\int_{-\infty}^{-d} I_x\, dt$ on Q_r.

(d) $|I_{xx}| \leq \dfrac{8M\,y}{(x-t)^4} + \dfrac{2M\,y}{(x-t)^4} \leq \dfrac{10M\,y}{(t+r)^4}$ for each $t \leq -d$

(e) $s_{xx} = \dfrac{1}{\pi}\displaystyle\int_{-\infty}^{-d} I_{xx}\, dt$ on Q_r

4. Show that if $v = \dfrac{1}{\pi}\displaystyle\int_{-d}^{d} I\, dt$, then

$$v_{xx} = \frac{1}{\pi}\int_{-d}^{d} I_{xx}\, dt \quad \text{on } Q_r.$$

Hint. For a fixed y, we see that I, I_x, and I_{xx} are continuous in the region $\{(x,t): |x| \leq r \text{ and } |t| < d\}$. Use results in advanced calculus.

5. Show that $u_{xx} = \dfrac{1}{\pi}\displaystyle\int_{-\infty}^{\infty} I_{xx}\, dt$ where u is given in (9.88).

Solution. Let s, v, and w be the functions given in Ex. 3, 4, and 1. Then

$$u = s + v + w.$$

Hence

$$u_{xx} = s_{xx} + v_{xx} + w_{xx}.$$

We now use Ex. 3(e), 4, and 2(d).

6. Verify each of the following.

(a) $I_y = \dfrac{[(t-x)^2 - y^2]\, f(t)}{[(t-x)^2 + y^2]^2}$

(b) $|I_y| \leq \dfrac{[(t-x)^2 + y^2]\, M}{[(t-x)^2 + y^2]^2} \leq \dfrac{M}{(t-x)^2}$

(c) For a fixed x in $[-r, r]$, the integral $\displaystyle\int_{d}^{\infty} I_y\, dt$ converges uniformly on $\left[\dfrac{1}{r}, r\right]$ with respect to y.

(d) If w is given in Ex. 1(d), then $w_y = \dfrac{1}{\pi} \displaystyle\int_d^\infty I_y\, dt$ on Q_r.

7. Prove each of the following.

(a) $I_{yy} = \dfrac{[2y^3 - 6y\,(t-x)^2]\,f(t)}{[(t-x)^2 + y^2]^3}$

(b) $|I_{yy}| \le \dfrac{2M\,r^3}{(t-x)^6} + \dfrac{6M\,r}{(t-x)^4}$

(c) For a fixed x in $[-r, r]$, the integral $\displaystyle\int_d^\infty I_{yy}\, dt$ converges uniformly on $\left[\dfrac{1}{r}, r\right]$ with respect to y.

(d) If w is given in Ex. 1(d), then $w_{yy} = \dfrac{1}{\pi} \displaystyle\int_d^\infty I_{yy}\, dt$ on Q_r.

8. Show that $u_{yy} = \dfrac{1}{\pi} \displaystyle\int_{-\infty}^\infty I_{yy}\, dt$ where u is given in (9.88).

9. Show that $\nabla^2 u = \dfrac{1}{\pi} \displaystyle\int_{-\infty}^\infty \nabla^2 I\, dt$ where u is given in (9.88).

Hint. Use Ex. 5 and 8.

10. Show that $\dfrac{y}{(t-x)^2 + y^2} = \mathcal{I}\left(\dfrac{1}{t-z}\right)$.

Remarks 9.4.9 Let f be a bounded real-valued function which is piecewise continuous on the interval $[-a, a]$ and which is continuous on the intervals $(-\infty, -a]$ and $[a, \infty)$ where a is some positive number. If $u(x, y)$ is given by (9.88), then u is harmonic in the upper half-plane H°. Furthermore, for each x at which f is continuous

$$\lim_{y \to 0} u(x, y) = f(x).$$

The proof is the same as the proof of 9.4.7, except that in 9.4.8, we take $d \ge a$ and we write $\int_{-d}^d I\, dt$ as the sum of a finite number of integrals over intervals on which f is continuous. Also in proving (9.89), we take a point x at which f is continuous.

9.5 Green's Function and the Dirichlet Problem

Remarks 9.5.1 (a) Let C be a parametrized curve given by

$$z = z(t) \quad \text{on } [a, b]. \tag{9.95}$$

If $z'(t_0)$ exists and is not 0, then the vector $z'(t_0)$ is tangent to C at the point $z(t_0)$. To see this, we observe that

$$z'(t_0) = \lim_{t \to t_0} \frac{z(t) - z(t_0)}{t - t_0}. \tag{9.96}$$

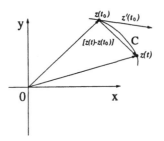

Figure 9.6: $z'(t_0) = \lim\limits_{t \to t_0} \frac{z(t)-z(t_0)}{t-t_0}$

Now as in Figure 9.6,

$$\frac{1}{t-t_0}\,[z(t) - z(t_0)]$$

is a scalar multiple of the vector $z(t) - z(t_0)$. Thus by definition, the limit in (9.96) is a tangent vector to C at $z(t_0)$.

(b) Suppose the parameter t in (9.95) is the arc length of C [measured from the initial point $p = z(0)$]. Then C is given by

$$z = z(s) \quad \text{on } [0, L]$$

where L is the length of C. Furthermore, suppose that C is piecewise smooth. From the calculus, at a point where $x'(s)$ and $y'(s)$ are continuous,

$$s = \int_0^s \sqrt{[x'(\sigma)]^2 + [y'(\sigma)]^2} \, d\sigma.$$

By a fundamental theorem of calculus,

$$1 = \frac{ds}{ds} = \sqrt{[x'(s)]^2 + [y'(s)]^2}.$$

Thus at a point where $z'(s)$ is continuous,

$$|z'(s)| = \sqrt{[x'(s)]^2 + [y'(s)]^2} = 1.$$

Hence $z'(s)$ is a unit tangent vector to C.

Definition 9.5.2 Let $f(x, y)$ be a real-valued function defined in a domain D such that f_x and f_y are continuous in D and let C be a piecewise smooth curve in D given by

$$z = z(s) = x(s) + iy(s) \quad \text{on } [0, L]$$

where s is arc length. Let

$$g(s) = f[x(s), y(s)] \quad \text{for each } s \text{ in } [0, L]. \tag{9.97}$$

If $g'(s)$ exists, then $g'(s)$ is called the **directional derivative** of f at $z(s)$ in the direction of C.

Remarks 9.5.3 **(a)** By the chain rule in calculus, it follows from (9.97) that the directional derivative $g'(s)$ is given by

$$g'(s) = f_x(z)\, x'(s) + f_y(z)\, y'(s) \quad \text{where } z = [x(s), y(s)]. \tag{9.98}$$

We observe by definition that the directional derivative of f at z along C is the rate of change of f (along C) with respect to **arc length**. [The rate of change of f along C with respect to an arbitrary parameter t is given by $f_x \frac{dx}{dt} + f_y \frac{dy}{dt} = f_x \frac{dx}{ds}\frac{ds}{dt} + f_y \frac{dy}{ds}\frac{ds}{dt} = g'(s)\frac{ds}{dt} \neq g'(s)$ unless $\frac{ds}{dt} = 1$ or $g'(s) = 0$.]

(b) In 9.5.1, we noticed that $z'(s) = x'(s) + iy'(s)$ is the unit tangent vector to C at z. Thus (9.98) shows that for a given f, its directional derivative at z along C is determined by the unit tangent vector to C at z (that is, by the "direction" of C). Hence f has the same directional derivative at z along all curves having the same direction at z [that is, the same unit tangent vector at the point $z(s)$].

(c) Since $z'(s) = x'(s) + iy'(s)$ is a unit vector, $iz'(s) = -y'(s) + ix'(s)$ is a unit vector perpendicular to $z'(s)$. Now the **normal derivative** $\frac{\partial f}{\partial n}$ of f at z along C is defined by

$$\begin{aligned} \frac{\partial f}{\partial n} &= -f_x(z)\, y'(s) + f_y(z)\, x'(s) \\ &= -f_x(z)\, \mathcal{I}(z'(s)) + f_y(z)\, \mathcal{R}(z'(s)). \end{aligned} \tag{9.99}$$

(d) By Parts (b) and (c), we see that $\frac{\partial f}{\partial n}$ at z along C is the directional derivative of f at z along any curve K whose unit tangent vector at z is $iz'(s)$ [where $z = z(s)$ is the parametrization of C with respect to its arc length]. Such a curve K is perpendicular to C at z.

As an example, we let C be the cco circle given by $|z| = a$ where $a > 0$. Parametrizing C in terms of its arc length s, we have

$$C: \quad z = a\cos\frac{s}{a} + ia\sin\frac{s}{a} \quad \text{on } [0, 2\pi a].$$

Now $x'(s) = -\sin\frac{s}{a}$ and $y'(s) = \cos\frac{s}{a}$. Thus by (9.99), the normal derivative of f at z along C is given by

$$\begin{aligned} \frac{\partial f}{\partial n} &= -f_x \cos\frac{s}{a} - f_y \sin\frac{s}{a} \\ &= -f_x \cos\theta - f_y \sin\theta \quad \text{where } \theta = \frac{s}{a}. \end{aligned}$$

Using polar coordinates, $x = r\cos\theta$ and $y = r\sin\theta$. Thus $f(x,y)$ is a function of r and θ, and we notice that

$$\begin{aligned} f_r &= f_x\, x_r + f_y\, y_r \\ &= f_x\,\cos\theta + f_y\,\sin\theta. \end{aligned}$$

Comparing the last expression for f_r and the last expression for $\frac{\partial f}{\partial n}$, we see that

$$\frac{\partial f}{\partial n} = -f_r \qquad \text{for this circle } C.$$

Thus the normal derivative of f at z along the circle C is the negative of the rate of change of f with respect to the distance r along the ray from the center of the circle through the point z. This ray is normal to C at the point z where both $\frac{\partial f}{\partial n}$ and f_r are evaluated.

(e) Let C be a contour given by

$$C: \quad z = z(s) \quad \text{on } [0, L]$$

where s is the arc length along C. Let $f(z) = f(x,y)$ be a real-valued function whose domain contains C. If the definite integral $\int_0^L f[x(s), y(s)]\, ds$ exists, then it is denoted by $\int_C f\, ds$ and is called the (line) integral along C of $f\, ds$. Thus

$$\int_C f\, ds = \int_0^L f\,[x(s), y(s)]\, ds \quad \text{if this integral exists.}$$

If in the integral $\int_C$ of $g\, ds$, a normal derivative $\frac{\partial f}{\partial n}$ appears in the expression for g, then it is understood that $\frac{\partial f}{\partial n}$ is the normal derivative at $z(s)$ along C.

(f) By a simple argument simular to the proofs of 3.5.11 and 3.5.12, we obtain an extension of Green's Theorem to a multiply connected domain. This extension states that if C denotes the oriented boundary B in 3.5.12, then

$$\int_C (P\, dx + Q\, dy) = \iint\limits_{I(C)} (Q_x - P_y)\, dx\, dy$$

where $I(C)$ stands for D° in 3.5.11 [provided P and Q satisfy the hypotheses of Green's Theorem on $\overline{I(C)}$].

Theorem 9.5.4 Let C be a cco simple closed contour. Let F and K be real-valued functions such that

$$F, F_x, F_y, K, K_x \text{ and } K_y \text{ are continuous in } \overline{I(C)}$$

and such that F_{xx}, F_{yy}, K_{xx}, and K_{yy} are continuous and bounded in $I(C)$. Then

$$\int_C \left(K \frac{\partial F}{\partial n} - F \frac{\partial K}{\partial n} \right) ds \ = \ \iint\limits_{I(C)} (F \ \nabla^2 K - K \ \nabla^2 F) \, dx \, dy. \qquad (9.100)$$

Proof. By (9.99),

$$\int_C K \frac{\partial F}{\partial n} \, ds \ = \ \int_C [K \, F_y \, dx - K \, F_x \, dy]$$

$$= \ - \iint\limits_{I(C)} [K \nabla^2 F + F_x \, K_x + F_y \, K_y] \, dx \, dy, \qquad (9.101)$$

by Green's Theorem with $P = K \, F_y$ and $Q = -K \, F_x$. Interchanging F and K in (9.101), we have

$$\int_C F \frac{\partial K}{\partial n} \, ds = - \iint\limits_{I(C)} [F \nabla^2 K + F_x \, K_x + F_y \, K_y] \, dx \, dy. \qquad (9.102)$$

Now (9.100) follows by subtracting the left and right members of (9.102) from the corresponding members of (9.101).

Remarks 9.5.5 (a) If $K \equiv 1$ on $\overline{I(C)}$ in 9.5.4, then

$$\int_C \frac{\partial F}{\partial n} \, ds = - \iint\limits_{I(C)} \nabla^2 F \, dx \, dy.$$

(b) If in 9.5.4, the functions F and K are harmonic in $I(C)$, then

$$\int_C \left(K \frac{\partial F}{\partial n} - F \frac{\partial K}{\partial n} \right) ds = 0. \qquad (9.103)$$

(c) The conclusion in 9.5.4 holds if the hypothesis that C is a simple closed contour is replaced by the hypothesis that C is the oriented boundary B of the region used in 9.5.3(f), since Green's Theorem is valid in this case. [If $C = \sum_{k=1}^n C_k$, then $\int_C = \sum_{k=1}^n \int_{C_k}$.] Hence Part (b) is still valid if C is the oriented boundary mentioned above.

Theorem 9.5.6 Let F and C satisfy the hypotheses of 9.5.4. Also, let F be harmonic in $I(C)$. Then for each a in $I(C)$,

$$F(a) = \frac{1}{2\pi} \int_C \left[\frac{\partial F}{\partial n} \operatorname{Log} r - F \frac{\partial}{\partial n} \operatorname{Log} r \right] ds \qquad (9.104)$$

where $r = |z - a|$.

Proof. Let $a \in I(C)$ and let C_1 be a cco circle given by $|z - a| = r_1$ where r_1 is sufficiently small that $\overline{I(C_1)} \subset I(C)$. By 9.1.6(c), we see that $\operatorname{Log} r$ is harmonic in $R^2 - \{a\}$. Thus by 9.5.5(b) and (c) with $K = \operatorname{Log} r$,

$$\int_{C-C_1} \left[\frac{\partial F}{\partial n} \operatorname{Log} r - F \frac{\partial}{\partial n} (\operatorname{Log} r) \right] ds = 0.$$

Hence

$$\int_C \left[\frac{\partial F}{\partial n} \operatorname{Log} r - F \frac{\partial}{\partial n} (\operatorname{Log} r) \right] ds$$

$$= \int_{C_1} \left[\frac{\partial F}{\partial n} \operatorname{Log} r - F \frac{\partial}{\partial n} (\operatorname{Log} r) \right] ds$$

$$= \operatorname{Log} r_1 \int_{C_1} \frac{\partial F}{\partial n} ds - \int_{C_1} \left(-\frac{1}{r_1} \right) F \, ds \quad \text{by 9.5.3(d)}$$

$$= 0 + \frac{1}{r_1} \int_0^{2\pi r_1} F(z) \, ds \quad \text{by 9.5.5(a)}$$

$$= \frac{1}{r_1} \int_0^{2\pi} F(z) \, r_1 \, d\theta \quad \text{where } s = r_1 \theta$$

$$= 2\pi F(a) \qquad (9.105)$$

by the mean-value property for harmonic functions. Thus (9.104) follows from (9.105).

Remarks 9.5.7 (a) Theorem 9.5.6 is valid if C and $I(C)$ are interpreted as in 9.5.3(f).

(b) If F and its normal derivative are known along C, then (9.104) may be used to find the values of F at interior points of C.

(c) However if $\frac{\partial F}{\partial n}$ along C is not given, then we consider the following. Suppose

$$K \in H(I(C)) \quad \text{and} \quad K \equiv \operatorname{Log} r \quad \text{on } C \qquad (9.106)$$

where $r = |z - a|$ for a fixed a in $I(C)$. Also, suppose F, K, and C satisfy the hypotheses of 9.5.4 where $F \in H(I(C))$. Then (9.103) and (9.104) hold.

Multiplying by $\frac{1}{2\pi}$ in (9.103) and subtracting from (9.104), we have for each a in $I(C)$,

$$
\begin{aligned}
F(a) &= \frac{1}{2\pi} \int_C \left[(\operatorname{Log} r - K) \frac{\partial F}{\partial n} - F \frac{\partial}{\partial n} (\operatorname{Log} r - K) \right] ds \\
&= \frac{1}{2\pi} \int_C F \frac{\partial}{\partial n} (K - \operatorname{Log} r) \, ds \quad \text{by 9.106.} \tag{9.107}
\end{aligned}
$$

(d) If K satisfies (9.106) and if K is continuous on $\overline{I(C)}$, then for a in $I(C)$, the function $G(z, a)$ given by

$$
G(z, a) = K(z) - \operatorname{Log} |z - a| \tag{9.108}
$$

is called **Green's function** for $I(C)$.

(e) By 9.1.11 for a given C such that $a \in I(C)$, we see that K is unique; and hence $G(z, a)$ is unique for $I(C)$. Also for a fixed a,

$$
G \in H[I(C) - \{a\}] \quad \text{and} \quad G \equiv 0 \quad \text{on } C. \tag{9.109}
$$

Also,

$$
G \text{ is continuous in } \overline{I(C)} - \{a\}. \tag{9.110}
$$

(f) Now $K(z) \to K(a)$ as $z \to a$. But $\operatorname{Log} |z - a| \to -\infty$ as $z \to a$. Thus

$$
G(z, a) \to +\infty \quad \text{as } z \to a. \tag{9.111}
$$

By (9.107) and 9.5.7(d), we have the following result.

Theorem 9.5.8 Let D be the set given in 3.5.11, let B be the oriented boundary of D, and let $a \in D^\circ$. If F and K are harmonic in D°, if $K \equiv \operatorname{Log} |z - a|$ on B, and if F and K satisfy the hypotheses of 9.5.4 with C replaced by B, then

$$
F(a) = \frac{1}{2\pi} \int_B F(z) \frac{\partial}{\partial n} G(z, a) \, ds
$$

where G is Green's function for $I(B)$.

Theorem 9.5.9 If $a \in I(C)$, then $G(z, a) > 0$ for each z in $I(C)$ such that $z \neq a$.

Proof. Let a be fixed. By (9.111), let C_1 be a circle centered at a such that $\overline{I(C_1)} \subset I(C)$ and such that

$$
G(z, a) > 0 \quad \text{for each } z \text{ on } C_1. \tag{9.112}
$$

Let

$$
D = I(C) - \overline{I(C_1)}.
$$

Then
$$G \in H(D) \quad \text{and} \quad G \in C(\overline{D}).$$

By 9.1.10, the function G assumes its minimum on $B(D) = C \cup C_1$. Thus by (9.109) and (9.112),
$$\min_{z \in \overline{D}} G(z, a) = 0.$$

Hence by 9.1.9, it follows that $G(z, a) > 0$ on D. Letting the radius of C_1 approach zero, we see that 9.5.9 is proved.

Theorem 9.5.10 Green's function for $I(C)$ is symmetric, that is, if a and b are different points in $I(C)$, then $G(a,b) = G(b,a)$.

   Proof. Let
$$g_1(z) = G(z, a) = K_1(z) - \text{Log}\,|z - a| \tag{9.113}$$

and let
$$g_2(z) = G(z, b) = K_2(z) - \text{Log}\,|z - b| \tag{9.114}$$

where K_1 and K_2 are harmonic in $I(C)$ and where
$$K_1 = \text{Log}\,|z - a| \quad \text{and} \quad K_2 = \text{Log}\,|z - b| \quad \text{on } C.$$

Also, let C_1 and C_2 be disjoint cco circles centered at a and b, respectively, such that $\overline{I(C_k)} \subset I(C)$ for $k = 1, 2$ and such that $\overline{I(C_1)} \cap \overline{I(C_2)} = \emptyset$. Since g_1 and g_2 are harmonic in $I(C) - \overline{I(C_1) \cup I(C_2)}$, we apply 9.5.5(b) and (c) to obtain
$$\int_{C-C_1-C_2} \left(g_2 \frac{\partial g_1}{\partial n} - g_1 \frac{\partial g_2}{\partial n} \right) ds = 0. \tag{9.115}$$

Since $g_1 \equiv 0 \equiv g_2$ on C, we conclude from (9.115) that
$$\int_{C_1} \left(g_2 \frac{\partial g_1}{\partial n} - g_1 \frac{\partial g_2}{\partial n} \right) ds = \int_{C_2} \left(g_1 \frac{\partial g_2}{\partial n} - g_2 \frac{\partial g_1}{\partial n} \right) ds. \tag{9.116}$$

Now
$$\int_{C_1} \left(g_2 \frac{\partial g_1}{\partial n} - g_1 \frac{\partial g_2}{\partial n} \right) ds$$
$$= \int_{C_1} \left[g_2 \frac{\partial}{\partial n}(K_1 - \text{Log}\,|z - a|) - (K_1 - \text{Log}\,|z - a|) \frac{\partial g_2}{\partial n} \right] ds \quad \text{by (9.113)}$$
$$= \int_{C_1} \left(g_2 \frac{\partial K_1}{\partial n} - K_1 \frac{\partial g_2}{\partial n} \right) ds + \int_{C_1} \left(-g_2 \frac{\partial}{\partial n} \text{Log}\,|z - a| + \frac{\partial g_2}{\partial n} \text{Log}\,|z - a| \right) ds$$
$$= 0 + \int_{C_1} \left[\frac{\partial g_2}{\partial n} \text{Log}\,|z - a| - g_2 \frac{\partial}{\partial n} \text{Log}\,|z - a| \right] ds \quad \text{by 9.5.5(b)}$$
$$= 2\pi\, g_2(a) = 2\pi\, G(a, b) \quad \text{by 9.5.6 and (9.114)}. \tag{9.117}$$

Similarly,

$$\int_{C_2} \left(g_1 \frac{\partial g_2}{\partial n} - g_2 \frac{\partial g_1}{\partial n} \right) ds = 2\pi G(b, a). \tag{9.118}$$

Thus $G(a, b) = G(b, a)$ by (9.116)–(9.118).

Exercises 9.5.11

1. Observe that the function K used in the definition of Green's function in 9.5.7(d) is the solution of Dirichlet's problem for $I(C)$ where $f(z) = \text{Log}\,|z - a|$ on C. [Compare the definition of Dirichlet's problem in 9.2.4 with the definition of K.]

2. (a) Show that if C is a circle with center at a and radius ρ, then

$$G(z, a) = \text{Log}\,\rho - \text{Log}\,|z - a| \quad \text{for each } z \text{ in } \overline{I(C)}.$$

   (b) In particular if $a = 0$ and $\rho = 1$, show that

$$G(z, 0) = -\text{Log}\,|z| \quad \text{for each } z \text{ in } \overline{I(C)} \text{ such that } z \neq 0.$$

Hint for Part (a). See 9.5.7(d) and note that a constant function is harmonic.

9.6 Green's Function and Transformations

Remarks 9.6.1 (a) Let D_1 be a bounded simply connected domain in R^2 which has a simple closed curve B_1 as its boundary. Suppose the solution of Dirichlet's problem is known for each continuous real-valued function defined on B_1. This solution may be used to solve Dirichlet's problem for an arbitrary bounded simply connected domain D and a given continuous real-valued function f on $B = B(D)$, provided B is a simple closed curve. For let

$$h \text{ be a one-to-one continuous mapping of } \overline{D} \text{ onto } \overline{D_1}$$
$$\text{such that } h \in A(D), \ h(D) = D_1, \text{ and } h^{-1} \in C(\overline{D_1}). \tag{9.119}$$

[Such a mapping h exists by 11.3.7 and Ex. 2 in 11.3.6.]

To solve Dirichlet's problem for D, we must find a function u such that

$$\text{(i) } u \in H(D), \quad \text{(ii) } u \in C(\overline{D}), \quad \text{and} \quad \text{(iii) } u = f \text{ on } B.$$

Let

$$f_1 = f \circ h^{-1} \quad \text{on } B_1. \tag{9.120}$$

By 2.2.15 and (9.119), we see that $f_1 \in C(B_1)$. Let

$$u_1 \in H(D_1) \text{ such that } u_1 \in C(\overline{D_1}) \text{ and } u_1 = f_1 \text{ on } B_1. \tag{9.121}$$

Now let

$$u = u_1 \circ h \quad \text{on } \overline{D}. \tag{9.122}$$

We shall see that u satisfies (i)–(iii). First we observe by (9.122) and 9.1.17, that $u \in H(D)$. Next $u \in C(\overline{D})$ by (9.119), (9.120), (9.122), and 2.2.15. Finally to see (iii), we observe that

$$
\begin{aligned}
f &= f_1 \circ h \quad \text{on } B \quad \text{by (9.120)} \\
&= u_1 \circ h \quad \text{on } B \quad \text{by (9.121)} \quad \text{since } h(B) = B_1 \\
&= u \quad \text{on } B \quad \text{by (9.122).}
\end{aligned}
$$

(b) In view of 9.2.5, we may take D_1 to be the unit disk $N_1(0)$.

Theorem 9.6.2 Let h be a mapping satisfying (9.119). If Green's function $G_1(w, b)$ is known for D_1, then Green's function $G(z, a)$ for D is given by

$$G(z, a) = G_1[h(z), h(a)] \quad \text{for each } a \text{ in } D \text{ and for each } z \text{ in } \overline{D} - \{a\}.$$

Proof. We shall let $B = B(D)$ and $B_1 = B(D_1)$. By 9.5.7(d) for a fixed b in D_1,

$$G_1(w, b) = K_1(w) - \operatorname{Log}|w - b| \quad \text{for each } w \text{ in } \overline{D_1} - \{b\}$$

where K_1 depends on b and where

$$K_1 \in H(D_1), \ K_1 \in C(\overline{D_1}), \text{ and } K_1(w) = \operatorname{Log}|w - b| \quad \text{on } B_1. \tag{9.123}$$

Then for a fixed a in D and for each z in $\overline{D} - \{a\}$, we have with $b = h(a)$,

$$
\begin{aligned}
G_1[h(z), h(a)] &= K_1(h(z)) - \operatorname{Log}|h(z) - h(a)| \\
&= \left[K_1(h(z)) - \operatorname{Log}\left| \frac{h(z) - h(a)}{z - a} \right| \right] - \operatorname{Log}|z - a| \\
&= K(z) - \operatorname{Log}|z - a| \tag{9.124}
\end{aligned}
$$

where $K(z)$ is defined by

$$K(z) = K_1(h(z)) - \operatorname{Log}|q(z)| \tag{9.125}$$

and where

$$
q(z) = \begin{cases}
\dfrac{h(z) - h(a)}{z - a} & \text{if } z \neq a \\[2mm]
h'(a) & \text{if } z = a.
\end{cases} \tag{9.126}
$$

Now by (9.123) and 9.1.17,

$$K_1(h(z)) \in H(D). \tag{9.127}$$

By Ex. 17 in 6.3.16, we have $q \in A(a)$ and hence $q \in A(D)$. By (9.119), (9.126), and 8.2.13, it follows that q is never zero in D. Thus by 9.1.6(c) and 9.1.17,

$$\text{Log} |q(z)| \text{ is harmonic in } D. \tag{9.128}$$

Hence by (9.125), (9.127), and (9.128), it follows that

$$K \in H(D). \tag{9.129}$$

By (9.125), 2.2.15, Ex. 2(a) in 2.4.9, and by Ex. 17 in 6.3.16,

$$K \in C(\overline{D}). \tag{9.130}$$

By (9.119), we see that if $z \in B$ then $h(z) \in B_1$. Hence by (9.123) and the first equality in (9.124), we see that $G_1[h(z), h(a)] = 0$ on B. Thus by (9.124),

$$K(z) = \text{Log} |z - a| \quad \text{on } B. \tag{9.131}$$

Hence by (9.129)–(9.131) and 9.5.7(d), it follows that $G_1[h(z), h(a)]$ is Green's function for D.

Theorem 9.6.3 Let U be the open unit disk $\{z : \ |z| < 1\}$ and let a be any point in U. Then Green's function for U is given by

$$G(z, a) = \text{Log} \left| \frac{\overline{a}z - 1}{z - a} \right| \quad \text{for each } z \text{ in } \overline{U} - \{a\}.$$

Proof. For each fixed a in U, let

$$h_a(z) = \frac{z - a}{\overline{a}z - 1} \quad \text{for each } z \text{ in } \overline{U}. \tag{9.132}$$

By 10.3.8, the function h_a is a one-to-one continuous mapping of $\overline{U}$ onto $\overline{U}$ such that $h_a \in A(U)$ and such that $h_a(U) = U$. Thus

$$
\begin{aligned}
G(z, a) &= G[h_a(z), h_a(a)] \quad \text{by 9.6.2} \\
&= G[h_a(z), 0] \quad \text{by (9.132)} \\
&= -\text{Log} |h_a(z)| \quad \text{by Ex. 2(b) in 9.5.11} \\
&= \text{Log} \left| \frac{\overline{a}z - 1}{z - a} \right| \quad \text{by (9.132).} \tag{9.133}
\end{aligned}
$$

In Chapter 11, we shall show that if D is a simply connected domain in R^2 such that $D \neq R^2$, then there is a one-to-one analytic mapping of D onto the unit disk U. (See 11.3.5.) We may use this result and 9.6.3 to obtain **Green's function for an arbitrary bounded simply connected domain D**. For (by Ex. 2 in 11.3.6 and by 11.3.7) let h be a one-to-one continuous mapping of $\overline{D}$ onto $\overline{U}$ such that $h \in A(D)$ and such that $h(D) = U$. Then Green's function G for D is given by

$$\begin{aligned} G(z,a) &= G_1[h(z), h(a)] \qquad \text{by 9.6.2} \\ &= \text{Log} \left| \frac{\overline{h(a)}\, h(z) - 1}{h(z) - h(a)} \right| \qquad \text{by 9.6.3} \end{aligned}$$

where G_1 is Green's function for the unit disk U.

Remarks 9.6.4 We may use 9.6.3 to show that for the unit disk, the formula in 9.5.8 reduces to Poisson's Formula given in 9.2.2. To do this, we shall need to compute $\frac{\partial G}{\partial n}$ along the cco unit circle $B = B(U)$ where G is Green's function for the unit disk U. By 9.6.3,

$$\begin{aligned} G(z,a) &= \text{Log} \left| \frac{\overline{a}z - 1}{z - a} \right| = \mathcal{R} \left[\log \frac{\overline{a}z - 1}{z - a} \right] \\ &= \mathcal{R} \left[\log \left(\overline{a}z - 1 \right) - \log \left(z - a \right) \right] \\ &= \mathcal{R} \left[f(z) \right] \end{aligned} \qquad (9.134)$$

for appropriate branches of the log terms in the neighborhood of a point of B at which calculations are to be made. By 5.4.9 for an analytic function $f = u + iv$,

$$f'(z) = e^{-i\theta}(u_r + i v_r) \qquad \text{where } z = re^{i\theta} \neq 0. \qquad (9.135)$$

Thus

$$u_r = \mathcal{R} \left[e^{i\theta} f'(z) \right]. \qquad (9.136)$$

Now along the unit circle B,

$$\begin{aligned} \frac{\partial G}{\partial n} &= -\frac{\partial G}{\partial r} \qquad \text{by 9.5.3(d)} \\ &= -\mathcal{R} \left[e^{i\theta} f'(z) \right] \qquad \text{by (9.134) and (9.136)} \\ &= -\mathcal{R} \left[\frac{z}{r} \left(\frac{\overline{a}}{\overline{a}z - 1} - \frac{1}{z - a} \right) \right] \\ &\qquad \text{by (9.135), (9.136), and Ex. 11 in 5.4.11} \\ &= \mathcal{R} \left[\frac{z}{z - a} + \frac{\overline{a}}{\overline{z} - \overline{a}} \right] \qquad \text{since } r = |z| = 1 \\ &= \mathcal{R} \left[K(a, z) \right] \qquad \text{by 9.2.3(c)} \\ &= K(a, z) \qquad \text{since } K(a, z) \text{ is real.} \end{aligned}$$

Also, on the unit circle, $s = \phi$ implies $ds = d\phi$. Therefore, the formula in 9.5.8 reduces to Poisson's Integral Formula in 9.2.2.

Chapter 10

Conformal Mapping

10.1 The Extended Complex Plane

The function $w = f(z) = \frac{1}{z}$ is a one-to-one continuous mapping of $R^2 - \{0\}$ onto $R^2 - \{0\}$. In order to extend the domain of f to all of R^2 we adjoin to R^2 an **ideal point** called the **point at infinity** and denoted by ∞.

For each $r > 0$, the set

$$\{z : \quad z = \infty \quad \text{or} \quad |z| > r\} \tag{10.1}$$

is called a **neighborhood of** ∞ and is denoted by $N_r(\infty)$. The system $R^2 \cup \{\infty\}$ with neighborhoods of ∞ given by (10.1) (and with the structure on R^2 given in Chapter 1) is called the **extended complex plane** and is denoted by R^2_∞ in this book. Now the function

$$w = f(z) = \begin{cases} \infty & \text{if } z = 0 \\ \frac{1}{z} & \text{if } z \in R^2 - \{0\} \\ 0 & \text{if } z = \infty \end{cases}$$

is a one-to-one mapping of R^2_∞ onto R^2_∞. We observe that the image of $N_r(0)$ is $N_{\frac{1}{r}}(\infty)$ and vice versa. See Figure 10.1. Thus if continuity in R^2_∞ is defined by using neighborhoods in the usual way, then f is continuous at $z = 0$ and $z = \infty$.

We find it convenient to represent the complex number system by the complex plane. However, to obtain a geometric representation of R^2_∞, it is convenient to use a sphere. This is possible since there is a **one-to-one mapping** of R^2_∞ onto a **sphere** which is continuous on R^2_∞ and whose inverse is continuous on the sphere.

Definition 10.1.1 Let the complex plane be tangent to the surface S of a sphere of unit diameter such that $z = 0$ is the south pole on S and let G be the north pole. See Figure 10.2. The **stereographic projection** of R^2_∞ onto S is the mapping f defined

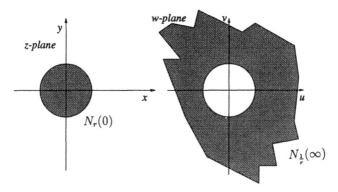

Figure 10.1: $f[N_r(0)] = N_{\frac{1}{r}}(\infty)$ and $f\left[N_{\frac{1}{r}}(\infty)\right] = N_r(0)$

as follows. For an arbitrary point P in R^2_∞,

$$f(P) = \begin{cases} Q & \text{if } P \neq \infty \quad \text{where } Q \text{ is the point of intersection of} \\ & \qquad\qquad\quad S - \{G\} \text{ and the line through } G \text{ and } P \\ G & \text{if } P = \infty. \end{cases}$$

Remarks 10.1.2 (a) A **line** in R^2_∞ (or **extended line**) is defined to be $L \cup \{\infty\}$ where L is a line in R^2. Thus each line in R^2_∞ "goes through" the point ∞.

(b) Let $H = L \cup \{\infty\}$ where L is any line in R^2. Then L and the north pole G determine a plane in space which intersects S in a circle K, which contains the point G. Clearly $f(H) = K$ where f is the stereographic projection in 10.1.1. If L goes through $z = 0$, then K is a great circle. Thus the stereographic projection maps each line in R^2_∞ onto a circle on S through G.

(c) Also, it is clear that the image of a circle in R^2 centered at $z = 0$ is a circle on S in a plane parallel to R^2. Theorem 10.1.3 shows that the image of each circle in R^2 is a circle on S.

(d) Clearly f is a one-to-one mapping of R^2_∞ **onto** S. Thus for each point Q on S, there is exactly one point P in R^2_∞ such that $f(P) = Q$. The mapping which associates with each Q on S, the unique P in R^2_∞ such that $f(P) = Q$ is called the inverse (mapping) of f and is denoted by f^{-1}. (This particular mapping f^{-1} is called the stereographic projection of S onto R^2_∞.)

(e) Clearly if P_1 is near P_2 in R^2, then $f(P_1)$ is near $f(P_2)$ on S. Conversely, if Q_1 is near Q_2 on $S - \{G\}$, then $f^{-1}(Q_1)$ is near $f^{-1}(Q_2)$ in R^2. Roughly, this means that f is continuous on R^2 and f^{-1} is continuous on $S - \{G\}$. This may be

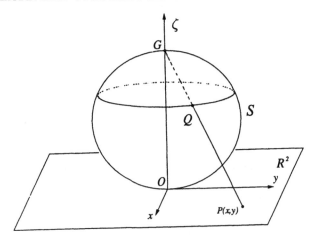

Figure 10.2: Sterographic Projection

proved by using (10.2) and (10.5). To see that f is continuous at ∞, we observe that if $U = N_\epsilon(G)$ is a neighborhood of G (other than S), then $U = f(V)$ where $V = N_r(\infty)$ is some neighborhood of ∞. Similarly f^{-1} is continuous at G. See Figure 10.3.

(f) By Part (e), we see that f is a one-to-one mapping of R^2_∞ onto S, that f is continuous on R^2_∞, and that f^{-1} is continuous on S. Thus R^2_∞ has any property possessed by S which is invariant under a one-to-one continuous mapping with a continuous inverse. Hence R^2_∞ and S are **equivalent** with respect to these properties, and we use S as a model to represent R^2_∞.

In 10.1.1, we introduce coordinate axes in space consisting of the x- and y-axes in R^2 and a ζ-axis perpendicular to the plane R^2 so that $z = 0$ is the origin and the north pole G is on the positive ζ-axis at $(0,0,1)$. If for each $P = (x,y)$ in R^2, the coordinates of $Q = f(P)$ are (ξ, η, ζ), then we have

$$\xi = \frac{x}{x^2 + y^2 + 1}, \quad \eta = \frac{y}{x^2 + y^2 + 1} \quad \text{and} \quad \zeta = \frac{x^2 + y^2}{x^2 + y^2 + 1}. \qquad (10.2)$$

To verify (10.2), we use Figure 10.4.

From the equation of the sphere S with center $\left(0, 0, \frac{1}{2}\right)$ and radius $\frac{1}{2}$, we have

$$\xi^2 + \eta^2 = \zeta(1 - \zeta). \qquad (10.3)$$

From similar triangles QBA and PCO and similar triangles QAG and POG, we have

$$\frac{\xi}{x} = \frac{\eta}{y} = \frac{\rho}{r} = \frac{1 - \zeta}{1} \qquad (10.4)$$

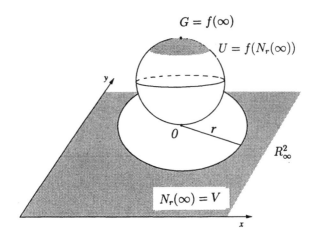

Figure 10.3: $f(N_r(\infty)) \subset f(N_r(\infty)) = U = N_\epsilon(f(\infty))$

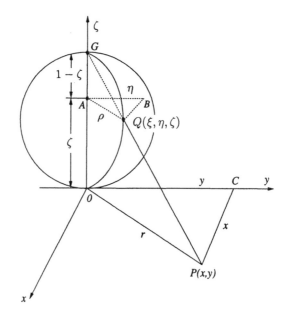

Figure 10.4: Coordinates (ξ, η, ζ) are given by (10.2).

where $r^2 = x^2 + y^2$ and $\rho^2 = \xi^2 + \eta^2$. From (10.4), we have

$$x = \frac{\xi}{1 - \zeta} \quad \text{and} \quad y = \frac{\eta}{1 - \zeta}. \tag{10.5}$$

From (10.5), we obtain

$$r^2 = x^2 + y^2 = \frac{\xi^2 + \eta^2}{(1 - \zeta)^2} = \frac{\zeta(1 - \zeta)}{(1 - \zeta)^2} = \frac{\zeta}{1 - \zeta} \quad \text{by (10.3).} \tag{10.6}$$

Solving (10.6) for ζ, we get

$$\zeta = \frac{r^2}{r^2 + 1} = \frac{x^2 + y^2}{x^2 + y^2 + 1} \quad \text{by (10.4).} \tag{10.7}$$

From (10.5), (10.7), and (10.4), we obtain

$$\xi = x\,(1 - \zeta) = \frac{x}{r^2 + 1} = \frac{x}{x^2 + y^2 + 1}$$

and

$$\eta = y\,(1 - \zeta) = \frac{y}{r^2 + 1} = \frac{y}{x^2 + y^2 + 1}.$$

[The preceding argument does not apply if $xy = 0$. But if $xy = 0$, then (10.2) and (10.5) may be verified by inspecting the cases: $x \neq 0$, $y = 0$; $x = 0$, $y \neq 0$; and $x = y = 0$.]

Theorem 10.1.3 The stereographic projection maps circles and lines in R_∞^2 onto circles on S and vice versa.

Proof. Let E be the circle (or line if $a = 0$) in R_∞^2 given by

$$ax^2 + ay^2 + bx + cy + d = 0 \quad \text{where the coefficients are real.} \tag{10.8}$$

If $P = (x, y) \in E$ and $(\xi, \eta, \zeta) = f(P)$, then by (10.5) and (10.8),

$$a\frac{\xi^2 + \eta^2}{(1 - \zeta)^2} + b\frac{\xi}{1 - \zeta} + c\frac{\eta}{1 - \zeta} + d = 0$$

or

$$\frac{a\zeta}{1 - \zeta} + \frac{b\xi}{1 - \zeta} + \frac{c\eta}{1 - \zeta} + d = 0 \quad \text{by (10.3)}$$

or

$$b\xi + c\eta + (a - d)\zeta + d = 0, \tag{10.9}$$

which is an equation of a plane, say K. Thus

$$f(E) \subset K. \tag{10.10}$$

But since f maps R^2_∞ onto S,

$$f(E) \subset S. \tag{10.11}$$

By (10.10) and (10.11),

$$f(E) \subset K \cap S = C \quad \text{where } C \text{ is a circle.}$$

Now by the geometry of the stereographic projection, $f(E)$ must be a closed curve on S. Hence $f(E)$ is the entire circle C.

Now let C be a circle on S. Then $C = S \cap K$ where K is some plane in space. Thus for each point $Q = (\xi, \eta, \zeta)$ on C, the coordinates (ξ, η, ζ) must satisfy an equation of the form (10.9). Substituting (10.2) into (10.9), we obtain (10.8), which means that $f^{-1}(C)$ is contained in the circle (or line if $a = 0$) given by (10.8). But for each $P = (x, y)$ satisfying (10.8), $f(P) = (\xi, \eta, \zeta)$ satisfies (10.9). Hence $f^{-1}(C)$ is the whole circle or line in R^2_∞.

Definitions 10.1.4 In R^2_∞, we define the following operations.

(i) If $z \in R^2$, then $z \pm \infty = \infty \pm z = \infty$, $\quad \dfrac{z}{\infty} = 0$, $\quad$ and $\quad \dfrac{\infty}{z} = \infty$.

(ii) If $z \in R^2$ and $z \neq 0$, then $z \cdot \infty = \infty \cdot z = \infty$ $\quad$ and $\quad \dfrac{z}{0} = \infty$.

(iii) $\infty + \infty = \infty \cdot \infty = \infty$.

10.2 Transformations

A function, or mapping, $w = f(z)$ may be regarded as a **transformation** of points from the domain of definition in the z-plane into points in the w-plane, which may be identical with the z-plane.

The Linear Transformation. If a and b are complex constants and $a \neq 0$, then the mapping given by

$$w = az + b \quad \text{for each } z \text{ in } R^2_\infty$$

is called the linear transformation.

Case 1. Let $w = z + b$ where $b = (b_1, b_2)$ is a fixed point of the z-plane and where the point $z = (x, y)$ varies over that plane. If the w-plane is identical to the z-plane, then the mapping $w = z + b$ translates any set of points in the z-plane through the vector b, that is, through the distance $|b|$ and in the direction $\arg b$.

Case 2. Let $w = az$ where $a \neq 0$. If $a = |a| e^{i\alpha}$ and $z = |z| e^{i\theta}$, then

$$w = |a| |z| e^{i(\theta + \alpha)}. \tag{10.12}$$

By (10.12), we see that the image of z is obtained by rotating the vector z through the angle α and magnifying (or contracting) the length of z by the factor $|a|$. Thus the

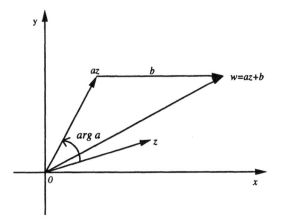

Figure 10.5: $w = az + b$

transformation $w = az$ is referred to as a rotation and magnification of the z-plane. Under this transformation, all distances from the origin are magnified by the factor $|a|$, while all figures (that is, sets of points) are rotated about the origin through the angle α. Hence all figures are mapped onto similar figures by this transformation.

Case 3. Let $w = az + b$ where $ab \neq 0$. If $w_1 = az = f(z)$ and $w_2 = z + b = g(z)$, then the transformation w is the composite f followed by g, that is, $w = g(f(z))$. Hence in the finite plane, w consists of a rotation of the z-plane through the angle $\arg a$ and a magnification by the factor $|a|$ followed by a translation through the vector b.

Remarks 10.2.1 All figures are mapped onto similar figures by a linear transformation. See Figure 10.5 for the image of an arbitrary vector. By 10.1.4, the point at ∞ is mapped onto the point at ∞ by the linear transformation.

Example 10.2.2 In view of 10.2.1, the transformation

$$w = \left(\sqrt{3} + i\right) z + (1 + 2i)$$

maps the rectangle ABCD and its interior onto the rectangle A'B'C'D' and its interior as shown in Figure 10.6. Here the angle of rotation is $\mathrm{Arg}\left(\sqrt{3} + i\right) = \frac{\pi}{6}$, the magnification factor is $\left|\sqrt{3} + i\right| = 2$ and the translation vector is $1 + 2i$.

The Reciprocal Transformation. The reciprocal transformation given by

$$w = \frac{1}{z} \tag{10.13}$$

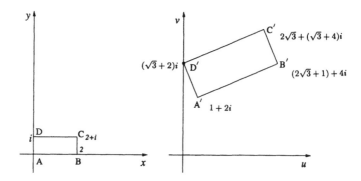

Figure 10.6: $w = \left(\sqrt{3} + i\right) z + (1+2i)$ maps the rectangle
ABCD onto the rectangle A'B'C'D'

is a one-to-one mapping of the extended plane onto itself (where $w = \infty$ if $z = 0$
and $w = 0$ if $z = \infty$). If $z = re^{i\theta}$ where $r > 0$, then

$$w = \frac{1}{z} = \frac{1}{r}e^{-i\theta}. \tag{10.14}$$

Now if

$$w_1 = f(z) = \frac{1}{r}e^{i\theta} \quad \text{and} \quad w_2 = g(z) = \overline{z}, \tag{10.15}$$

then the mapping in (10.14) is the composite function given by $w = g[f(z)]$. The
first transformation in (10.15) is an **inversion** with respect to the unit circle γ given
by $|z| = 1$. That is, the point $w_1 = \frac{1}{r}e^{i\theta}$ lies on the ray from the origin through the
point z, and $|w_1| = \frac{1}{r}$. [If $z \in E(\gamma)$, then $w_1 = f(z) \in I(\gamma)$, whereas if $z \in I(\gamma)$, then
$w_1 \in E(\gamma)$.] Now the second transformation in (10.15) is the reflection in the real axis
(since $\overline{z}$ is the symmetric image of z with respect to the x-axis). Thus the reciprocal
transformation in (10.14) is the result of an inversion in the circle γ followed by the
reflection in the real axis. Points on the unit circle are reflected by (10.14) in the real
axis into points on the unit circle. See Figure 10.7.

Let $z = x + iy \neq 0$ and let $w = f(z) = \frac{1}{z} = u + iv$. Then

$$u + iv = w = \frac{1}{z} = \frac{1}{x+iy} = \frac{x}{x^2+y^2} - i\frac{y}{x^2+y^2}.$$

Thus

$$u = \frac{x}{x^2+y^2} \quad \text{and} \quad v = -\frac{y}{x^2+y^2}. \tag{10.16}$$

By (10.16), we have

$$u^2 + v^2 = \frac{x^2+y^2}{(x^2+y^2)^2} = \frac{1}{x^2+y^2}. \tag{10.17}$$

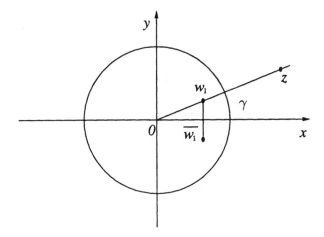

Figure 10.7: $w = \frac{1}{z} = \frac{1}{|z^2|} \, \bar{z} = \frac{1}{r} e^{-i\theta}$

Thus by (10.16) and (10.17),

$$x = u\left(x^2 + y^2\right) = \frac{u}{u^2 + v^2}. \tag{10.18}$$

Similarly,

$$y = -\frac{v}{u^2 + v^2}. \tag{10.19}$$

Let G be a circle or an extended line given by

$$a\left(x^2 + y^2\right) + bx + cy + d = 0 \tag{10.20}$$

where a, b, c, and d are real constants. Let $z = (x, y)$ be any nonzero point in R^2 which is on G. Dividing (10.20) by $x^2 + y^2$, we see from (10.16) and (10.17) that the image $w = (u, v) = \frac{1}{z}$ must satisfy the equation

$$d\left(u^2 + v^2\right) + bu - cv + a = 0 \tag{10.21}$$

which determines a circle (if $d \neq 0$) or an extended line (if $d = 0$). We let H be this circle or extended line in the w-plane. Again using (10.16) and (10.17), we see that if $w = (w, v) \neq (0, 0)$ satisfies (10.21), then $z = \frac{1}{w}$ satisfies (10.20). Thus we see that the transformation $w = \frac{1}{z}$ maps the set of all nonzero points on $G \cap R^2$ **onto** the set of all nonzero points on $H \cap R^2$. [Now $z = 0$ is on G iff $D = 0$, and $D = 0$ iff $w = \frac{1}{0} = \infty$ is on H since (10.21) represents a line when $D = 0$ (and since ∞ is on each line). Furthermore, $z = \infty$ is on G iff $a = 0$, and $a = 0$ iff $w = \frac{1}{\infty} = 0$ is on H.] Thus we have the following result.

The transformation $w = \frac{1}{z}$ maps circles and lines

onto circles or lines, where "line" means "extended line."   (10.22)

Exercises 10.2.3

1. Find the image of the semi-infinite strip $1 < x < 2$, $y > 0$ under the transformation $w = 2iz$

   (a) by finding its image under a rotation and magnification,

   (b) by writing $w = u + iv = 2ix - 2y$ and finding the restrictions on u and v.
   $$u < 0, \quad 2 < v < 4$$

2. Find in two ways the image of the set $1 < x < 2$, $y > 0$ under $w = iz + 3i$.
   $$u < 0, \quad 4 < v < 5$$

3. Under the transformation $w = \dfrac{1}{z}$, what may be said about the image of

   (a) a line through the origin,

   (b) a line not through the origin,

   (c) a circle through the origin,

   (d) a circle not through the origin?

   Hint. See (10.20) and (10.21) with $a = d = 0$, $a = 0$ and $d \neq 0$, $a \neq 0$ and $d = 0$, and $ad \neq 0$.

4. Let c be a positive constant. Under the transformation $w = f(z) = \dfrac{1}{z}$, find the image of

   (a) the (extended) line determined by $y = c$,

   (b) the extended x-axis,

   (c) the strip $0 < y < c$,

   (d) the half-plane $y > c$,

   (e) the half-plane $y < 0$.

   Solution for (a). By (10.20)–(10.22), the function f maps the extended line $y = c$ onto the circle
   $$c(u^2 + v^2) + v = 0 \quad \text{or} \quad u^2 + \left(v + \frac{1}{2c}\right)^2 = \frac{1}{4c^2}.$$

   Ans. (b) the extended line $v = 0$

   (c) $v < 0$ and $u^2 + \left(v + \dfrac{1}{2c}\right)^2 > \dfrac{1}{4c^2}$

   (d) the disk $u^2 + \left(v + \dfrac{1}{2c}\right)^2 < \dfrac{1}{4c^2}$

   (e) the half-plane $\{(u, v): \ v > 0\}$

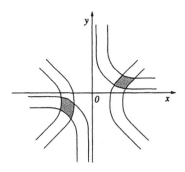

Figure 10.8: $w = f(z) = z^2$

5. (a) Find the image of the quadrant $y > 0$, $x > c$ (where $c > 0$) under $w = \dfrac{1}{z}$.

   the half-disk $\{(u, v) : \left(u - \dfrac{1}{2c}\right)^2 + v^2 < \dfrac{1}{4c^2}$ and $v < 0\}$

   (b) What set in the z-plane is mapped onto the hyperbola $u^2 - v^2 = 1$ under $w = \dfrac{1}{z}$?   the Lemniscate of Bernoulli given by $r^2 = \cos 2\theta$

6. Find the image of the circular sector $0 < \theta < \dfrac{\pi}{3}$, $0 < r < 2$ under the transformation $w = z^2 = r^2 e^{2i\theta}$.

$$0 < \phi < \dfrac{2\pi}{3},\ 0 < \rho < 4 \quad \text{where } w = \rho e^{i\phi}$$

7. Show that $w = z^2$ maps the line $x = c$ (where $c \neq 0$) **onto** the parabola $u = c^2 - \dfrac{v^2}{4c^2}$. **Hint.** $u + iv = x^2 - y^2 + 2xyi$

8. Let $w = f(z) = z^2$. Find the set of all points z such that $f(z)$ is in the rectangular region $2 < u < 4$, $2 < v < 4$ in the w-plane. See Figure 10.8.
$$\{(x, y) : \ 2 < x^2 - y^2 < 4 \text{ and } 1 < xy < 2\}$$

The Linear Fractional (Bilinear or Möbius) Transformations. The linear fractional transformation is given by

$$w = f(z) = \frac{az + b}{cz + d} \quad \text{where } \begin{vmatrix} a & b \\ c & d \end{vmatrix} \neq 0 \tag{10.23}$$

and where a, b, c, and d are complex constants.

Let $c = 0$. Then $ad - bc \neq 0$ implies $ad \neq 0$. Thus

$$w = \frac{a}{d}z + \frac{b}{d}, \tag{10.24}$$

which is a linear transformation.

Let $c \neq 0$. By division,

$$w = A + B\,\frac{1}{cz + d} \quad \text{where} \quad A = \frac{a}{c} \quad \text{and} \quad B = \frac{bc - ad}{c}. \tag{10.25}$$

Thus if

$$g(z) = cz + d, \quad h(z) = \frac{1}{z} \quad \text{and} \quad k(z) = A + Bz, \tag{10.26}$$

then the transformation in (10.25) is given by $w = k(h[g(z)])$, which is the composite g followed by h and then by k. Since [by (10.22) and 10.2.1] each transformation in (10.26) maps circles and lines onto circles or lines, the same is true for the composite in (10.25). Thus [since this is also true for the mapping in (10.24)], we have the following result.

> The linear fractional transformation maps
>
> circles and lines onto circles or lines.

Remarks 10.2.4 (a) Equation (10.23) may be written as

$$c\,(wz) + dw - az - b = 0,$$

which is linear in z and linear in w. Hence transformation (10.23) is also called the **bilinear transformation**.

(b) The condition $ad - bc \neq 0$ in (10.23) holds if and only if w is not constant. To prove the "Only If Part," suppose

$$w = \frac{az + b}{cz + d} = k.$$

Then $az + b = kcz + kd$ which implies

$$a = kc \quad \text{and} \quad b = kd,$$

and hence $ad - bc = kcd - kcd = 0$, a contradiction.

To prove the "If Part," assume $ad - bc = 0$. Then $ad = bc$. Not both c and d are zero. If $c \neq 0$, then $b = \frac{ad}{c}$. Hence

$$w = \frac{az + \frac{ad}{c}}{cz + d} = \frac{a}{c},$$

which is a constant. Similarly, if $d \neq 0$, then $w = \frac{b}{d}$, which is a constant. Thus (in either case) w is constant, a contradiction.

It follows that a constant function is not a bilinear transformation.

(c) Observe that the inverse in (10.23) is given by

$$z = \frac{-dw + b}{cw - a} \quad \text{and} \quad \begin{vmatrix} -d & b \\ c & -a \end{vmatrix} = ad - bc \neq 0.$$

(d) In (10.23), it is understood that

$$f\left(-\frac{d}{c}\right) = \infty \quad \text{and} \quad f(\infty) = \frac{a}{c}.$$

To explain why we say $f(\infty) = \frac{a}{c}$, we first notice that

$$\lim_{z \to \infty} \frac{az + b}{cz + d} = \lim_{z \to \infty} \frac{a + \frac{b}{z}}{c + \frac{d}{z}} = \frac{a}{c}.$$

Next, we notice that there is no z in R^2 such that $f(z) = \frac{a}{c}$ (since $ad - bc \neq 0$). Thus we complete the rule for determining the image of z under the bilinear map f in (10.23) by prescribing that $f(\infty) = \frac{a}{c}$. Hence we may record the explicit rule for the bilinear map f as

$$f(z) = \begin{cases} \dfrac{az + b}{cz + d} & \text{if } z \in R^2 \\[2mm] \dfrac{a}{c} & \text{if } z = \infty. \end{cases}$$

On the other hand, $f\left(-\frac{d}{c}\right)$ is determined from the rule in (10.23) by using the operations defined in 10.1.4.

(e) In view of (c) and (d), each bilinear transformation is a one-to-one mapping of R_∞^2 onto R_∞^2.

(f) If $c = 0$ in (10.23), then $w = \frac{a}{d} z + \frac{b}{d}$ (a linear transformation). But if $c \neq 0$, then

$$w = \frac{a}{c} - \frac{1}{c} \frac{ad - bc}{cz + d}.$$

Thus any bilinear transformation may be written in one of the forms

$$w = Az + B \quad \text{where } A \neq 0 \tag{10.27}$$

or

$$w = A + B \frac{1}{z + C} \quad \text{where } B \neq 0. \tag{10.28}$$

Theorem 10.2.5 If z_1, z_2, and z_3 are distinct points in R^2 and if w_1, w_2, and w_3 are distinct points in R^2, then there is a unique bilinear transformation f such that $f(z_k) = w_k$ for $k = 1, 2, 3$.

Proof. Let $w = f(z)$ be the transformation determined by the equation

$$\frac{w - w_1}{w_2 - w_1} \frac{z - z_3}{z_2 - z_3} = \frac{w - w_3}{w_2 - w_3} \frac{z - z_1}{z_2 - z_1}. \tag{10.29}$$

By putting $z = z_k$ in (10.29), we see that

$$f(z_k) = w_k \quad \text{for } k = 1, 2, 3. \tag{10.30}$$

[In particular, if $z = z_2$, we may cancel the common factors in (10.29) and obtain

$$(w - w_1)(w_2 - w_3) = (w - w_3)(w_2 - w_1),$$

which holds iff $w = w_2$.] Solving (10.29) for w, we obtain the form

$$w = \frac{az + b}{cz + d} = f(z).$$

Since w_1, w_2, and w_3 are distinct, (10.30) implies that $f(z)$ is not constant. Hence $ad - bc \neq 0$ by 10.2.4(b). Thus $w = f(z)$ is one bilinear transformation which maps z_k onto w_k.

Now suppose $w = g(z)$ is any bilinear transformation such that $g(z_k) = w_k$ for $k = 1, 2, 3$. Then $g(z)$ has the form of equation (10.27) or the form of equation (10.28). In case equation (10.28) holds, we have

$$w_k = A + \frac{B}{z_k + C} \quad \text{for } k = 1, 2, 3. \tag{10.31}$$

By (10.28) and (10.31), we have the following four equations.

$$w - w_1 = -\frac{B(z - z_1)}{(z + C)(z_1 + C)} \qquad w_2 - w_3 = -\frac{B(z_2 - z_3)}{(z_2 + C)(z_3 + C)}$$

$$\tag{10.32}$$

$$w - w_3 = -\frac{B(z - z_3)}{(z + C)(z_3 + C)} \qquad w_2 - w_1 = -\frac{B(z_2 - z_1)}{(z_2 + C)(z_1 + C)}$$

By (10.32), we have

$$\frac{w - w_1}{w_2 - w_1} \cdot \frac{w_2 - w_3}{w - w_3} = \frac{z - z_1}{z_2 - z_1} \cdot \frac{z_2 - z_3}{z - z_3}. \tag{10.33}$$

Now the last equation implies (10.29). Similarly, if $w = g(z)$ has the form of equation (10.27), we have the following four equations.

$$w - w_1 = A(z - z_1) \qquad w_2 - w_3 = A(z_2 - z_3)$$
$$w - w_3 = A(z - z_3) \qquad w_2 - w_1 = A(z_2 - z_1)$$

These equations give (10.33), which implies (10.29). Thus any bilinear transformation which maps z_k onto w_k for $k = 1, 2, 3$, must be the one given by (10.29).

Example 10.2.6 Find the bilinear transformation which maps 0, i, and $i+1$ onto $2i$, -1, and 0, respectively.

Solution. Substituting into (10.33), we have

$$\frac{w-2i}{w-0} \cdot \frac{-1-0}{-1-2i} = \frac{z-0}{z-(i+1)} \cdot \frac{i-(i+1)}{i-0}.$$

Solving for w, we obtain

$$w = \frac{-2z+(2+2i)}{(3i+1)z+(1-i)}. \tag{10.34}$$

To verify that the mapping in (10.34) is the correct one, we substitute the given values of z and obtain the corresponding values of w.

Remarks 10.2.7 If in R^2_∞, the points z_1, z_2, and z_3 are distinct and the points w_1, w_2, and w_3 are distinct, then there is a unique bilinear transformation $w = f(z)$ such that $f(z_k) = w_k$ for $k = 1, 2, 3$. Observe that

$$\lim_{z_3 \to \infty} \frac{z-z_1}{\frac{z}{z_3}-1} \cdot \frac{\frac{z_2}{z_3}-1}{z_2-z_1} = \frac{z-z_1}{z_2-z_1}.$$

Hence if $z_3 = \infty$ while w_1, w_2, and w_3 are finite, then equation (10.33) is replaced by

$$\frac{w-w_1}{w-w_3} \cdot \frac{w_2-w_3}{w_2-w_1} = \frac{z-z_1}{z_2-z_1}.$$

Similarly, if $w_k = \infty$ (or $z_k = \infty$), we obtain the transformation from (10.33) by excluding those factors involving w_k (or z_k). The uniqueness of the transformation follows just as in the proof of 10.2.5.

Example 10.2.8 Find the bilinear transformation which maps 0, 1, ∞ onto 1, ∞, 0, respectively.

Solution. Excluding the appropriate factors in (10.33), we have

$$\frac{w-1}{w-0} = \frac{z-0}{1-0} \quad \text{which implies} \quad w = \frac{1}{1-z}.$$

Exercises 10.2.9

1. Find the bilinear transformation which maps, respectively,

   (a) 0, $1+i$, -1 onto -1, $1-2i$, 0;

   (b) -1, ∞, 1 onto 2, 1, 0;

   (c) $2i$, ∞, i onto 4, 2, ∞.

$$\text{Ans. (a)} \quad w = \frac{z+1}{z-1} \quad \text{(b)} \quad w = \frac{z-1}{z} \quad \text{(c)} \quad w = \frac{2z}{z-i}$$

2. (a) Find the inverse transformation of $w = \dfrac{z+1}{z-1}$ $z = \dfrac{w+1}{w-1}$

   (b) Let $w = \frac{z-1}{z+1} = f(z)$ and let L be the line in the z-plane through $z = 0$
        and $z = 1 + i$. Find the image of L under f. **Hint.** First find the inverse
        transformation $z = f^{-1}(w)$, and then express x and y in terms of u and v.
$$u^2 + (v+1)^2 = 2$$

3. A point z is called a **fixed point** (or **invariant point**) under a transformation
    f if and only if $f(z) = z$. Find all fixed points of the transformation $w = \frac{z+1}{z-1}$
$$z = 1 \pm \sqrt{2}$$

Prove each statement in Ex. 4–6.

4. Every bilinear transformation has at least one fixed point in R^2_∞.

5. The identity mapping $w = z$ is the only bilinear mapping with more than two
    fixed points.

6. Any bilinear transformation having both zero and ∞ as fixed points, must be
    of the form $w = az$.

7. Use Ex. 6 to find the bilinear transformation which has zero and ∞ as fixed
    points and which maps $1 + i$ onto $2 + 3i$.

8. When is the transformation (10.23) the same as its inverse, which is given in
    10.2.4(c)?
$$d = -a \quad \text{or} \quad b = c = 0 \text{ and } d = \pm a \neq 0$$

10.3 Special Bilinear Transformations and Inverse Points

Theorem 10.3.1 A bilinear transformation $w = f(z)$ maps the extended half-plane
$H = \{z : \; \mathcal{I}(z) \geq 0 \text{ or } z = \infty\}$ onto the disk $D = \{w : \; |w| \leq 1\}$ if and only if the
function f has the form

$$f(z) = e^{i\alpha} \frac{z - p}{z - \overline{p}} \quad \text{where} \quad \alpha \in R \quad \text{and} \quad \mathcal{I}(p) > 0. \tag{10.35}$$

Proof of "If Part." Suppose (10.35) holds. Then for each real number $z = x$,
we have

$$|f(z)| = \frac{|x - p|}{|x - \overline{p}|} = 1 \qquad \text{since } \overline{x - p} = x - \overline{p}.$$

Thus the extended x-axis is mapped **onto** the circle $|w| = 1$.

If $z = (x, y)$ such that $y > 0$, then with $p = (p_1, p_2)$,

$$|f(z)| = \frac{\sqrt{(x - p_1)^2 + (y - p_2)^2}}{\sqrt{(x - p_1)^2 + (y + p_2)^2}} < 1 \quad \text{since } y > 0 \text{ and } p_2 > 0.$$

Thus the upper half-plane $y > 0$ is mapped **into** the interior of the unit circle $|w| = 1$.

Similarly, if $\mathcal{I}(z) = y < 0$, then $|f(z)| > 1$. Hence the lower half-plane $y < 0$ is mapped **into** the exterior of the circle $|w| = 1$.

By 10.2.4(e), we know that f maps **onto** R_∞^2. Since the lower half-plane goes into the exterior of $|w| = 1$, it follows that H must be mapped **onto** D.

Proof of "Only If Part." Let

$$w = f(z) = \frac{az + b}{cz + d} \tag{10.36}$$

be any bilinear transformation which maps H onto D and let L be the extended x-axis. Since $L \subset H$,

$$f(L) \subset f(H) = D. \tag{10.37}$$

Now $f(L)$ is a line or circle, which must be contained in the disk D by (10.37). But D contains no line. Thus

$$f(L) \text{ is some circle } K' \text{ contained in } D. \tag{10.38}$$

Now if K' were (partly) in the interior of D, then by the continuity of f, some points near L in the lower half-plane $y < 0$ would be mapped onto points in D. But this is impossible, since $f(H) = D$ and since f is one-to-one by 10.2.4(e). See Figure 10.9. Thus K' is not (partly) interior to D. Hence by (10.38),

$$f(L) = K \quad \text{where } K \text{ is the circle } |w| = 1. \tag{10.39}$$

Since $\infty \in L$, (10.39) shows that

$$1 = |f(\infty)| = \left|\frac{a}{c}\right| \quad \text{by 10.2.4(d).} \tag{10.40}$$

Thus $\frac{a}{c} = e^{i\alpha}$ for some α in R, and $ac \neq 0$. Hence equation (10.36) may be written as

$$w = f(z) = \frac{a}{c} \frac{z + \frac{b}{a}}{z + \frac{d}{c}} = e^{i\alpha} \frac{z - p}{z - q} \tag{10.41}$$

where

$$p = -\frac{b}{a} \quad \text{and} \quad q = -\frac{d}{c}.$$

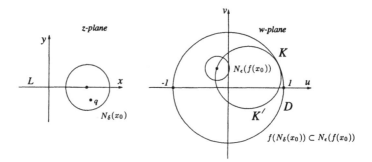

Figure 10.9:   If $f(x_0)$ were in D° for some real x_0 on L, then $N_\epsilon(f(x_0)) \subset D^\circ$ for some $\epsilon > 0$. Then $f(N_\delta(x_0)) \subset N_\epsilon(f(x_0))$ for some $\delta > 0$. So $f(q) \in N_\epsilon(f(x_0)) \subset D^\circ$ for some q below the x-axis in $N_\delta(x_0)$, contrary to the fact that $f(H) = D$.

By (10.39) and (10.41) (since $0 \in L$), we have $1 = |f(0)| = \left|\frac{p}{q}\right|$. Therefore,

$$|q| = |p|. \tag{10.42}$$

Also by (10.39) and (10.41) (since $1 \in L$), we have $1 = |f(1)| = \frac{|1-p|}{|1-q|}$. Thus $(1 - p)(1 - \overline{p}) = |1 - p|^2 = |1 - q|^2 = (1 - q)(1 - \overline{q})$. This gives

$$1 - 2\mathcal{R}(p) + |p|^2 = 1 - 2\mathcal{R}(q) + |q|^2,$$

and hence

$$\mathcal{R}(p) = \mathcal{R}(q) \quad \text{by (10.42)}. \tag{10.43}$$

Now by (10.42) and (10.43), it follows that $q = \overline{p}$.

Comparing (10.41) and (10.35), we see the proof will be complete if we show that $\mathcal{I}(p) > 0$. By hypothesis, f maps H **onto** D. Thus if z is below the x-axis, then $f(z)$ is not a point in D (since f is one-to-one). By (10.41),

$$f(p) = 0 \in D.$$

Hence p is not below the x-axis. So $\mathcal{I}(p) \geq 0$. But p cannot be on the x-axis since $f(L) = K$ by (10.39). Thus $\mathcal{I}(p) > 0$, and the proof is complete.

Exercises 10.3.2

1. Prove the bilinear transformation in (10.23) maps the extended real axis onto the extended real axis iff a, b, c, and d may be chosen to be real. **Hint.** To prove the "Only If Part," let z, z_2, z_3, w_1, w_2, and w_3 be real in 10.2.5 and use (10.33).

2. Prove if a bilinear transformation maps the half-plane $\mathcal{I}(z) > 0$ onto itself, then f maps $R \cup \{\infty\}$ onto itself. **Hint.** Let $x_0 \in R$ and let $w_0 = f(z_0)$. Suppose $\mathcal{I}(w_0) \neq 0$ and use the continuity of f to reach a contradiction.

3. Prove if in (10.23), the coefficients a, b, c, and d are real, then

$$\mathcal{I}(w) = \frac{(ad - bc)\, y}{(cx + d)^2 + c^2 y^2} \quad \text{where } z = x + iy.$$

4. Prove the bilinear transformation (10.23) maps the half-plane $\mathcal{I}(z) > 0$ onto the half-plane $\mathcal{I}(w) > 0$ iff a, b, c, and d may be taken to be real and $ad - bc > 0$. **Hints.** To prove the "Only If Part," use Ex. 1, 2, and 3 in that order. To prove the "If Part," use Ex. 3.

5. Let f be any bilinear transformation. Prove $f \in C(R^2_\infty)$, f is one-to-one, and $f^{-1} \in C(R^2_\infty)$. **Hints.** f^{-1} is given in 10.2.4(c). This shows that f is one-to-one. See 2.2.13 and Section 10.1.

6. Show that the determinant associated with the transformation (10.35) is not zero.

7. Find the point on the circle $|w| = 1$ which is the image of $z = \infty$ under the transformation (10.35).

Definition 10.3.3 Let K be the circle with center at z_0 and radius $r > 0$ and let p and q be points in R^2. Then p and q are **inverse points** with respect to K (or **inverse in** K) iff

(i) $\operatorname{Arg}(p - z_0) = \operatorname{Arg}(q - z_0)$ and

(ii) $|p - z_0||q - z_0| = r^2$.

Also, the center z_0 and ∞ are inverse points with respect to K (or inverse in K).

Remarks 10.3.4 (a) Finite points p and q are inverse with respect to a circle iff they are on a ray with endpoint at the center of the circle and the product of their distances from this center is equal to the square of the radius of the circle.

(b) We see that (i) and (ii) hold iff

$$p = z_0 + \lambda e^{i\alpha} \quad \text{and} \quad q = z_0 + \frac{r^2}{\lambda} e^{i\alpha} \tag{10.44}$$

for some $\lambda > 0$ and some α in R.

(c) Thus finite points p and q are inverse points with respect to the circle

$$|z - z_0|^2 = r^2 \quad \text{or} \quad (z - z_0)(\overline{z} - \overline{z_0}) = r^2 \tag{10.45}$$

if and only if

$$(p - z_0)(\overline{q} - \overline{z_0}) = r^2. \tag{10.46}$$

[Clearly, (10.44) implies (10.46). Conversely, suppose (10.46) holds. Let $p - z_0 = \lambda e^{i\alpha}$ where $\lambda > 0$. Then by (10.46), we see that $\overline{q} - \overline{z_0} = \frac{r^2}{\lambda} e^{-i\alpha}$ and hence $q - z_0 = \frac{r^2}{\lambda} e^{i\alpha}$. Thus (10.46) implies (10.44).]

(d) It is clear that a point p is its own inverse in a circle K iff p is a point on K.

Now (10.45) may be written as

$$z\overline{z} - \overline{z_0} z - z_0 \overline{z} + |z_0|^2 - r^2 = 0$$

or as

$$z\overline{z} + \overline{b}z + b\overline{z} + c = 0, \tag{10.47}$$

where

$$b = -z_0 \quad \text{and} \quad c = |z_0|^2 - r^2 < |z_0|^2 = |b|^2. \tag{10.48}$$

Conversely, if $c < |b|^2$, then (10.47) is an equation of the circle with center at $-b$ and radius $\sqrt{|b|^2 - c}$. We see that (10.46) holds iff

$$p\overline{q} + \overline{b}p + b\overline{q} + c = 0 \tag{10.49}$$

where b and c are given in (10.48). Thus from 10.3.4(c), we have the following result.

Theorem 10.3.5 Finite points p and q are inverse points with respect to the circle given by equation (10.47) iff equation (10.49) holds [which is the result of replacing z by p and $\overline{z}$ by $\overline{q}$ in (10.47)].

Theorem 10.3.6 Let $w = f(z)$ be a bilinear transformation which maps a circle K onto a circle K'. If p and q are inverse points with respect to K, then $f(p)$ and $f(q)$ are inverse points with respect to K'.

Proof. By 10.2.4(f), any bilinear transformation is a composite of transformations of the following types.

(i)   translation $w = z + B$
(ii) magnification and rotation $w = Az$ where $A \neq 0$
(iii) reciprocal transformation $w = \frac{1}{z}$

[To justify the last statement, we may also see the discussion which includes the displayed equations (10.24) – (10.26).] Thus the proof of Theorem 10.3.6 will be complete when it is proved that the second statement in the theorem holds for all transformations of types (i)–(iii). See 10.3.7.

Suppose f is of type (i). Then f preserves distances, since $|(z+B)-(z_0+B)| = |z-z_0|$. Let z_0 be the center of K. Then $f(z_0)$ is the center of K' (since f preserves distances), and $f(\infty) = \infty$. Now $|f(p)-f(z_0)| = |p-z_0|$ and $|f(q)-f(z_0)| = |q-z_0|$. Also, $\text{Arg}\,[(p+B)-(z_0+B)] = \text{Arg}\,(p-z_0)$ and $\text{Arg}\,[f(q)-f(z_0)] = \text{Arg}\,(q-z_0)$. Thus by the definition of inverse points, the second statement in 10.3.6 holds for each translation f.

Now suppose f is of the type (iii). Thus

$$w = f(z) = \frac{1}{z} \tag{10.50}$$

and let K be given by (10.47). By hypothesis, K' (the image of K) is a circle, not a line. Thus, by using equations (10.20) and (10.21), we see that K cannot go through the origin. Hence $c \neq 0$ in (10.47). Now $z = \frac{1}{w}$ by (10.50). Thus, substituting $\frac{1}{w}$ for z in (10.47), an equation of K' is

$$w\,\overline{w} + \frac{b}{c}\,w + \frac{\overline{b}}{c}\,\overline{w} + \frac{1}{c} = 0, \tag{10.51}$$

and in view of (10.48), the center of K' is

$$-\frac{\overline{b}}{c} \quad \text{(the negative of the coefficient of } \overline{w}\text{).} \tag{10.52}$$

Case 1. Suppose p and q are finite and $pq \neq 0$. Since p and q are inverse points with respect to K, we have

$$p\overline{q} + \overline{b}p + b\overline{q} + c = 0 \quad \text{by 10.3.5.}$$

Dividing by $p\overline{q}c$, we obtain

$$\frac{1}{c} + \frac{\overline{b}}{c}\,\overline{\left(\frac{1}{q}\right)} + \frac{b}{c}\frac{1}{p} + \frac{1}{p}\,\overline{\left(\frac{1}{q}\right)} = 0.$$

Comparing the last equation with the equation of the circle K' given in (10.51), we see that $\frac{1}{p}$ and $\frac{1}{q}$ are inverse points with respect to K' by 10.3.5.

Case 2. Suppose $pq = 0$. [Not both p and q are 0 by 10.3.4(d) since 0 is not on K.] Let us name the points so that $q = 0$ and hence $p \neq 0$. Then by 10.3.5, we put $\overline{q} = q = 0$ in (10.49) and obtain

$$p = -\frac{c}{\overline{b}}.$$

Thus $f(p) = \frac{1}{p} = -\frac{\overline{b}}{c}$, which is the center of K' by (10.52). But $f(q) = f(0) = \infty$. Thus $f(p)$ and $f(q)$ are inverse points with respect to K' by Definition 10.3.3.

Case 3. Suppose $pq = \infty$. Then at least one of the points p or q must be ∞. We choose the labeling so that $p = \infty$. Then q must be the center of the circle K by the last part of 10.3.3. Hence $q \neq \infty$. In our present case, $pq = \infty \cdot q = \infty$. Thus $q \neq 0$ since $\infty \cdot 0$ is not defined in 10.1.4.

Since q is the center of K, we have $b = -q$ by (10.48). Thus the equation of K' given in (10.51) becomes

$$w\,\overline{w} - \frac{q}{c}\,w - \frac{\overline{q}}{c}\,\overline{w} + \frac{1}{c} = 0. \tag{10.53}$$

Now $f(p) = f(\infty) = 0$ and $f(q) = \frac{1}{q}$. We see that (10.53) holds with $w = 0$ and $\overline{w} = \overline{\left(\frac{1}{q}\right)}$. Thus 0 and $\frac{1}{q}$ are inverse points with respect to K' by 10.3.5.

Case 4. The only remaining possibility is that one of the points is ∞ and the other is 0. We choose the labeling so that $p = \infty$ and $q = 0$ then 0 is the center of K by Definition 10.3.3. Thus K is the circle given by $|z| = r$ for some $r > 0$. Hence $K' = f(K)$ in the circle given by $|w| = \frac{1}{r}$. So the center of K' is the point $0 = f(\infty) = f(p)$. Also, $f(q) = f(0) = \frac{1}{0} = \infty$. Thus $f(p)$ and $f(q)$ are inverse points with respect to K' by the definition of inverse points.

The proof for a magnification and rotation is left as an exercise.

Remarks 10.3.7 In the proof of 10.3.6, it is important to observe that transformations of types (i) and (ii) always map circles onto circles and always map lines onto lines. However, $w = \frac{1}{z}$ maps some circles onto lines and vice versa. But any bilinear transformation is a composite of transformations of types (i), (ii), and (iii) where **at most one** is the reciprocal transformation. In 10.3.6, we are given by hypothesis that $f(k)$ is a circle K'. If an intermediate reciprocal transformation were to map an image of K onto a line L, then the succeeding transformations would map L onto a line—not a circle K'. Thus in our proof for $w = \frac{1}{z}$, we need only consider the case where a circle maps onto a circle.

Theorem 10.3.8 A bilinear transformation maps the interior of the circle $|z| = r$ onto the interior of the circle $|w| = r$ iff it is of the form

$$w = r^2\, e^{i\alpha}\, \frac{z - p}{\overline{p}z - r^2} \tag{10.54}$$

where α is real and $|p| < r$.

Proof of "Only If Part." Let $w = f(z)$ be any bilinear transformation which maps the interior of the circle

$$K = \{z : \ |z| = r\}$$

onto the interior of the circle

$$K' = \{w : \ |w| = r\}.$$

Let $z \in K$. Since f is one-to-one, $f(z) \notin I(K')$. Since $f \in C(R_\infty^2)$, it follows that $f(z) \notin E(K')$, where it is understood that ∞ is considered to be in the exterior of each circle. Thus $f(z) \in K'$, and hence $f(K) = K'$.

Now let $p \in I(K)$ such that

$$f(p) = 0, \quad \text{the center of } K'. \tag{10.55}$$

Case 1. Suppose $p \neq 0$. Let q be the inverse of p in K. By (10.46), we have $q = \frac{r^2}{\bar{p}}$. Thus $f\left(\frac{r^2}{\bar{p}}\right)$ must be inverse to $f(p)$ in K' by 10.3.6. Hence by (10.55), we have

$$f\left(\frac{r^2}{\bar{p}}\right) = \infty. \tag{10.56}$$

Now

$$
\begin{aligned}
f(z) &= \frac{az+b}{cz+d} \quad \left[\begin{array}{ll} a \neq 0; \text{ for if } a = 0, \text{ then } f(\infty) = 0 = f(p) & \text{by (10.55)} \\ c \neq 0; \text{ for if } c = 0, \text{ then } f(\infty) = \infty = f\left(\frac{r^2}{\bar{p}}\right) & \text{by (10.56)} \end{array}\right] \\
&= \frac{a}{c} \frac{z-p}{z - \frac{r^2}{\bar{p}}} \quad \text{by (10.55) and (10.56)} \\
&= B \frac{z-p}{\bar{p}z - r^2} \quad \text{where } B = \frac{\bar{p}a}{c}. \tag{10.57}
\end{aligned}
$$

Since $r \in K$ and $f(K) = K'$, we have $|f(r)| = r$. Hence by (10.57),

$$r = |f(r)| = \left| B \frac{r-p}{\bar{p}r - r^2} \right| = \frac{|B|}{r} \frac{|r-p|}{|\bar{p}-r|} = \frac{|B|}{r}$$

or $|B| = r^2$. Thus $B = r^2 e^{i\alpha}$ for some real number α. Hence equation (10.57) has the form in (10.54).

Case 2. Suppose $p = 0$. Now $p = 0$ and ∞ are inverse in K. Thus $f(p) = 0$ and $f(\infty)$ are inverse in K' by 10.3.6 and (10.55). Hence $f(\infty) = \infty$. Thus by Ex. 6 in 10.2.9, we know that f must have the form

$$f(z) = Az. \tag{10.58}$$

Since $r \in K$ and $f(K) = K'$, $r = |f(r)| = |Ar|$. Thus $|A| = 1$, that is, $A = e^{i\beta}$. Hence equation (10.58) is in the form of equation (10.54).

Proof of "If Part." Suppose equation (10.54) holds. If $|z| = r$, then

$$|w| = r^2 \left| \frac{z-p}{\bar{p}z - \bar{z}z} \right| = \frac{r^2}{|-z|} \left| \frac{z-p}{\bar{z} - \bar{p}} \right| = r \left| \frac{z-p}{\bar{z} - \bar{p}} \right| = r.$$

Thus the circle $K = \{z : \ |z| = r\}$ is mapped onto the circle $K' = \{w : \ |w| = r\}$. By hypothesis, $|p| < r$. Thus $p \in I(K)$. Since $f(p) = 0 \in I(K')$, it follows that $f[I(K)] = I(K')$ by Ex. 3 in 10.3.9.

Exercises 10.3.9

1. Let $f(z) = Az$ where $A \neq 0$ and let K be a circle with center at z_0.

   (a) Using the definition of a circle, prove that $f(K)$ is a circle with center at $Az_0 = f(z_0)$.

   (b) Using Definition 10.3.3, prove directly that if p and q are inverse in K, then $f(p)$ and $f(q)$ are inverse in $f(K)$.

2. Let K be the circle or line given by (10.20) where $a \geq 0$. The complementary domains of K are defined to be

   (i) $I(K)$ and $E(K)$ if K is a circle, where it is understood that $\infty \in E(K)$,

   (ii) the two domains (half-planes) whose union is $R^2 - K$ if K is a line.

   Let $f(z) = \frac{1}{z}$ and let $K' = f(K)$ as given in (10.21). Prove that f maps one of the complementary domains of K **onto** one of the complementary domains of K' and that the other complementary domain of K is mapped **onto** the remaining complementary domain of K'. **Hint. Case 1.** $a > 0, d < 0$. Then $a\left(x^2 + y^2\right) + bx + cy + d < 0$ represents $I(K)$. By (10.18) and (10.19), it follows that $d\left(u^2 + v^2\right) + bu - cv + a < 0$ or $u^2 + v^2 + \frac{b}{d}u - \frac{c}{d}v + \frac{a}{d} > 0$ (since $d < 0$). Thus by (10.21), we see that $f[I(K)] \subset E(K')$. Similarly $f[E(K)] \subset I(K')$. Hence $f[I(K)] = E(K')$ and $f[E(K)] = I(K')$ since f maps onto R^2_∞. The other cases are similar. In fact all cases may be considered together.

3. Let K and K' be given as in Ex. 2 and let $f(z)$ be any bilinear transformation. Prove f maps one of the complementary domains of K **onto** one of the complementary domains of K' and the other complementary domain of K is mapped onto the remaining complementary domain of K'. **Hint.** Clearly the statement is true for a translation and a magnification and rotation. But f is a composite of transformations of types (i), (ii), and (iii) in the proof of 10.3.6.

4. Let $p, q \in R^2$ such that $p \neq q$ and let L be the perpendicular bisector of the line joining p and q. Prove that an equation of L is

   (A) $\overline{\beta}z + \beta\overline{z} + \gamma = 0$ where $\beta = q - p$ and $\gamma = |p|^2 - |q|^2$.

   Hint. L is given by $|z - p|^2 = |z - q|^2$ or by

(B) $(z - p)(\bar{z} - \bar{p}) = (z - q)(\bar{z} - \bar{q})$.

5. Let L be the (extended) line given by

(C) $\bar{b}z + b\bar{z} + c = 0$ where c is real and $b \neq 0$.

Points p and q are said to be inverse in L iff p and q are symmetric with respect to L (that is, $p = q \in L$, or L is the perpendicular bisector of the segment from p to q). Prove that finite points p and q are inverse in L iff

(D) $\bar{b}p + b\bar{q} + c = 0$.

Hint. Suppose p and q are inverse in L. If $p = q \in L$, then (D) is the result of putting $z = p$ in (C). If $p \neq q$, then (A) is an equation of L. So (C) is a "constant multiple" of (A). Thus observe that (B) holds if we replace z by p and replace $\bar{z}$ by $\bar{q}$. Thus (A), and hence (C) holds, if we replace z by p and $\bar{z}$ by $\bar{q}$. This gives (D). Conversely, let (D) hold. Let p^* be inverse to p in L. By the foregoing proof,

(E) $\bar{b}p + b\overline{p^*} + c = 0$.

Now (D) and (E) imply $q = p^*$.

6. Let $w = f(z) = \dfrac{1}{z}$, and let K be the circle (or line if $\alpha = 0$) given by

(F) $\alpha z\bar{z} + \bar{\beta}z + \beta\bar{z} + \gamma = 0$ where α and γ are real.

Prove that if p and q are inverse in K, then $f(p)$ and $f(q)$ are inverse in $f(K)$.
Hint.

Case 1. Suppose p and q are finite and $pq \neq 0$. Since p and q are inverse in K, we have

(G) $\alpha p\bar{q} + \bar{\beta}p + \beta\bar{q} + \gamma = 0$ $\begin{cases} \text{by 10.3.5} & \text{if } \alpha \neq 0 \\ \text{by Ex. 5} & \text{if } \alpha = 0. \end{cases}$

Now (G) implies $\alpha + \bar{\beta}\dfrac{1}{\bar{q}} + \beta\dfrac{1}{p} + \gamma\dfrac{1}{p}\cdot\dfrac{1}{\bar{q}} = 0$. This implies $f(p)$ and $f(q)$ are inverse in $f(K)$ by 10.3.5 if $\gamma \neq 0$ (by Ex. 5 if $\gamma = 0$) since an equation of $f(K)$ is

(H) $\alpha + \bar{\beta}\bar{w} + \beta w + \gamma w\bar{w} = 0$.

Case 2. Suppose $p = 0$ and q is finite and $q \neq 0$. Then $f(0) = \infty$ and we must show that $f(K)$ is a circle with center at $f(q)$. Since p and q are inverse in K, by (F), 10.3.5, and Ex. 5,

$$\beta\bar{q} + \gamma = 0, \quad \text{that is,} \quad \bar{q} = -\frac{\gamma}{\beta} \text{ and } \beta \neq 0.$$

(If $\beta = 0$, then $\gamma = 0$ and K would not be a circle or a line.) Hence

$$f(q) = \frac{1}{q} = -\frac{\overline{\beta}}{\gamma}.$$

Now see (10.52) and (H).

Case 3. Suppose $p = q$. Then $p, q \in K$ and $f(p) = f(q) \in f(K)$. By definition, $f(p)$ and $f(q)$ are inverse points in $f(K)$.

Case 4. Suppose $p = \infty$ and q is finite. Then K must be a circle centered at q. Thus by (10.47) and (10.48), K is given by

$$z\overline{z} - \overline{q}z - q\overline{z} + c = 0.$$

Thus $f(K)$ is given by

$$cw\overline{w} - \overline{q}\,\overline{w} - qw + 1 = 0.$$

Now suppose $q \neq 0$. Then $f(p) = f(\infty) = 0$ and $f(q) = \frac{1}{q}$. Replacing w by $f(p) = 0$ and $\overline{w}$ by $\overline{f(q)} = \frac{1}{\overline{q}}$, we have $-\overline{q} \cdot \frac{1}{\overline{q}} + 1 = 0$. Thus by 10.3.5 (if $c \neq 0$) or by Ex. 5 (if $c = 0$), we see that $f(p)$ and $f(q)$ are inverse in $f(K)$.

But if $q = 0$, then K is a circle centered at $z = 0$ and $f(K)$ is a circle centered at $w = 0$. Also $f(q) = \infty$ and $f(p) = 0$, the center of $f(K)$. Hence $f(p)$ and $f(q)$ are inverse in $f(K)$.

7. Let $w = f(z)$ be any bilinear transformation and let K be a circle or a line. Prove if p and q are inverse in K, then $f(p)$ and $f(q)$ are inverse in $f(K)$.

8. Let $w = F(z) = z^m$ where $0 < m < 1$. Prove that F is a one-to-one mapping of the upper half-plane $H = \{z : \ \mathcal{I}(z) > 0\}$ onto the angular region $A = \{w : 0 < \text{Arg}\, w < m\pi\}$. **Hint.** $F(z) = |z|^m e^{im\theta}$ where $0 < \theta = \text{Arg}\, z < \pi$.

9. Let $F(z) = \text{Log}\, z$ and let $H = \{z : \ \mathcal{I}(z) > 0\}$. Show that $F(H) = \{w : 0 < \mathcal{I}(w) < \pi\}$. **Hint.** $\text{Log}\, z = \text{Log}\, |z| + i \,\text{Arg}\, z$.

10.4 Transcendental Mappings

A function $w = f(z)$ is called an **algebraic function** if it is a solution of an equation of the form

$$P(z, w) = 0$$

where $P(z, w)$ is a polynomial in z and w of degree at least **one** in w. Here, $P(z, w)$ may be written in the form

$$P(z, w) = \sum_{k=0}^{n} [a_k(z)] w^k,$$

where $n \geq 1$, each $a_k(z)$ is a polynomial in z, and $a_n(z)$ is not identically 0.

A function which is not algebraic is said to be **transcendental**.

We see that each rational function is algebraic. Also if $n \in N$, then the function $w = r^{\frac{1}{n}} e^{\frac{i\theta}{n}}$ (where $0 \leq \theta < 2\pi$) is algebraic since $w^n - z = 0$.

The Exponential Function e^z. The function e^z is transcendental. (See Ex. 8 in 10.4.1.) Let

$$w = f(z) = e^z = e^x e^{iy} \quad \text{for each } z \text{ in } R^2. \tag{10.59}$$

For this function,

$$\rho = |w| = e^x \quad \text{and} \quad \phi = \arg w = y. \tag{10.60}$$

Hence

$$\text{the line } x = a \text{ is mapped onto the circle } |w| = e^a \tag{10.61}$$

and

$$\text{the line } y = b \text{ is mapped onto the ray } \phi = b \text{ with } \rho > 0. \tag{10.62}$$

In particular the x-axis ($y = 0$) is mapped **onto** the ray $\phi = 0$ with $\rho > 0$ while the line $y = \pi$ is mapped **onto** the ray $\phi = \pi$ with $\rho > 0$. Now as b increases from 0 to π, the ray $\phi = b$ rotates through the upper half-plane. This means that the horizontal strip

$$-\infty < x < +\infty \quad \text{and} \quad 0 \leq y \leq \pi$$

is mapped **onto** the upper half-plane (with the origin deleted)

$$\rho > 0 \quad \text{and} \quad 0 \leq \phi \leq \pi.$$

See Figure 10.10.

Similiarly, the strip $-\infty < x < +\infty$ and $-\pi < y < 0$ is mapped **onto** the lower half-plane $\rho > 0$ and $-\pi < \phi < 0$. Thus f maps the strip

$$\{z : \quad -\pi < y \leq \pi\} \tag{10.63}$$

onto $R^2 - \{0\}$.

Now for each integer k, we have

$$w = f(z + 2k\pi i) = e^z \cdot e^{2k\pi i} = e^z = f(z) \quad \text{for each } z \text{ in } R^2.$$

Thus for each integer k, the strip

$$\{z : \quad (2k-1)\pi < y \leq (2k+1)\pi\} \tag{10.64}$$

is mapped **onto** $R^2 - \{0\}$ in the same way as the strip in (10.63). For this reason, the strip in (10.63) is called the **fundamental strip**.

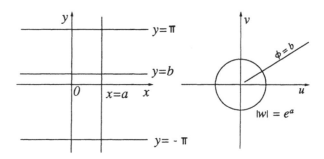

Figure 10.10: The line $x = a$ is mapped onto the circle $|w| = e^a$.
The line $y = b$ is mapped onto the ray $\phi = b$ with
$\rho > 0$.

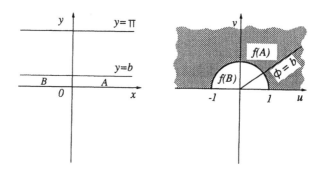

Figure 10.11: $f(A) = \{w : \ |w| \geq 1 \text{ and } \mathcal{I}(w) \geq 0\}$
$f(B) = \{w : \ 0 < |w| < 1 \text{ and } \mathcal{I}(w) \geq 0\}$

Exercises 10.4.1

In Ex. $1-7$, let $w = f(z) = e^z$ for each z in R^2.

1. If $z = (x, b)$, what part of the ray $\phi = b$ is traversed by $f(z)$ when x varies

   (a) from zero to $+\infty$,

   (b) from zero to $-\infty$?

   Hint. $|f(z)| = e^x$.

2. Use Ex. 1 to find $f(A)$ and $f(B)$ where A and B are the semi-infinite strips given by

$$A = \{(x,y): \ 0 \le x < +\infty \ \text{ and } \ 0 \le y \le \pi\}$$
$$B = \{(x,y): \ -\infty < x < 0 \ \text{ and } \ 0 \le y \le \pi\}.$$

$$\text{Ans.} \ f(A) = \{w: \ |w| \ge 1 \text{ and } \mathcal{I}(w) \ge 0\}$$
$$f(B) = \{w: \ |w| < 1, \ \mathcal{I}(w) \ge 0 \text{ and } w \ne 0\}$$

   Hint. Let b vary from 0 to π. See Figure 10.11.

3. Find $f(S)$ where S is the segment $\ x = a \ $ and $\ 0 \le y \le \pi$.

4. In Ex. 3, let a vary from zero to $-\infty$ to find $f(B)$ where B is given in Ex. 2.

5. Find $f(C)$ where C is the strip $-\infty < x < +\infty \ $ and $\ c \le y \le d$. [See Ex. 1 and let b vary from c to d.]

6. Find $f(D)$ where D is the strip $a \le x \le b \ $ and $\ -\infty < y < +\infty$. **Hint.** See (10.61) and let x vary from a to b.

7. Use Ex. 5 and 6 to find $f(E)$ where E is the rectangular region

$$a \le x \le b \ \text{ and } \ c \le y \le d.$$

   See Figure 10.12.

8. Prove the function $w = f(z) = e^z$ is transcendental.

9. Let $w = F(z) = z^m$ where $0 < m < 1$. Prove F is a one-to-one mapping of the upper half-plane $y > 0$ onto the angular region $0 < \text{Arg}\, w < m\pi$. **Hint.** $F(z) = |z|^m \, e^{im\theta}$ where $0 < \theta = \text{Arg}\, z < \pi$.

10. Let $F(z) = \text{Log}\, z$ and let H be the half-plane $y > 0$. Show that $F(H) = \{w: \ 0 < \mathcal{I}(w) < \pi\}$. **Hint.** $\text{Log}\, z = \text{Log}\, |z| + i\, \text{Arg}\, z$.

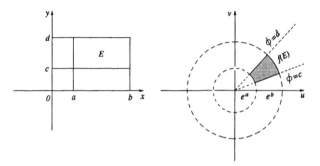

Figure 10.12: $f(E) = \{w : \ e^a \le |w| \le e^b \text{ and } c \le \phi \le d\}$

Solution of Ex. 8. Let $P(z, w)$ be **any** polynomial of the form

$$P(z, w) = a_0(z) + a_1(z)\, w + \cdots + a_n(z)\, w^n$$

where $n \ge 1$, $a_n(z) \not\equiv 0$, and where $a_k(z)$ is a polynomial in z for $k = 0, 1, \cdots, n$. Let z_0 be such that $a_n(z_0) \ne 0$. Then $P(z_0, w)$ is a polynomial in w of degree $n \ge 1$, and hence has at most n zeros by 8.2.9. Thus we let w_0 be such that $w_0 \ne 0$ and such that $P(z_0, w_0) \ne 0$. So $P(z, w_0)$ is a polynomial in z which is not identically zero. Hence $P(z, w_0)$ has only a finite number of zeros by 8.2.9. Since $f(z) = e^z$ maps each strip in (10.46) **onto** $R^2 - \{0\}$, there are infinitely many distinct values of z, say $z_1, z_2, z_3, \cdots$, such that

$$e^{z_j} = w_0 \qquad \text{for } j = 1, 2, 3, \cdots.$$

Since $P(z, w_0)$ has only a finite number of zeros, it follows that there is some z_m such that $P(z_m, w_0) = P(z_m, e^{z_m}) \ne 0$. Thus $P(z, e^z) \not\equiv 0$. Since $P(z, w)$ was **just any** polynomial of the type described, this means that $f(z) = e^z$ is not algebraic. Therefore, the exponential function is transcendental.

The Sine and Cosine Functions. Since $\sin(z + 2k\pi) = \sin z$, it is clear that $w = \sin z$, is not algebraic and hence is transcendental. See Ex. 8 in 10.4.1.

Example 10.4.2 Let $w = f(z) = \sin z$ and let S be the semi-infinite strip $|x| \le \frac{\pi}{2}$ and $y \ge 0$. Show that $f(S)$ is the half-plane H given by $v \ge 0$.

Solution. By 5.2.9,

$$w \ = \ f(z) \ = \ \sin x \ \cosh y \ + \ i \cos x \ \sinh y,$$

so that

$$u \ = \ \sin x \ \cosh y \quad \text{and} \quad v \ = \ \cos x \ \sinh y. \qquad (10.65)$$

From (10.65), the segment S_0 given by

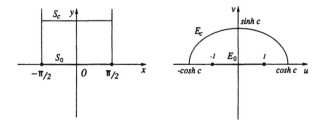

Figure 10.13: As c varies over $[0, +\infty)$, the image $f(S_c)$ generates H.

$$|x| \leq \frac{\pi}{2} \quad \text{and} \quad y = 0$$

is mapped onto the segment E_0 given by

$$|u| \leq 1 \quad \text{and} \quad v = 0.$$

See Figure 10.13. Also by (10.65), the segment S_c given by

$$|x| \leq \frac{\pi}{2} \quad \text{and} \quad y = c \quad \text{where } c > 0$$

is mapped onto the semiellipse E_c given by

$$\frac{u^2}{\cosh^2 c} + \frac{v^2}{\sinh^2 c} = 1 \quad \text{and} \quad v \geq 0. \tag{10.66}$$

Now as c varies over the positive reals, S_c sweeps out the strip

$$S = \{(x, y) : \ |x| \leq \frac{\pi}{2} \ \text{and} \ 0 < y < +\infty\}$$

and E_c sweeps out the half-plane $H - E_0$. To see this, let

$$(u_0, v_0) \in (H - E_0).$$

Then there is a semiellipse

$$\frac{u^2}{a^2} + \frac{v^2}{b^2} = 1 \quad \text{with} \quad v \geq 0 \tag{10.67}$$

through (u_0, v_0) with foci at $(\pm 1, 0)$. From analytic geometry,

$$a > 1 \quad \text{and} \quad b = \sqrt{a^2 - 1}.$$

Since $a > 1$, let $c_0 > 0$ such that $\cosh c_0 = a$. But $\cosh^2 c_0 - \sinh^2 c_0 = 1$ and $c_0 > 0$, so that

$$\sinh c_0 = \sqrt{\cosh^2 c_0 - 1} = b.$$

Thus the semiellipse in (10.67) is E_{c_0} in (10.66). Hence $f(S) = H$.

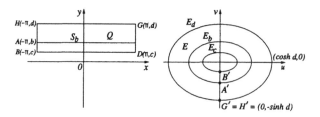

Figure 10.14: $f(S_b) = E_b$

Example 10.4.3 Let $w = f(z) = \sin z$, let $0 < c < d$, and let Q be the closed rectangular region $-\pi \le x \le \pi$ and $c \le y \le d$. Show that $f(Q)$ is the closed region E bounded by the confocal ellipses

$$\frac{u^2}{\cosh^2 c} + \frac{v^2}{\sinh^2 c} = 1 \quad \text{and} \quad \frac{u^2}{\cosh^2 d} + \frac{v^2}{\sinh^2 d} = 1. \qquad (10.68)$$

See Figure 10.14.

Solution. Let $z \in S_b$ where S_b is the segment given by $y = b$ and $-\pi \le x \le \pi$. Then by (10.65),

$$u = \cosh b \, \sin x \quad \text{and} \quad v = \sinh b \, \cos x \quad \text{with} -\pi \le x \le \pi, \qquad (10.69)$$

which are parametric equations of an ellipse E_b. When $z \in S_b$, we see by (10.69) that u and v vary as indicated in the following table.

as x increases from	u	v
$-\pi$ to $-\frac{\pi}{2}$	decreases from 0 to $-\cosh b$	increases from $-\sinh b$ to 0
$-\frac{\pi}{2}$ to 0	increases from $-\cosh b$ to 0	increases from 0 to $\sinh b$
0 to $\frac{\pi}{2}$	increases from 0 to $\cosh b$	decreases from $\sinh b$ to 0
$\frac{\pi}{2}$ to π	decreases from $\cosh b$ to 0	decreases from 0 to $-\sinh b$

Thus as $z = x + ib$ moves along $y = b$ from $x = -\pi$ to $x = \pi$, its image $\sin z$ traces the complete ellipse E_b clockwise from $A' = (0, -\sinh b)$. Now as b varies from c to d, the segment S_b sweeps out the rectangular region Q and E_b sweeps out the elliptic region E. Thus $f(Q) = E$. See Figure 10.14.

Remarks 10.4.4 By (10.65), we have the following results for $\sin z$.

as z moves along	$\sin z$ traces
side BD from B to D	ellipse E_c clockwise
side DG from D to G	segment B′ G′ from B′ to G′
side GH from G to H	ellipse E_d counterclockwise
side HB from H to B	segment H′ B′ from H′ to B′

Furthermore, the interior of rectangle BDGH is mapped onto the set consisting of the open elliptic ring bounded by the ellipses E_c and E_d given in (10.68) with segment B′G′ deleted.

The transformation

$$w = \cos z = \sin\left(z + \frac{\pi}{2}\right)$$

is the composite

$$w = f[g(z)] \quad \text{where} \quad g(z) = z + \frac{\pi}{2} \quad \text{and} \quad f(z) = \sin z. \tag{10.70}$$

Thus $w = \cos z$ consists of a translation of the z-plane $\frac{\pi}{2}$ units to the right followed by the transformation $f(z) = \sin z$.

Similarly the transformation $w = \sinh z = -i \sin iz$ may be studied as the composite $h \circ g \circ f$ where

$$f(z) = iz, \quad g(z) = \sin z, \quad \text{and} \quad h(z) = -iz.$$

We notice that f is a rotation through $90°$ and h is a rotation though $-90°$.

Exercises 10.4.5

1. Let $|c| \leq \frac{\pi}{2}$.

   (a) Use (10.65) to show that $w = f(z) = \sin z$ maps the line L given by $x = c$ onto the curve

   $$u = \sin c \cosh y \quad \text{and} \quad v = \cos c \sinh y \quad \text{where} \quad -\infty < y < +\infty.$$

   (b) Show that $f(L)$ is the set described as follows.

   the right-hand half of the hyperbola
   $$\frac{u^2}{\sin^2 c} - \frac{v^2}{\cos^2 c} = 1 \qquad \text{if } 0 < c < \frac{\pi}{2}$$
   the left-hand half of this hyperbola $\quad$ if $-\frac{\pi}{2} < c < 0$
   the v-axis $\qquad\qquad\qquad\qquad\qquad$ if $c = 0$
   the ray $\quad v = 0 \quad$ and $\quad u \geq 1 \qquad$ if $c = \frac{\pi}{2}$
   the ray $\quad v = 0 \quad$ and $\quad u \leq -1 \qquad$ if $c = -\frac{\pi}{2}$

2. Let $0 < a < b < \frac{\pi}{2}$ and let $0 < c < d$. Find the image under $w = \sin z$ of the rectangular region $a \le x \le b$ and $c \le y \le d$.

3. Use (10.70) and 10.4.3 to find the image under $w = \cos z$ of the rectangular region $-\frac{3\pi}{2} \le x \le \frac{\pi}{2}$ and $1 \le y \le 4$.

4. Let $f(z) = \cosh z$ and let A be the rectangular region $1 \le x \le 4$ and $-\frac{\pi}{2} \le y \le \frac{3\pi}{2}$. Find $f(A)$. **Hint.** Since $\cosh z = \cos iz$, we have $f(z) = g[h(z)]$ where $h(z) = iz$ and $g(z) = \cos z$. Note that $h(A)$ is the rectangular region given in Ex. 3.

10.5   Conformal Mapping

Definitions 10.5.1 (a) Let z be a point of intersection of arcs C_1 and C_2 in R^2 with tangents T_1 and T_2 at z, respectively. Then the angle from T_1 to T_2 is referred to as the **angle from C_1 to C_2 at z** and is denoted by $\angle(C_1, C_2, z)$.

(b) Let $z \in D$ where D is a domain in R^2 and let f be a mapping of D into R^2. The mapping f is **conformal at z** iff there is some neighborhood U of z such that if C_1 and C_2 are smooth arcs in U intersecting at z, then $f(C_1)$ and $f(C_2)$ are arcs with tangents at $f(z)$ such that

$$\angle[f(C_1), f(C_2), f(z)] = \angle(C_1, C_2, z).$$

(c) A mapping f is **conformal in a domain D** iff f is conformal at each point of D.

Theorem 10.5.2 If f is analytic in a domain D, then f is conformal at each point z in D at which $f'(z) \ne 0$.

Proof. Let $z \in D$ such that $f'(z) \ne 0$. By 8.2.12, let $N(z)$ be a neighborhood of z contained in D such that f is one-to-one on $N(z)$. For $k = 1, 2$, we let C_k be a smooth arc in $N(z)$ through the point z given by

$$C_k : \qquad z_k = z_k(t) \quad \text{on } [a_k, b_k]$$

where $z_k'(t)$ is never zero in $[a_k, b_k]$ and where $z_1(t_1) = z_2(t_2) = z$. Then $f(C_1)$ and $f(C_2)$ are given by

$$f(C_k) : \qquad w_k = f[z_k(t)] \quad \text{on } [a_k, b_k] \quad \text{for } k = 1, 2.$$

Now

$$w_k'(t_k) = f'(z)\, z_k'(t_k) \ne 0.$$

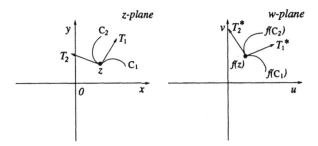

Figure 10.15: $w_k'(t_k) = f'(z) z_k'(t_k) \neq 0$

Since $w_k'(t_k) \neq 0$, it follows that $f(C_k)$ has a tangent at $f(z)$ and its direction is given by

$$\arg w_k'(t_k) = \arg f'(z) + \arg z_k'(t_k) \quad \text{for } k = 1, 2.$$

Thus the angle from $f(C_1)$ to $f(C_2)$ at $f(z)$ is given by

$$
\begin{aligned}
\arg w_2'(t_2) - \arg w_1'(t_1) &= [\arg f'(z) + \arg z_2'(t_2)] - [\arg f'(z) + \arg z_1'(t_1)] \\
&= \arg z_2'(t_2) - \arg z_1'(t_1)
\end{aligned}
$$

which is the angle from C_1 to C_2 at z if the appropriate arguments are used. See Figure 10.15. [The curve $f(C_k)$ is an arc since $w_k = f[z_k(t)]$ is a composite of one-to-one continuous functions.]

Remark 10.5.3 Let f be analytic in a domain D and suppose f' is never zero in D. If C is a smooth arc in D given by

$$C: \quad z = z(t) \quad \text{on } [a, b],$$

then its image $f(C)$ is given by

$$f(C): \quad w = f[z(t)] \quad \text{on } [a, b]. \tag{10.71}$$

As in the proof of 10.5.2,

$$\arg w'(t) = \arg f'[z(t)] + \arg z'(t). \tag{10.72}$$

Thus f rotates the tangent to C at z through the angle $\arg f'(z)$.

Example 10.5.4 Let $f(z) = -iz$, let C_1 be the line segment from $z = 0$ to $z = 2$, and let C_2 be the line segment from $z = 0$ to $z = 2 + 2i$. Then $f(C_1)$ is the line segment from $w = 0$ to $w = -2i$ and $f(C_2)$ is the line segment from $w = 0$ to $w = 2 - 2i$. Thus $\angle(C_1, C_2, 0) = \frac{\pi}{4}$ and $\angle(f(C_1), f(C_2), 0) = \arg(2 - 2i) - \arg(-2i) = \frac{7\pi}{4} - \frac{3\pi}{2} = \frac{\pi}{4} = \angle(C_1, C_2, 0)$. Note $f'(0) = -i \neq 0$. Hence the linear transformation $w = -iz$ is conformal at $z = 0$. We see that $\arg f(z) = \arg(-i) + \arg z = \frac{3\pi}{2} + \arg z$ and $|f(z)| = |z|$. Thus f rotates all points through 270° about the origin.

Remark 10.5.5 If $f'(z_0)$ exists, then

$$\lim_{z \to z_0} \left| \frac{f(z) - f(z_0)}{z - z_0} \right| = |f'(z_0)|.$$

Proof. Let $\epsilon > 0$. Then there is a $\delta > 0$ such that $0 < |z - z_0| < \delta$ implies

$$\epsilon > \left| \frac{f(z) - f(z_0)}{z - z_0} - f'(z_0) \right| \quad \text{by 2.3.1}$$

$$\geq \left| \left| \frac{f(z) - f(z_0)}{z - z_0} \right| - |f'(z_0)| \right| \quad \text{by 1.3.1(c).}$$

Definitions 10.5.6 Let f be a function on D into R^2 where $D \subset R^2$. Now $|z - z_0|$ is the distance between the points z and z_0 in the z-plane, and $|f(z) - f(z_0)|$ is the distance between their images in the w-plane. The ratio

$$\frac{|f(z) - f(z_0)|}{|z - z_0|}$$

may be called the **magnification** of the segment from the point z_0 to the point z due to f. Thus it is natural to call

$$\lim_{z \to z_0} \frac{|f(z) - f(z_0)|}{|z - z_0|}$$

the **magnification at z_0** due to f (if the limit exists).

In view of 10.5.5, if $f'(z_0)$ exists, then $|f'(z_0)|$ is the magnification at z_0 due to f. Furthermore, if z is near z_0, then

$$|f(z) - f(z_0)| \approx |z - z_0| \, |f'(z_0)|.$$

Exercises 10.5.7

1. Show that the linear transformation is conformal at each point z in R^2.

2. Show that the bilinear transformation

$$T(z) = \frac{az + b}{cz + d}$$

   is conformal in $R^2 - \left\{ -\dfrac{d}{c} \right\}$.

3. Show that if $a \neq 0$, then $w = az^2 + bz + c$ is conformal at each point z_0 in R^2 such that $z_0 \neq -\frac{b}{2a}$.

4. Show that if $a \neq 0$ then $w = az^3 + bz^2 + cz + d$ is conformal in $R^2 - E$ where E consists of at most two points and where the coefficients are complex numbers.

5. Prove if $f \in A(z_0)$ and $f'(z_0) = 0$, then f is not conformal at z_0.

6. Use Ex. 3 and 5 to show that $w = az^2 + bz + c$ is not conformal at exactly one point, if $a \neq 0$.

7. Show that if $a \neq 0$, then $w = az^3 + bz^2 + cz + d$ is not conformal at some point.

8. Prove that $f(C_1)$ is a smooth arc, where C_1 is given in the proof of 10.5.2. **Hint.** See 8.2.13 and use the chain rule.

Solution of Ex. 5. Case 1. Suppose $f^{(k)}(z_0) = 0$ for each k in N. Then $f(z) = f(z_0)$, a constant, in some neighborhood of z_0. Thus f is not conformal at z_0 by 10.5.1(b).

Case 2. Suppose $f^{(k)}(z_0) \neq 0$ for some positive integer k. Let n be the least value of k in N such that $f^{(k)}(z_0) \neq 0$. Then $n \geq 2$ and for each z in some neighborhood $N(z_0)$,

$$f(z) = f(z_0) + (z - z_0)^n [a_n + a_{n+1}(z - z_0) + \cdots]$$

where $a_n, a_{n+1}, \cdots$ are the coefficients in the Taylor series for f. Thus for each z in $N(z_0)$,

$$\arg[f(z) - f(z_0)] = n \arg(z - z_0) + \arg[a_n + a_{n+1}(z - z_0) + \cdots].$$

Now (for $k = 1, 2$) we choose a smooth arc C_k in $N(z_0)$ which is a straight line segment with an endpoint at z_0. We let $\gamma = \angle(C_1, C_2, z_0) = \alpha_2 - \alpha_1$ where α_k is the angle of inclination of C_k. We choose C_1 and C_2 so that

$$n\gamma \text{ and } \gamma \text{ do not differ by an integral multiple of } 2\pi. \tag{10.73}$$

This is possible since we need only to take γ so that γ is not a multiple of $\frac{2\pi}{n-1}$. See Figure 10.16.

Letting z approach z_0, first along C_2 and then along C_1, we obtain

$$\alpha_2' = n\alpha_2 + \arg a_n \quad \text{and} \quad \alpha_1' = n\alpha_1 + \arg a_n \tag{10.74}$$

where α_1' and α_2' indicate the angles of inclination within multiples of 2π of the tangents T_1' and T_2' to $f(C_1)$ and $f(C_2)$, respectively, at $f(z_0)$. By (10.74),

$$\gamma' = \alpha_2' - \alpha_1' = n(\alpha_2 - \alpha_1) = n\gamma$$

where γ' indicates the angle $\angle[f(C_1), f(C_2), f(z_0)]$ within a multiple of 2π. Thus $\angle[f(C_1), f(C_2), f(z_0)] \neq \gamma = \angle(C_1, C_2, z_0)$ since $n\gamma$ and γ do not differ by a multiple of 2π by (10.73). Hence f is not conformal at z_0.

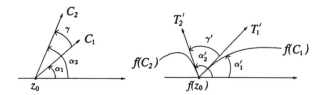

Figure 10.16: $\gamma = \alpha_2 - \alpha_1$ $\gamma' = n(\alpha_2 - \alpha_1) = n\gamma$

Remark 10.5.8 Let $f \in A(z_0)$. Then f is conformal at z_0 iff $f'(z_0) \neq 0$. This follows by 10.5.2 and Ex. 5 in 10.5.7.

Theorem 10.5.9 Let

$$f, f_x, f_y \in C(D) \quad \text{where } D \text{ is a domain.} \tag{10.75}$$

(a) If f is conformal in D, then $f \in A(D)$.

(b) Let $z_0 \in D$. If the magnification at z_0 due to f exists, then either $f'(z_0)$ exists or $(\overline{f})'(z_0)$ exists.

(c) Let $z_0 \in D$ such that $f_x(z_0) \neq 0$. If the magnification at each point of D due to f exists, then either $f \in A(z_0)$ or $\overline{f} \in A(z_0)$.

Proof of (a). Let $z_0 \in D$ and let C be a smooth arc in D given by

$$z = z(t) \quad \text{on } [a, b]$$

where $z(a) = z_0$. Then $f(C)$ is given by

$$w = f[z(t)] \quad \text{on } [a, b].$$

By (10.75), $w'(a)$ exists and is given by

$$w'(a) = f_x(z_0)\, x'(a) + f_y(z_0)\, y'(a). \tag{10.76}$$

Putting

$$x' = \frac{1}{2}(z' + \overline{z'}) \quad \text{and} \quad y' = \frac{i}{2}(\overline{z'} - z')$$

in (10.76), we have $w'(a) = \frac{1}{2}(f_x - if_y)\, z' + \frac{1}{2}(f_x + if_y)\, \overline{z'}$ and hence

$$\frac{w'(a)}{z'(a)} = \frac{1}{2}(f_x - if_y) + \frac{1}{2}(f_x + if_y)\frac{\overline{z'}}{z'}, \tag{10.77}$$

where z' is computed at $t = a$ and f_x and f_y are computed at $z = z_0$.

Now since f is conformal at z_0, it follows that

$$\arg \frac{w'(a)}{z'(a)} \text{ is independent of } \arg z'(a), \tag{10.78}$$

that is, independent of the direction of C at z_0. To verify (10.78) let

$$C_1: \qquad z = z_1(t) \quad \text{on } [a_1, b_1]$$

be another smooth arc in D such that $z_1(a_1) = z_0$. Then $f(C_1)$ is given by

$$f(C_1): \qquad w_1(t) = f[z_1(t)] \quad \text{on } [a_1, b_1].$$

Now

$$
\begin{aligned}
\arg z_1'(a_1) - \arg z'(a) &= \angle (C, C_1, z_0) \\
&= \angle [f(C), f(C_1), f(z_0)] \quad \text{by 10.5.1(b)} \\
&= \arg w_1'(a_1) - \arg w'(a). \tag{10.79}
\end{aligned}
$$

Thus by transposing terms in (10.79),

$$\arg w'(a) - \arg z'(a) = \arg w_1'(a_1) - \arg z_1'(a_1),$$

that is,

$$\arg \frac{w'(a)}{z'(a)} = \arg \frac{w_1'(a_1)}{z_1'(a_1)}. \tag{10.80}$$

Now (10.80) implies (10.78).

We observe that (10.77) is of the form

$$\frac{w'(a)}{z'(a)} = \alpha(z_0) + \beta(z_0)e^{-2i\theta}, \tag{10.81}$$

where

$$\alpha = \frac{1}{2}[f_x(z_0) - if_y(z_0)] \quad \text{and} \quad \beta = \frac{1}{2}[f_x(z_0) + if_y(z_0)] \tag{10.82}$$

and where $z'(a) = re^{i\theta}$. If we allow the $\arg z'(a)$ to vary from $\theta = 0$ to $\theta = 2\pi$, the point $\frac{w'(a)}{z'(a)}$ in (10.81) describes a circle, with center at α and radius $|\beta|$. Thus $\arg \frac{w'(a)}{z'(a)}$ would vary unless $\beta = 0$. Hence by (10.78) and (10.82),

$$0 = \beta = \frac{1}{2}[f_x(z_0) + if_y(z_0)],$$

which implies

$$f_x = -if_y. \tag{10.83}$$

Equation (10.83) is called the **complex form** of the Cauchy-Riemann equations since it is equivalent to them. For (10.83) implies $u_x + iv_x = -i(u_y + iv_y)$, which implies

$$u_x = v_y \quad \text{and} \quad u_y = -v_x$$

[and conversely the last equations imply (10.83) but we do not use this converse implication in our present proof.] Thus $f'(z_0)$ exists by (10.75) and 2.3.11 and hence $f \in A(D)$.

Proof of (b). Let C and $f(C)$ be given as in the proof of Part (a). The magnification at z_0 due to f is given by

$$\lim_{t \to a} \frac{|f[z(t)] - f(z_0)|}{|z(t) - z_0|} = \lim_{t \to a} \frac{\left|\frac{w(t)-f(z_0)}{t-a}\right|}{\left|\frac{z(t)-z_0}{t-a}\right|} = \frac{|w'(a)|}{|z'(a)|}.$$

Thus by hypothesis, $\left|\frac{w'(a)}{z'(a)}\right|$ has a constant value as the direction of C at z_0 varies from 0 to 2π. But by (10.81), the point $\frac{w'(a)}{z'(a)}$ traces a circle with center at α and radius $|\beta|$ as $\arg z'(a)$ varies from 0 to 2π. Now if the modulus on a circle is constant, then its radius is zero or its center is at the origin. Thus

$$\beta = 0 \quad \text{or} \quad \alpha = 0.$$

If $\beta = 0$, then (10.82) implies (10.83) which with (10.75) implies that $f'(z_0)$ exists. Now suppose $\alpha = 0$. Then at z_0,

$$f_x = i f_y \quad \text{by (10.82)}.$$

Thus $\overline{f_x} = -i\,\overline{f_y}$. Hence

$$(\overline{f})_x = -i(\overline{f})_y \quad \text{by Ex. 2 in 10.5.10.} \tag{10.84}$$

The last equation is the complex form of the Cauchy-Riemann equations for $\overline{f}$ at z_0. Thus $(\overline{f})'(z_0)$ exists in case $\alpha = 0$ by (10.75) and 2.3.11.

Proof of (c). The hypothesis that $f_x(z_0) \neq 0$ implies by (10.82) that not both $\alpha = 0$ and $\beta = 0$. Suppose $\beta \neq 0$. By (10.75) and (10.82),

$$\beta(z) = \frac{1}{2}[f_x(z) + if_y(z)] \quad \text{is continuous in D}.$$

Since $\beta = \beta(z_0) \neq 0$, we let $\rho > 0$ such that $N_\rho(z_0) \subset D$ and such that

$$\beta(z) \text{ is never zero in } N_\rho(z_0).$$

See the proof of (c) in 2.2.6. By the argument used earlier (applied to z instead of z_0), we have

$$\alpha(z) = \frac{1}{2}[f_x(z) - if_y(z)] = 0 \quad \text{for each } z \text{ in } N_\rho(z_0). \tag{10.85}$$

Now (10.85) implies that (10.84) holds on $N_\rho(z_0)$. Thus $(\overline{f})'(z)$ exists on $N_\rho(z_0)$ and hence $\overline{f} \in A(z_0)$. Similarly, if $\alpha \neq 0$ at z_0, then $f \in A(z_0)$.

Exercises 10.5.10

1. Let $f(z) = \begin{cases} z & \text{if } \mathcal{I}(z) \geq 0 \\ \overline{z} & \text{if } \mathcal{I}(z) < 0. \end{cases}$

   In Parts (a)–(c), find the indicated set.

   (a) $\{z : \ f'(z) \text{ exists or } (\overline{f})'(z) \text{ exists }\}$ $R^2 - R$

   (b) $\{z : \ f \text{ is conformal at } z\}$ $\{z : \ \mathcal{I}(z) > 0\}$

   (c) $\{z : \text{ the magnification at } z \text{ due to } f \text{ exists }\}$ R^2

   (d) By the answer to Part (c), the magnification at z due to f exists for each z on the real axis. By the answer to Part (a), neither f' nor $(\overline{f})'$ exists at any point on the real axis. Why does this not contradict 10.5.9(b)?

   (e) Along the real axis, $f_x \equiv 1 \neq 0$ and the magnification at each point on the x-axis due to f exists. But f is not analytic at any point on the x-axis. Why is there no contradiction to 10.5.9(c)?

2. Prove if f_x exists, then $\overline{f_x} = (\overline{f})_x$, where $f(x,y) = u + iv$. State this result in words.

10.6   The Schwarz-Christoffel Transformation

In this section, we discuss the problem of finding a one-to-one analytic mapping of the upper half-plane

$$H = \{z : \ \mathcal{I}(z) > 0\}$$

onto the interior of a polygon. [A polygon is a simple closed contour consisting of line segments. A closed polygonal line is a closed contour (which may cross itself) consisting of line segments.] We set forth here some conditions and labeling to be used in our next theorem. For each integer $j = 1$ to $n + 1$, we let x_j and k_j be such that

$$x_1 < x_2 < \cdots < x_n, \quad x_{n+1} = \infty, \quad -1 < k_j < 1, \quad \text{and}$$

$$\sum_{j=1}^{n+1} k_j = 2. \tag{10.86}$$

We let $D = \{z : \ z \in R^2 \text{ and } -\frac{\pi}{2} \text{ is not an argument of } z - x_j \text{ for any } j = 1 \text{ to } n\}$ and for each z in D we let $g(z)$ be given by

$$g(z) = (z - x_1)^{-k_1}(z - x_2)^{-k_2} \cdots (z - x_n)^{-k_n} \tag{10.87}$$

where

$$(z - x_j)^{-k_j} = |z - x_j|^{-k_j} e^{-i k_j \theta_j} \quad \text{with } -\frac{\pi}{2} < \theta_j = \arg(z - x_j) < \frac{3\pi}{2}. \tag{10.88}$$

Finally, for some fixed point z_0 in D, we let

$$f(z) = \begin{cases} \int_{z_0}^z g(t)\,dt & \text{if } z \in D \\ w_j & \text{if } z = x_j, \end{cases}$$

where

$$w_j = \lim_{z \to x_j} f(z) \quad \text{for } j = 1 \text{ to } n$$

and

$$w_{n+1} = \lim_{\substack{z \to \infty \\ z \in H^-}} f(z).$$

We shall show that these limits exist in R^2.

Theorem 10.6.1 Let $H, D, f, g, k_j, x_j,$ and w_j be as given above and for $j = 1$ to n, let $k_j \neq 0$.

(a) The function f is analytic in D, and $f' = g$ on D.

(b) For $j = 1$ to n, the limit w_j exists in R^2.

(c) The limit w_{n+1} also exists in R^2.

(d) For $j = 1$ to n, the function f maps the segment $[x_j, x_{j+1}]$ onto the segment $\overrightarrow{w_j\,w_{j+1}}$ and f maps the segment $[x_{n+1}, x_1]$ onto the segment $\overrightarrow{w_{n+1}\,w_1}$, where it is understood that $[x_n, x_{n+1}] = [x_n, \infty] = \{x : \ x \geq x_n \ \text{ or } \ x = \infty\}$ and $[x_{n+1}, x_1] = \{x : \ x \leq x_1 \ \text{ or } \ x = \infty\}$.

(e) Let P be the closed polygonal line consisting of the segments $\overrightarrow{w_j\,w_{j+1}}$ for $j = 1$ to n and the segment $\overrightarrow{w_{n+1}\,w_1}$. Then f maps the extended real axis onto P. See Figure 10.17. If

$$\text{no two ``non-consecutive'' sides of } P \text{ intersect,} \tag{10.89}$$

then P is a polygon.

(f) The mapping f is one-to-one on $[x_j, x_{j+1}]$ for $j = 1$ to n and on $[x_{n+1}, x_1]$.

(g) If (10.89) holds, then f is a one-to-one mapping of the upper half-plane H onto $I(P)$, the interior of P.

Proof of (a). Part (a) follows from 3.6.3.

Proof of (b). To prove (b) observe that

$$g(z) = (z - x_1)^{-k_1} h(z) \tag{10.90}$$

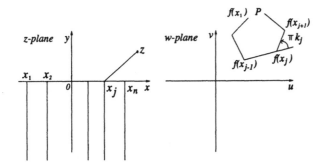

Figure 10.17:   The function f maps the extended real axis onto the polygonal line P consisting of the segments shown in the figure.

where $h(z)$ is the product of the remaining factors in $g(z)$. Since $h \in A(x_1)$, we may write by 6.1.6,

$$h(z) = \sum_{n=0}^{\infty} a_n(z - x_1)^n \quad \text{on some neighborhood } N(x_1) \text{ of } x_1, \tag{10.91}$$

where the numbers a_n are the Taylor coefficients. Let $p \in D \cap N(x_1)$. Then for each z in $D \cap N(x_1)$, we have by (10.90) and (10.91),

$$
\begin{aligned}
f(z) &= \int_{z_0}^{p} g(t)\, dt + \int_{p}^{z} \sum_{n=0}^{\infty} a_n(t - x_1)^{n-k_1}\, dt \\
&= \int_{z_0}^{p} g(t)\, dt + (z - x_1)^{1-k_1} \sum_{n=0}^{\infty} \frac{a_n}{n - k_1 + 1} (z - x_1)^n \\
&\quad - \sum_{n=0}^{\infty} \frac{a_n}{n - k_1 + 1} (p - x_1)^{n-k_1+1}
\end{aligned}
\tag{10.92}
$$

by 4.4.3, 4.4.9, and Ex. 7 of 4.4.11. [To see that $(t - x_1)^{-k_1} \in C(D)$, recall 5.4.7 and Ex. 11 in 5.4.11.]

Now by (10.92),

$$
\begin{aligned}
\lim_{z \to x_1} f(z) &= \int_{z_0}^{p} g(t)\, dt - \sum_{n=0}^{\infty} \frac{a_n}{n - k_1 + 1} (p - x_1)^{n-k_1+1} \\
&= \text{a constant,}
\end{aligned}
$$

since $(z - x_1)^{1-k_1} \to 0$ as $z \to x_1$ and since $\sum_{n=0}^{\infty} \frac{a_n}{n-k_1+1} (z - x_1)^n$ is continuous at x_1. Similarly,

$$\lim_{z \to x_j} f(z) \text{ exists for each } j = 2, 3, \cdots, n.$$

Proof of (c). If $|z| > 2|x_j|$, then

(i) $|z - x_j| \geq |z| - |x_j| > |z| - \frac{|z|}{2} = \frac{|z|}{2}$.

If $|z| > |x_j|$, then

(ii) $|z - x_j| \leq |z| + |x_j| < 2|z|$.

Hence for $|z| > 2|x_j|$,

$$|z - x_j|^{-k_j} < \begin{cases} 2^{k_j}|z|^{-k_j} & \text{if } k_j > 0 \quad \text{by (i)} \\ \\ 2^{-k_j}|z|^{-k_j} & \text{if } k_j < 0 \quad \text{by (ii)}. \end{cases} \qquad (10.93)$$

Let

$$r > \max\{2|x_j| : \quad j = 1, 2, \cdots, n\}.$$

Now by (10.93), if $|z| > r$, then

$$|g(z)| < \frac{M}{|z|^s} \quad \text{where } s = \sum_{j=1}^{n} k_j \qquad (10.94)$$

and where M is some positive number. Now $k_{n+1} < 1$. Hence $-k_{n+1} > -1$. Thus by (10.86),

$$s = 2 - k_{n+1} = 2 + (-k_{n+1}) > 2 + (-1) = 1. \qquad (10.95)$$

Now for $|z| > r$, we have

$$f(z) = \int_{z_0}^{r} g(t)\, dt + \int_{r}^{|z|} g(x)\, dx + \int_{C} g(t)\, dt \qquad (10.96)$$

where C is the circular arc from $|z|$ to z given by $|t| = |z|$. See Figure 10.18. The first integral in (10.96) is a constant. To show that the second integral has a limit as $|z| \to \infty$, we see that

$$\left| \int_{r}^{|z|} g(x)\, dx \right| \leq M \int_{r}^{|z|} x^{-s}\, dx \quad \text{by (10.94)}$$

$$= \frac{M}{1-s} \left[\frac{1}{|z|^{s-1}} - \frac{1}{r^{s-1}} \right]$$

$$\to \frac{M}{(s-1)r^{s-1}} \quad \text{as } z \to \infty \quad \text{by (10.95).} \qquad (10.97)$$

Finally, the 3rd integral in (10.96) approaches zero as $z \to \infty$, since

$$\left| \int_{C} g(t)\, dt \right| \leq \frac{M}{|z|^s}(2\pi|z|) \quad \text{by (10.94) and 3.4.10(g)}$$

$$= \frac{2\pi M}{|z|^{s-1}} \to 0 \quad \text{as } z \to \infty. \qquad (10.98)$$

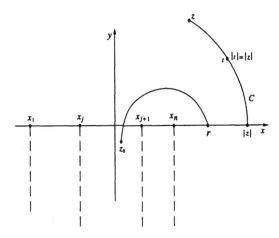

Figure 10.18: $f(z) = \int_{z_0}^r g(t)\,dt + \int_r^{|z|} g(x)\,dx + \int_C g(t)\,dt$

Now (c) follows by (10.96) – (10.98).

Proof of (d). We first observe that if x and x_j are real and if $\alpha_j(x) = \mathrm{Arg}\,(x - x_j)$, then

$$\alpha_j(x) = \begin{cases} \pi & \text{when } x - x_j < 0 \\ 0 & \text{when } x - x_j > 0 \end{cases} \tag{10.99}$$

By Part (a) and (10.87),

$$\begin{aligned} \beta_x &= \arg f'(x) = -k_1\alpha_1(x) - k_2\alpha_2(x) - \cdots - k_{j-1}\alpha_{j-1}(x) \\ &\quad -k_j\alpha_j(x) - \cdots - k_n\alpha_n(x) \\ &= -k_1\cdot 0 - \cdots - k_{j-1}\cdot 0 - k_j\cdot\pi - \cdots - k_n\pi \end{aligned} \tag{10.100}$$

if $x_{j-1} < x < x_j$. Hence as $z = x$ moves along the x-axis from x_{j-1} to x_j, β_x is a constant (which means that the tangent vector to the curve $w = f(x)$ has a constant direction). See Ex. 1 in 10.6.9. Thus $w = f(x)$ moves in a fixed direction along a line as x moves from x_{j-1} to x_j. Hence f maps $[x_{j-1}, x_j]$ onto the segment $\overrightarrow{w_{j-1}\,w_j}$. But β_x increases (or decreases) abruptly by $|k_j\pi|$ as $z = x$ passes through x_j since the term $-k_j\pi$ will be replaced by 0 in (10.100) if $x_j < x < x_{j+1}$. Also by the first line in (10.100), we see that β_x is constant for $x < x_1$ and is constant for $x > x_n$. Thus, f maps $[x_n, \infty]$ onto the segment $\overrightarrow{w_n w_{n+1}}$ and f maps $[x_{n+1}, x_1]$ onto the segment $\overrightarrow{w_{n+1}w_1}$, and hence Part (d) is proved.

Proof of (e). The first assertion in Part (e) follows from Part (d), and the second assertion is clear.

Note. Suppose (10.89) holds. If w_n, w_{n+1} and w_1 are collinear, then w_{n+1} is not a vertex of P, and P has n sides where n is the number of factors in $g(t)$. But if w_n, w_{n+1} and w_1 are not collinear, then w_{n+1} is a vertex and P has $n+1$ sides. Thus P has n vertices or $n+1$ vertices [assuming (10.89) holds]. Also, by (10.100), the exterior angle of P at w_j is $k_j\pi$.

Proof of (f). If $x_{j-1} < x < x' < x_j$ for a fixed j, then $\theta = \arg g(t)$ is constant as t varies over the interval $[x, x']$, and we have

$$
\begin{aligned}
f(x') - f(x) &= \int_x^{x'} g(t)\, dt \\
&= e^{i\theta} \int_x^{x'} |g(t)|\, dt \\
&\neq 0 \quad \text{since } g(t) \neq 0 \text{ in } [x, x'].
\end{aligned}
$$

Thus $f(x') \neq f(x)$, and f is one-to-one on $[x_{j-1}, x_j]$. Similarly, f is one-to-one on $[x_n, x_{n+1}]$ and on $[x_{n+1}, x_1]$.

Proof of (g). Let C be the cco simple closed contour consisting of the semicircle S_r and the n semicircles S_{r_j}, joined by segments of the x-axis as shown in Figure 10.19. Let w_0 be a point not on P. By 8.2.7, the "number n_C of zeros" of $f(z) - w_0$ in $I(C)$ is given by

$$
n_C = \frac{1}{2\pi i} \int_C Q(z)\, dz \quad \text{where } Q(z) = \frac{f'(z)}{f(z) - w_0} \tag{10.101}
$$

where the radii of the semicircles are chosen so that $f(z) - w_0$ is never zero on C. Now [with $t, \tau, f(t)$, and $f(\tau)$ as shown in Figure 10.19]

$$
\int_t^\tau Q(x)\, dx = \int_{f(t)}^{f(\tau)} \frac{dw}{w - w_0}, \tag{10.102}
$$

where the last integral is along a side of P [since $w = f(x)$ is a parametrization of the segment from $f(t)$ to $f(\tau)$]. By 3.6.3 and Part (b),

$$
\begin{aligned}
\lim_{\substack{t \to x_j \\ \tau \to x_{j+1}}} \int_{f(t)}^{f(\tau)} \frac{dw}{w - w_0} &= \lim_{\tau \to x_{j+1}} \int_{w_j}^{f(\tau)} \frac{dw}{w - w_0} - \lim_{t \to x_j} \int_{w_j}^{f(t)} \frac{dw}{w - w_0} \\
&= \int_{w_j}^{w_{j+1}} \frac{dw}{w - w_0} - \int_{w_j}^{w_j} \frac{dw}{w - w_0} \\
&= \int_{w_j}^{w_{j+1}} \frac{dw}{w - w_0}. \tag{10.103}
\end{aligned}
$$

Thus by (10.102) and (10.103),

$$
\lim_{\substack{t \to x_j \\ \tau \to x_{j+1}}} \int_t^\tau Q(x)\, dx = \int_{w_j}^{w_{j+1}} \frac{dw}{w - w_0}. \tag{10.104}
$$

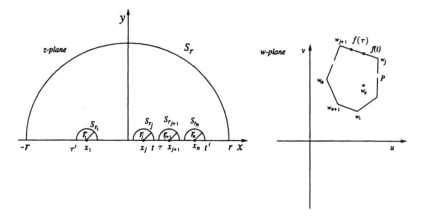

Figure 10.19: The function f is a one-to-one mapping of H onto $I(P)$ if (10.89) holds.

Similarly,

$$\lim_{\substack{t' \to x_n \\ r \to \infty}} \int_{t'}^{r} Q(x)\,dx = \int_{w_n}^{w_{n+1}} \frac{dw}{w - w_0} \tag{10.105}$$

and

$$\lim_{\substack{\tau' \to x_1 \\ r \to \infty}} \int_{-r}^{\tau'} Q(x)\,dx = \int_{w_{n+1}}^{w_1} \frac{dw}{w - w_0}. \tag{10.106}$$

By (10.101),

$$2\pi i n_C = \int_{S_r} Q(z)\,dz + \sum_{j=1}^{n} \int_{S_{r_j}} Q(z)\,dz + \int_{-r}^{\tau'} Q(x)\,dx$$

$$+ \cdots + \int_{t}^{\tau} Q(x)\,dx + \cdots + \int_{t'}^{r} Q(x)\,dx. \tag{10.107}$$

By (10.104)–(10.107), as $r \to \infty$ and each $r_j \to 0$,

$$n_C \to \frac{1}{2\pi i} \int_{P} \frac{dw}{w - w_0} = \begin{cases} 1 & \text{if } w_0 \in I(P) \quad \text{by 3.6.4} \\ 0 & \text{if } w_0 \in E(P) \quad \text{by 3.5.10.} \end{cases} \tag{10.108}$$

Now (10.108) gives the following conclusions (i) and (ii) directly.

(i) If $w_0 \in I(P)$, there is exactly one z in H such that $f(z) = w_0$.

(ii) If $w_0 \in E(P)$, then there is no z in H such that $f(z) = w_0$.

By 8.2.14, the set $f(H)$ is open. Thus we also have the following conclusion.

(iii) If w_0 is a point on P then there is no z in H such that $f(z) = w_0$.

[For suppose $f(z_0) = w_0 \in P$ where $z_0 \in H$. Since $f(H)$ is open, there is a $r > 0$ such that $N_r(w_0) \subset f(H)$. Hence there are points in $N_r(w_0) \cap E(P)$ which are images of points of H, contradicting (ii).]

Thus by (i) – (iii), we conclude that f maps H, one-to-one onto $I(P)$, and (g) is proved.

Remarks 10.6.2 (a) A theorem in geometry states that the sum of the radian measures of the exterior angles of a polygon is 2π. From the note following the proof of 10.6.1(e), the exterior angle of P at w_j has measure $k_j\pi$. This explains the significance of hypothesis (10.86).

(b) If $\sum_{j=1}^{n} k_j = 2$, then $k_{n+1} = 0$. So, the exterior angle measure $\pi k_{n+1} = 0$, and hence in this case, w_{n+1} is not a vertex of P but is a point on the side of P from w_n to w_1. Thus if $\sum_{j=1}^{n} k_j = 2$, then P has n sides [assuming (10.89)].

(c) If $\sum_{j=1}^{n} k_j \neq 2$ (that is, if $k_{n+1} \neq 0$), then $w_{n+1} = f(\infty)$ is a vertex of P, and P has $n+1$ sides [assuming (10.89)].

(d) Suppose (10.89) holds. Then the function f given in 10.6.1 is a one-to-one mapping of the extended real axis onto a polygon P and is a one-to-one analytic mapping of H onto $I(P)$.

(e) In geometry, two polygons are said to be similar iff there is a correspondence between their vertices such that corresponding angles are congruent and corresponding sides are proportional. A theorem states that if two triangles have their corresponding angles congruent, then they are similar. Thus if two angles of a triangle are congruent to two angles of another triangle, then they are similar (since the sum of the measures of the angles of any triangle is π radians).

Remarks 10.6.3 (a) Let Q be a triangle with vertices w_1', w_2' and w_3' and let πk_j denote its exterior angle at w_j'. Let

$$w = F(z) = \int_{z_0}^{z} (t - x_1)^{-k_1} (t - x_2)^{-k_2} \, dt \qquad (10.109)$$

where $x_1 < x_2$, where $z_0 \in D$ and where D is given in 10.6.1. Then (10.86) is satisfied, and $k_3 \neq 0$. Let $P = F(L)$ where L is the extended real axis. By 10.6.1(e), P is a closed polygonal line, and (10.89) holds (since P has no non-consecutive sides). Thus P is a triangle with vertices $w_1 = F(x_1), w_2 = F(x_2)$ and $w_3 = F(\infty)$. By the note following the proof of 10.6.1(e), the measure of the exterior angle of P at w_1 is πk_1, which is the measure of the exterior angle of Q at w_1'. Hence the interior angle of P at w_1 is congruent to the interior angle of Q at w_1'. Likewise, the interior angle of P at w_2 is congruent to the interior angle of Q at w_2'. Thus triangle P is similar to triangle Q. Hence there is a one-to-one analytical mapping F of the form f in 10.6.1, which maps the extended real axis onto a triangle P similar to the triangle Q and which maps H onto $I(P)$. (Note that x_1 and x_2 may be chosen arbitrarily in (10.109) such that $x_1 < x_2$.)

(b) Let

$$G(w) = Bw + E \quad \text{where } B, E \in R^2. \tag{10.110}$$

The constant B in (10.110) has the effect of rotating and stretching (or shrinking) the triangle P, whereas the constant E in (10.110) has the effect of translating P. Since P is similar to Q, we may choose B and E so that G maps $P \cup I(P)$ onto $Q \cup I(Q)$. Thus B and E may be chosen such that

$$h(z) = BF(z) + E \tag{10.111}$$

maps $\overline{H} \cup \{\infty\}$ onto $Q \cup I(Q)$ and such that $h \in A(H)$, where F is given in (10.109).

Example 10.6.4 For an equilateral triangle, $k_1 = k_2 = \frac{2}{3}$. Then (10.109) becomes

$$F(z) = \int_a^z (t + a)^{-\frac{2}{3}}(t - a)^{-\frac{2}{3}} \, dt$$

if we take $z_0 = a$, $x_1 = -a$, and $x_2 = a$ where $a > 0$. Then

$$w_2 = F(a) = \int_a^a (t + a)^{-\frac{2}{3}}(t - a)^{-\frac{2}{3}} \, dt = 0.$$

To obtain $w_1 = F(-a)$, we take our contour to be the segment of the x-axis from $x = a$ to $x = -a$. Since $\text{Arg}(x - x_j)$ is π or 0 according as x is to the left or right of x_j on the real axis, it follows by (10.88) that for $-a < x < a$,

$$(x + a)^{-\frac{2}{3}} = |x + a|^{-\frac{2}{3}} e^{0i} = (a + x)^{-\frac{2}{3}} \quad \text{since } x + a > 0,$$

and

$$(x - a)^{-\frac{2}{3}} = |x - a|^{-\frac{2}{3}} e^{-\frac{2}{3}\pi i} = (a - x)^{-\frac{2}{3}} e^{-\frac{2}{3}\pi i} \quad \text{since } x - a < 0.$$

Thus

$$
\begin{aligned}
w_1 = F(-a) &= \int_a^{-a} (x + a)^{-\frac{2}{3}}(x - a)^{-\frac{2}{3}} \, dx \\
&= -\int_{-a}^a (a + x)^{-\frac{2}{3}} \left[(a - x)^{-\frac{2}{3}} e^{-\frac{2}{3}\pi i} \right] \, dx \\
&= e^{\pi i} \int_{-a}^a (a^2 - x^2)^{-\frac{2}{3}} e^{-\frac{2}{3}\pi i} \, dx \quad \text{since } -1 = e^{\pi i} \\
&= 2e^{\frac{\pi i}{3}} \int_0^a \frac{dx}{(a^2 - x^2)^{\frac{2}{3}}}. \tag{10.112}
\end{aligned}
$$

By 10.6.2(c), since $k_1 + k_2 = \frac{4}{3} \neq 2$, it follows that $w_3 = F(\infty)$ is a vertex of our triangle. Now in view of 10.6.1(c),

$$w_3 = \int_a^\infty (x+a)^{-\frac{2}{3}}(x-a)^{-\frac{2}{3}}\, dx = \int_a^\infty \frac{dx}{(x^2-a^2)^{\frac{2}{3}}}, \qquad (10.113)$$

which implies that w_3 is on the positive u-axis since $\dfrac{1}{(x^2-a^2)^{\frac{2}{3}}} > 0$. But again by 10.6.1(c),

$$
\begin{aligned}
w_3 &= \int_a^{-\infty} (x+a)^{-\frac{2}{3}}(x-a)^{-\frac{2}{3}}\, dx \\
&= w_1 + \int_{-a}^{-\infty} (x+a)^{-\frac{2}{3}}(x-a)^{-\frac{2}{3}}\, dx \qquad \text{by (10.112)} \\
&= w_1 + \int_{-a}^{-\infty} \left[|x+a|^{-\frac{2}{3}} e^{-\frac{2}{3}\pi i}\right]\left[|x-a|^{-\frac{2}{3}} e^{-\frac{2}{3}\pi i}\right] dx \\
&= w_1 + e^{-\frac{4}{3}\pi i} \int_{-a}^{-\infty} \frac{dx}{(x^2-a^2)^{\frac{2}{3}}} \\
&= w_1 + e^{\frac{-\pi i}{3}} \int_a^\infty \frac{dr}{(r^2-a^2)^{\frac{2}{3}}}
\end{aligned}
$$

since $e^{-\frac{4}{3}\pi i} = -e^{-\frac{\pi i}{3}}\, dr = -dx$. By the last equation and (10.113), we have $w_3 = w_1 + e^{\frac{-\pi i}{3}} w_3$. Thus

$$
\begin{aligned}
w_3 &= \frac{w_1}{1-e^{\frac{-\pi i}{3}}} = \frac{e^{\frac{\pi i}{3}}}{1-e^{\frac{-\pi i}{3}}} \int_{-a}^a \frac{dx}{(a^2-x^2)^{\frac{2}{3}}} \qquad \text{by (10.112)} \\
&= \int_{-a}^a \frac{dx}{(a^2-x^2)^{\frac{2}{3}}} \qquad \text{since } \frac{e^{\frac{\pi i}{3}}}{1-e^{\frac{-\pi i}{3}}} = 1.
\end{aligned}
$$

Hence F maps the extended real axis onto the equilateral triangle with vertices

$$w_2 = 0, \quad w_3 = \int_{-a}^a \frac{dx}{(a^2-x^2)^{\frac{2}{3}}}, \quad \text{and} \quad w_1 = e^{\frac{\pi i}{3}} w_3.$$

Furthermore F maps the upper half-plane onto the interior of this triangle. See Figure 10.20. We observe that

$$|w_2 - w_3| = |w_3 - w_1| = |w_1 - w_2| = w_3.$$

Example 10.6.5 To map the extended real axis onto a rectangle, we choose $k_j = \frac{1}{2}$ and we choose

$$x_1 = -b, \quad x_2 = -a, \quad x_3 = a, \quad \text{and} \quad x_4 = b$$

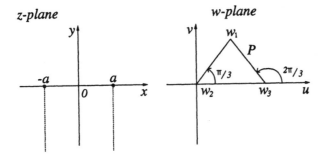

Figure 10.20: $F(H) = I(P)$

where $0 < a < b$. Let

$$F(z) = \int_0^z g(t)\,dt \tag{10.114}$$

where by (10.87),

$$g(t) = (t+b)^{-\frac{1}{2}}(t+a)^{-\frac{1}{2}}(t-a)^{-\frac{1}{2}}(t-b)^{-\frac{1}{2}}$$

and where (10.88) holds. Then

$$
\begin{aligned}
w_3 &= F(x_3) = F(a) = \int_0^a g(x)\,dx \\
&= \int_0^a (x+b)^{-\frac{1}{2}}(x+a)^{-\frac{1}{2}} \left[|x-a|^{-\frac{1}{2}}e^{-\frac{\pi i}{2}}\right]\left[|x-b|^{-\frac{1}{2}}e^{-\frac{\pi i}{2}}\right]\,dx
\end{aligned}
$$

by (10.88) and (10.99). Thus

$$w_3 = -\int_0^a \frac{dx}{\sqrt{(a^2-x^2)(b^2-x^2)}} < 0. \tag{10.115}$$

Now

$$
\begin{aligned}
w_2 &= F(x_2) = F(-a) = \int_0^{-a} g(x)\,dx \\
&= \int_0^{-a} (x+b)^{-\frac{1}{2}}(x+a)^{-\frac{1}{2}} \left[|x-a|^{-\frac{1}{2}}e^{-\frac{\pi i}{2}}\right]\left[|x-b|^{-\frac{1}{2}}e^{-\frac{\pi i}{2}}\right]\,dx \\
&= -\int_0^{-a} \frac{dx}{\sqrt{(a^2-x^2)(b^2-x^2)}} \\
&= \int_0^a \frac{dx}{\sqrt{(a^2-x^2)(b^2-x^2)}} \qquad \text{(replacing } x \text{ by } -x\text{).}
\end{aligned}
$$

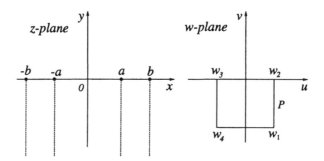

Figure 10.21: $F(H) = I(P)$

Thus

$$w_2 = -w_3 > 0 \quad \text{by (10.115)}.$$

Similarly,

$$
\begin{aligned}
w_4 = F(b) &= \int_0^a g(x)\,dx + \int_a^b g(x)\,dx \\
&= w_3 + \int_a^b (x+b)^{-\frac{1}{2}}(x+a)^{-\frac{1}{2}}|x-a|^{-\frac{1}{2}}|x-b|^{-\frac{1}{2}}e^{-\frac{\pi i}{2}}\,dx \\
&= w_3 - ip \quad \text{where } p = \int_a^b \frac{dx}{\sqrt{(x^2-a^2)(b^2-x^2)}}.
\end{aligned}
$$

Finally,

$$
\begin{aligned}
w_1 &= F(-b) = \int_0^{-b} g(x)\,dx = \int_0^{-a} g(x)\,dx + \int_{-a}^{-b} g(x)\,dx \\
&= w_2 + \int_{-a}^{-b} (x+b)^{-\frac{1}{2}}\left[|x+a|^{-\frac{1}{2}}e^{-\frac{\pi i}{2}}\right]\left[|x-a|^{-\frac{1}{2}}e^{-\frac{\pi i}{2}}\right]|x-b|^{-\frac{1}{2}}e^{-\frac{\pi i}{2}}\,dx.
\end{aligned}
$$

Hence

$$w_1 = w_2 - ip.$$

See Figure 10.21.

Thus the tranformation (10.114) maps the extended real axis and the upper half-plane onto $P \cup I(P)$ where P is the rectangle with vertices w_1, w_2, w_3, and w_4. (We note that P is determined by a and b.)

Remarks 10.6.6 (a) In 10.6.3, we saw that if Q is a given triangle, then $F(z)$ in (10.109) maps $\overline{H} \cup \{\infty\}$ onto $P \cup I(P)$, where P is similar to Q if $k_1\pi$ and $k_2\pi$ are measures of two exterior angles of Q. We observed that x_1 and x_2 were arbitrary in (10.109).

(b) Now given a rectangle Q, do there exist numbers a and b such that $F(z)$ in (10.114) maps the extended real axis onto a rectangle P which is similar to Q? It can be shown that the answer is "yes," and that a can be chosen arbitrarily. It is clear that not both a and b are arbitrary, unlike the case for x_1 and x_2 for the triangle in 10.6.3. (Two rectangles are not necessarily similar when their respective angles are congruent.)

(c) The question now arises: Given a polygon Q, is there a transformation of the form f in 10.6.1 which maps the extended real axis onto a polygon P similar to Q? The answer is "yes," but a proof is not easy. A proof may be based on the Riemann Mapping Theorem. (See Chapter 11.)

Let the given polygon Q in 10.6.6(c) have $n + 1$ vertices w_j' with exterior angles $m_j \pi$. To find a function f given in 10.6.1 which maps the extended real axis onto a polygon P similar to Q, we must determine appropriate numbers x_j and k_j. In order for P to be similar to Q, their corresponding angles must be congruent. Hence we take $k_j = m_j$. As stated in 10.6.6(c), there are numbers x_j such that the transformation

$$w = f(z) = \int_{z_0}^{z} (t - x_1)^{-m_1} \cdots (t - x_n)^{-m_n} \, dt \qquad (10.116)$$

maps the extended real axis onto a polygon P similar to Q.

Let $w_j = f(x_j), j = 1, 2, \cdots, (n + 1)$, be the vertices of the polygon P which is to be similar to Q. Clearly two of the points w_j, say w_1 and w_2, are arbitrary. But once $w_1 = f(x_1)$ and $w_2 = f(x_2)$ are determined, the remaining points w_j must be determined so that

$$\frac{|w_3 - w_2|}{|w_3' - w_2'|} = \cdots = \frac{|w_{n+1} - w_n|}{|w_{n+1}' - w_n'|} = \frac{|w_1 - w_{n+1}|}{|w_1' - w_{n+1}'|} = r,$$

where

$$r = \frac{|w_2 - w_1|}{|w_2' - w_1'|}.$$

This suggests that x_1 and x_2 may be chosen arbitrarily (which is in fact the case). By using n factors in (10.116) we have already determined that $f(\infty) = w_{n+1}$ is a vertex of P since $\sum_{j=1}^{n} m_j \neq 2$. [See 10.6.2(c).] Thus three points on the extended real axis, which are to map onto vertices of P, may be chosen arbitrarily.

To obtain a transformation of the form f in 10.6.1 which does not map the point ∞ onto a vertex of P, we write

$$w = f(z) = \int_{z_0}^{z} (t - x_1)^{-m_1} \cdots (t - x_{n+1})^{-m_{n+1}} \, dt \qquad (10.117)$$

and then determine the numbers x_j so that P is similar to the given polygon Q. [Here, $f(\infty)$ is not a vertex since $\sum_{j=1}^{n+1} m_j = 2$. See 10.6.2(b).]

If the polygon P is similar to Q, then P may be mapped onto Q by a linear transformation of the form

$$h(w) = Bw + E \qquad (10.118)$$

[that is, a rotation and stretching (or shrinking) followed by a translation]. Thus if the transformation f in (10.116) or (10.117) maps the extended real axis onto a polygon P similar to Q, then f may be followed by the transformation h in (10.118) to obtain

$$F(z) = h[f(z)] = Bf(z) + E. \qquad (10.119)$$

Then F in (10.119) maps the extended real axis onto Q and maps the upper half-plane H onto $I(Q)$ analytically and one-to-one.

The transformation

$$w = F(z) = B \int_{z_0}^{z} g(t)\, dt + E, \qquad (10.120)$$

where g is given in 10.6.1, is called the **Schwarz-Christoffel transformation**.

Remark 10.6.7 If we drop the hypothesis (10.86) in 10.6.1, then the image of the extended real axis may not be a closed polygonal line but may be a "degenerate polygon." We shall see an example of this in 10.6.8 where (10.86) does not hold for $-1 < k_3 < 1$.

Example 10.6.8 (Degenerate polygon). Find a Schwarz-Christoffel transformation which maps

$$\{z:\ \mathcal{I}(z) \geq 0\} \quad \text{onto} \quad \{w:\ \mathcal{I}(w) \geq 0 \text{ and } |\mathcal{R}(w)| \leq b\}$$

where $b > 0$.

Solution. Let $w_1 = -b$, $w_2 = b$ and $w_3 = b + iy$ where $y > 0$ and let T be the right triangle with vertices at these three points. The infinite half-strip is the limiting region of the closure $\overline{I(T)}$ of the interior of this triangle as $w_3 \to \infty$. See Figure 10.22.

The exterior angle at w_1 has measure $k_1 \pi$ which approaches $\frac{\pi}{2}$ as $w_3 \to \infty$. Thus $k_1 \to \frac{1}{2}$. In view of (10.120), this suggests that we should let F be given by

$$w = F(z) = B \int_0^z (t+1)^{-\frac{1}{2}}(t-1)^{-\frac{1}{2}}\, dt + E \qquad (10.121)$$

where B and E are to be determined so that $F(-1) = -b$ and $F(1) = b$. Thus

$$
\begin{aligned}
b = w_2 = F(1) &= B \int_0^1 |t+1|^{-\frac{1}{2}} \left[|t-1|^{-\frac{1}{2}} e^{-\frac{\pi i}{2}} \right] dt + E \\
&= -Bi \int_0^1 \frac{dt}{\sqrt{1-t^2}} + E. \\
&= -Bi \left[\operatorname{Arcsin} t \right]_0^1 + E
\end{aligned}
$$

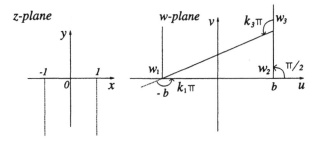

Figure 10.22: $F(\overline{H}) = \{w : \mathcal{I}(w) \geq 0 \text{ and } |\mathcal{R}(w)| \leq b\}$

Thus

$$b = -\frac{Bi\pi}{2} + E. \qquad (10.122)$$

(We should recognize that the last real integral from 0 to 1 is an improper integral. But since the real function Arcsin x on $(0,1)$ is continuous at $x = 1$, it is an easy mental exercise to justify the step involving $[\text{Arcsin}\, t]_0^1$.) Also by (10.121),

$$
\begin{aligned}
-b = w_1 = F(-1) &= B \int_0^{-1} |t+1|^{-\frac{1}{2}} \left[|t-1|^{-\frac{1}{2}} e^{-\frac{\pi i}{2}} \right] dt + E \\
&= -Bi \int_0^{-1} \frac{dt}{\sqrt{1-t^2}} + E
\end{aligned}
$$

or

$$-b = \frac{Bi\pi}{2} + E. \qquad (10.123)$$

By (10.122), and (10.123), we find that $E = 0$ and $B = \frac{2bi}{\pi}$. Thus (10.121) becomes

$$
\begin{aligned}
w = F(z) &= \frac{2bi}{\pi} \int_0^z (t+1)^{-\frac{1}{2}} (t-1)^{-\frac{1}{2}} dt \\
&= \frac{2bi}{\pi} \int_0^z \frac{dt}{(1+t)^{\frac{1}{2}} i(1-t)^{\frac{1}{2}}} \\
&= \frac{2b}{\pi} \int_0^z \frac{dt}{(1-t^2)^{\frac{1}{2}}}.
\end{aligned}
$$

Thus

$$w = F(z) = \frac{2b}{\pi} \text{Arcsin}\, z. \qquad (10.124)$$

See Ex. 4 in 10.8.3. Solving (10.124) for z, we have

$$z = \sin \frac{\pi w}{2b}. \qquad (10.125)$$

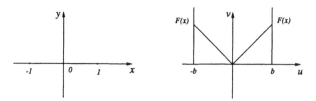

Figure 10.23: As x traces the real axis, $F(x)$ traces the boundary
of $\{w : \ \mathcal{I}(w) \geq 0$ and $|\mathcal{R}(w)| \leq b\}$.

By 10.4.2, we see that (10.125) maps the infinite half-strip.

$$\{w : \ |\mathcal{R}(w)| \leq b \text{ and } \mathcal{I}(w) \geq 0\} \tag{10.126}$$

onto the closed upper half-plane $\overline{H}$ given by $y \geq 0$.

It is interesting to observe that we can use (10.121) to show that the real axis
in the z-plane maps onto the boundary of the semi-infinite strip (10.126). For when
$x > 1$, it follows by (10.121) with $B = \frac{2bi}{\pi}$ and $E = 0$, that

$$
\begin{aligned}
F(x) &= B \int_0^1 g(t)\, dt + B \int_1^x g(t)\, dt \\
&= b + \frac{2bi}{\pi} \int_1^x |t+1|^{-\frac{1}{2}} |t-1|^{-\frac{1}{2}}\, dt \\
&= b + \frac{2bi}{\pi} \int_1^x \frac{dt}{\sqrt{t^2-1}}
\end{aligned}
$$

or

$$F(x) = b + \frac{2bi}{\pi} \ln\left(x + \sqrt{x^2-1}\right).$$

Thus as x varies along the positive real axis from 1 to ∞, we see that $F(x)$ moves
from the point $(b,0)$ to the point ∞ along the ray $u = b$ and $v \geq 0$. Similarly, by
Ex. 9 in 10.6.9, as x varies along the **negative real axis** from -1 to ∞, it follows
that $F(x)$ varies along the ray $u = -b$, $v \geq 0$ from $(-b,0)$ to ∞. See Figure 10.23.
Finally, as $z = x$ varies from -1 to 1, it follows from (10.124) that $w = F(x)$ varies
from $u = -b$ to $u = b$ along $v = 0$ (since Arcsin x varies from $-\frac{\pi}{2}$ to $\frac{\pi}{2}$).

Exercises 10.6.9

1. Let $f(x) = \int_{z_0}^x g(t)\, dt$ for each x in $[x_{j-1}, x_j]$ where g is given in (10.87). Prove
   each of the following statements.

   (a) The function $f(x) = u(x) + iv(x)$ with $x_{j-1} \leq x \leq x_j$ is a parametric
   representation of a curve with parameter x, and $f'(x) = g(x)$ represents a
   tangent vector to the curve at the point $[u(x), v(x)]$ for each x in (x_{j-1}, x_j).
   Hint. See 3.3.1(a) and 10.6.1(a) and (b).

(b) An argument of $f'(x)$ is constant on the open interval (x_{j-1}, x_j) and hence the curve traced by $f(x)$ is a line segment.

Solution of (b). Let $\mathrm{Arg}\,(u' + iv') = \alpha$ (a constant) on (x_{j-1}, x_j).

Case 1. Suppose u' is never zero in (x_{j-1}, x_j). Now $c = \tan \alpha = \frac{v'}{u'} = \frac{dv}{du}$ is constant. Thus $v = cu + b$ on (x_{j-1}, x_j) which means that $f(x)$ traces out a line segment.

Case 2. Suppose $u'(x) = 0$ for some x in (x_{j-1}, x_j). Then $v'(x) \neq 0$, since $u'(x) + iv'(x) = g(x)$, which is never zero in (x_{j-1}, x_j). Thus the tangent line is vertical at $f(x)$. Since the inclination α is constant, the tangent line remains vertical. Thus $u'(x) \equiv 0$ so that $u = k$ (a constant), which again means that $f(x)$ traces out a line segment.

In Ex. 2–4, let P be the polygon given in 10.6.1(e), let f and w_j be given in 10.6.1 and let $w_0 \in I(P) \cup E(P)$.

2. Prove the following statements.

(a) There is a number $r_0 > 0$ such that if $|z| > r_0$ and $z \in D$, then $|f(z) - w_0| > \frac{1}{2}|w_{n+1} - w_0|$.

(b) There is a number $\delta > 0$ such that if $z \in D \cap N_\delta(x_j)$, then $|f(z) - w_0| > \frac{1}{2}|w_0 - w_j|$.

Proof of (a). Since $w_{n+1} \in P$ and $w_0 \notin P$, we know that $|w_0 - w_{n+1}| > 0$. Let $\epsilon = \frac{1}{2}|w_0 - w_{n+1}|$. By 10.6.1(c), let r_0 be such that if $z \in D$ and $|z| > r_0$, then $|f(z) - w_{n+1}| < \epsilon = \frac{1}{2}|w_0 - w_{n+1}|$. Thus if $z \in D$ and $|z| > r_0$, then

$$
\begin{aligned}
|w_0 - f(z)| &= |(w_0 - w_{n+1}) - [f(z) - w_{n+1}]| \\
&\geq |w_0 - w_{n+1}| - |f(z) - w_{n+1}| \\
&> |w_0 - w_{n+1}| - \frac{1}{2}|w_0 - w_{n+1}| = \frac{1}{2}|w_0 - w_{n+1}|.
\end{aligned}
$$

3. Let S_r be the cco semicircle $|z| = r$, where $\mathcal{I}(z) \geq 0$. Prove

$$
\int_{S_r} \frac{f'(z)\,dz}{f(z) - w_0} \to 0 \quad \text{as } r \to \infty.
$$

Hint. By 3.4.10(g), (10.94), and Ex. 2(a), for $r > r_0$ in Ex. 2(a) and for $r > \max 2|x_j|$,

$$
\left| \int_{S_r} \frac{g(z)\,dz}{f(z) - w_0} \right| \leq \frac{2M\pi r}{|w_0 - w_{n+1}|\, r^s} \to 0 \quad \text{as } r \to \infty \text{ by (10.95).}
$$

4. Let S_{r_j} be the clockwise oriented semicircle $|z - x_j| = r_j$ where $\mathcal{I}(z) \geq 0$. Prove

$$\int_{S_{r_j}} \frac{f'(z)\,dz}{f(z) - w_0} \to 0 \quad \text{as } r_j \to 0.$$

Hint. Note that $g(z) = (z - x_j)^{-k_j} h(z)$ where $h(z)$ is the product of the remaining factors in $g(z)$. Let E be a closed half-disk $|z - x_j| \leq \rho$ with $\mathcal{I}(z) \geq 0$ such that $x_k \notin E$ if $k \neq j$. Since $h \in C(E)$, let $K = \max|h(z)|$ on E. For $r_j < \min\{\rho, \delta\}$, where δ is given in Ex. 2(b),

$$\left| \int_{S_{r_j}} \frac{g(z)\,dz}{f(z) - w_0} \right| = \left| \int_{S_{r_j}} \frac{(z - x_j)^{-k_j} h(z)\,dz}{f(z) - w_0} \right|$$

$$\leq \frac{2 r_j^{-k_j} K \pi r_j}{|w_0 - w_j|} \to 0 \quad \text{as } r_j \to 0.$$

5. In 10.6.4, show that $w_1 = e^{\frac{\pi i}{3}} \beta\left(\frac{1}{2}, \frac{1}{3}\right)$ if $a = 1$, where β is defined by (7.63).
   Hint. In (10.112), let $s = x^2$.

6. In 10.6.4 show that $F(0) = \dfrac{w_3}{2} e^{\frac{\pi i}{3}}$. **Hint.** $F(0) = \int_a^0 (x + a)^{-\frac{2}{3}} (x - a)^{-\frac{2}{3}}\,dx = -\int_0^a (x + a)^{-\frac{2}{3}} (a - x)^{-\frac{2}{3}} e^{-\frac{2}{3}\pi i}\,dx$

7. Show that the Schwarz-Christoffel transformation

$$w = F(z) = \int_0^z (t + 1)^{-\frac{1}{2}} t^{-\frac{1}{2}} (t - 1)^{-\frac{1}{2}}\,dt$$

   maps H onto $I(P)$ where P is the square with vertices

$$w_1 = \frac{1}{2} \beta\left(\frac{1}{4}, \frac{1}{2}\right), \quad w_2 = 0, \quad w_3 = -i w_1, \quad \text{and} \quad w_4 = w_1 + w_3.$$

8. (a) What can be said about $f(H)$ if $f(z) = B\,F(z) + E$ where B and E are complex constants and where F is given in Ex. 7?

   (b) Find B and E such that the vertices of $F(H)$ are

$$w_1' = \frac{i}{2} \beta\left(\frac{1}{4}, \frac{1}{2}\right), \quad w_2' = 0, \quad w_3' = -i w_1', \quad \text{and} \quad w_4' = w_3' + w_1'.$$

Ans. (a) $f(H)$ is the interior of a square.
     (b) $B = i \quad E = 0$

9. Show that in 10.6.8, as x varies along the **negative real axis** from -1 to ∞, the point $F(x)$ varies along the ray $u = -b$, $v \geq 0$ from the point $(-b, 0)$ to ∞.

   Hint.
   $$F(x) = F(-1) + \frac{2bi}{\pi} \int_{-1}^{x} e^{-\frac{\pi i}{2}} |t + 1|^{-\frac{1}{2}} e^{-\frac{\pi i}{2}} |t - 1|^{-\frac{1}{2}} \, dt$$
   $$= -b - \frac{2bi}{\pi} \ln \left| x + \sqrt{x^2 - 1} \right|$$
   $$= -b - \frac{2bi}{\pi} \ln \frac{1}{\left| x - \sqrt{x^2 - 1} \right|}$$
   $$= -b + \frac{2bi}{\pi} \ln \left| x - \sqrt{x^2 - 1} \right|.$$

10. Find a Schwarz-Christoffel transformation which maps H onto the semi-infinite strip $S = \{w : \ \mathcal{R}(w) > 0 \text{ and } 0 < \mathcal{I}(w) < b\}$. **Hint.** S is obtained by rotating the strip in 10.6.8 through $-90°$, shrinking it by a factor $\frac{1}{2}$ and translating the result by $\frac{bi}{2}$. Thus $w = -\frac{i}{2}F(z) + \frac{bi}{2}$ where F is given in (10.124). Solving for z, we obtain $z = \cosh \frac{\pi w}{b}$.

11. Solve Ex. 10 directly by using the Schwarz-Christoffel transformation (10.121). **Hint.** S is the limiting region of the interior of a triangle with vertices
    $$w_1 = bi, \quad w_2 = 0, \quad \text{and} \quad w_3 > 0 \text{ as } w_3 \to \infty.$$

Now substituting into (10.121),
$$bi = w_1 = B \int_0^{-1} (t + 1)^{-\frac{1}{2}} (1 - t)^{-\frac{1}{2}} e^{-\frac{\pi i}{2}} \, dt + E \quad \text{and}$$
$$0 = w_2 = B \int_0^{1} (t + 1)^{-\frac{1}{2}} (1 - t)^{-\frac{1}{2}} e^{-\frac{\pi i}{2}} \, dt + E.$$

12. Find a Schwarz-Christoffel transformation which maps the upper half-plane H onto the infinite strip $T = \{w : \ 0 < \mathcal{I}(w) < b\}$.

    Solution. Let P be a triangle with vertices $w_1 = -a + bi$, $w_2 = 0$, and $w_3 = a + bi$ where $a > 0$ and with exterior angles πk_1 at w_1 and πk_2 at w_2. Now T is the limiting region of $I(P)$ as w_1 and w_3 approach ∞. See Figure 10.24. Then $\pi k_1 \to \pi$ and $\pi k_2 \to 0$. Thus $k_1 \to 1$ and $k_2 \to 0$. This suggests that we try a transformation of the form
    $$w = F(z) = B \int_1^z (t - 0)^{-1} (t - 1)^0 \, dt + E$$
    where we have chosen $x_1 = 0$ and $x_2 = 1$. Thus $w_2 = F(x_2) = F(1) = 0$. This implies $E = 0$. Thus $w = F(z) = B \, \text{Log} \, z$. Now $F(x) = B[\text{Log} |x| + \pi i]$ if $x < 0$. Thus F maps the negative x-axis onto the line $v = B\pi$. Hence we take $B = \frac{b}{\pi}$. By Ex. 10 in 10.4.1, we see that the function $F(z) = \frac{b}{\pi} \, \text{Log} \, z$ maps H onto T.

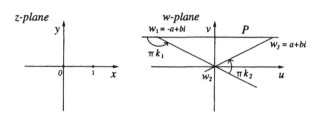

Figure 10.24: $I(P) \to T$ as $a \to \infty$.

13. Find a Schwarz-Christoffel transformation which maps H onto the interior of an isosceles right triangle.

Solution. In view of 10.6.3, we take $x_1 = -1$ and $x_2 = 1$ and we write

$$w = F(z) = \int_0^z (t+1)^{-\frac{3}{4}}(t-1)^{-\frac{3}{4}}\, dt.$$

Since $-1 < k_3 = 2 - \frac{3}{4} - \frac{3}{4} = \frac{1}{2} < 1$, it follows by 10.6.2(c) that F maps the extended real axis onto a closed polygon P with three sides. Also, since the interior angles of P are $\frac{\pi}{4}, \frac{\pi}{4}$, and $\frac{\pi}{2}$, it follows that P is an isosceles right triangle and 10.6.1 implies $F(H) = I(P)$. We may write $F(z)$ as

$$
\begin{aligned}
F(z) &= \int_0^z |t+1|^{-\frac{3}{4}} e^{-\frac{3}{4}\theta_1 i} |t-1|^{-\frac{3}{4}} e^{-\frac{3}{4}\theta_2 i}\, dt \\
&= \int_0^z |t^2-1|^{-\frac{3}{4}} e^{-\frac{3}{4}(\theta_1+\theta_2)i}\, dt \\
&= \int_0^z \frac{dt}{(t^2-1)^{\frac{3}{4}}}
\end{aligned}
$$

where $-\frac{\pi}{2} < \theta_j < \frac{3\pi}{2}$ by (10.88).

14. Find the vertices of triangle P in the solution of Ex. 13. See Figure 10.25.

Hint.
$$
\begin{aligned}
w_1 &= F(-1) = \int_0^{-1} (t+1)^{-\frac{3}{4}}(1-t)^{-\frac{3}{4}} e^{-\frac{3}{4}\pi i}\, dt = \frac{1+i}{\sqrt{2}} \int_0^1 \frac{dx}{(1-x^2)^{\frac{3}{4}}} \\
w_2 &= F(1) = \int_0^1 (t+1)^{-\frac{3}{4}}(1-t)^{-\frac{3}{4}} e^{-\frac{3}{4}\pi i}\, dt = -w_1 \\
w_3 &= w_2 + b = w_1 - bi \quad \text{where } 0 < b = \int_1^\infty \frac{dx}{(x^2-1)^{\frac{3}{4}}}
\end{aligned}
$$

15. Show that the transformation

$$w = F(z) = \int_0^z (t+1)^{-\frac{1}{2}} t^{-\frac{1}{2}}(t-1)^{-\frac{1}{2}}\, dt$$

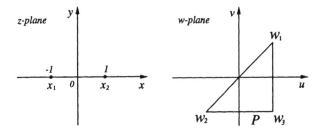

Figure 10.25: $F(H) = I(P)$, where P is an isosceles right triangle.

maps the upper half-plane H onto the interior of a square. **Hint.** Since $k_1 = k_2 = k_3 = \frac{1}{2}$, it follows that F maps the extended real axis onto a rectangle. Show that $|w_3 - w_2| = |w_2 - w_1|$. Now $w_2 = F(x_2) = F(0) = 0$.

$$w_1 = \int_0^{-1} (t+1)^{-\frac{1}{2}} (-t)^{-\frac{1}{2}} e^{-\frac{\pi i}{2}} (1-t)^{-\frac{1}{2}} e^{-\frac{\pi i}{2}} \, dt$$

$$= \int_0^1 \frac{dx}{\sqrt{x(1-x^2)}} > 0.$$

Also, $w_3 = F(1) = -iw_1$.

16. What is the image of the upper half-plane H under the transformation

$$w = F(z) = \int_0^z t^{-\frac{2}{3}} (t-1)^{-\frac{5}{6}} \, dt ?$$

Ans. $F(H)$ is the interior of a 30–60° right triangle with vertices $w_1 = 0$ and $w_2 = -\dfrac{\sqrt{3}+i}{2} \displaystyle\int_0^1 \dfrac{dx}{x^{\frac{2}{3}}(1-x)^{\frac{5}{6}}}$. See Figure 10.26.

17. Sketch the image of the extended real axis under the transformation

$$w = F(z) = \int_0^z (t+1)^{-\frac{1}{2}} t^{-\frac{1}{3}} (t-1)^{-\frac{2}{3}} \, dt.$$

Hint. $w_2 = 0 < w_1 = \displaystyle\int_0^1 \dfrac{dx}{x^{\frac{1}{3}}(1-x)^{\frac{1}{2}}(1+x)^{\frac{2}{3}}}$ and $w_3 = -\dfrac{1+\sqrt{3}i}{2} \displaystyle\int_0^1 (x+1)^{-\frac{1}{2}} x^{-\frac{1}{3}} (1-x)^{-\frac{2}{3}} \, dx$ See Figure 10.27.

18. Find a Schwarz-Christoffel transformation which maps H onto the angular region $\{w : \ 0 < \operatorname{Arg} w < m\pi\}$ where $0 < m < 1$. See Figure 10.28.

Solution. Since the degenerate polygon has only one vertex (or turning point), we expect only one x_j so we take $x_1 = 0$. See the note following the proof of

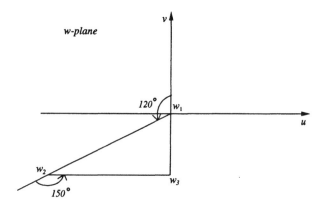

Figure 10.26: $F(H)$ is the interior of a $30 - 60°$ right triangle.

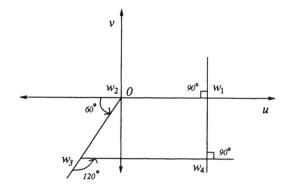

Figure 10.27: $F(z) = \int_0^z (t+1)^{-\frac{1}{2}} t^{-\frac{1}{3}} (t-1)^{-\frac{2}{3}} \, dt$

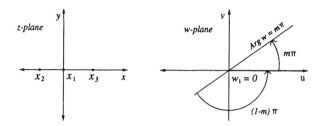

Figure 10.28: $F(H) = \{w : \ 0 < \ \mathrm{Arg}\, w < m\pi\}$

10.6.1(e). Now as $z = x'$ moves along the negative x-axis towards x_1, its image $F(z)$ must move along the ray $\text{Arg}\, w = m\pi$ towards $w_1 = 0$ as in the proof of 10.6.1(d). But as $z = x''$ passes through the point x_1 and moves to the right, the point $F(z)$ must make an abrupt turn through an angle $(1 - m)\pi$. Thus we try the transformation $w = B \int_0^z t^{m-1}\, dt = z^m$, taking $B = m$. By Ex. 9 in 10.4.1, this is the desired transformation.

10.7 Multivalued Functions

Definition 10.7.1 A multivalued function on X to Y is a correspondence f which associates with each x in X at least one member $y = f(x)$ in Y and which associates at least two members of Y with some x in X.

Note. A multivalued function is not a function. To distinguish between a multivalued function and a function, we sometimes refer to a function as a **single-valued function**.

If we write $w = (z-a)^{\frac{1}{2}}$ and indicate that the notation on the right stands for each square root of $z - a$, then the correspondence determined by this rule is an example of a multivalued function. In this example, if $z - a = re^{i\theta}$ with $r \geq 0$, then

$$w = \sqrt{r}\, e^{i\left(\frac{\theta}{2} + k\pi\right)} \quad \text{where } k = 0, 1. \tag{10.127}$$

Now for a real constant θ_0 and any complex constant a, we let

$$w_1 = \sqrt{r}e^{\frac{i\theta}{2}} \tag{10.128}$$

where $r = |z - a|$ and where $\theta = \arg(z - a)$ is in the interval $[\theta_0, \theta_0 + 2\pi)$. Then w_1 is a single-valued function whose domain is R^2. Here, $\sqrt{r}$ is still understood to be the non-negative square root of the non-negative real number r. See Figure 10.29. As in the figure, we shall let L_a be the ray from a which is given by

$$L_a = \{z : \ \theta_0 \text{ is an argument of } (z - a)\}.$$

Also, we shall let D be the domain given by

$$D = R^2 - L_a. \tag{10.129}$$

Remarks 10.7.2 (a) The single-valued function w_1 given in (10.128) is analytic in D. To show this, we use the polar form of the Cauchy-Riemann equations in 5.4.9 with

$$u = \sqrt{r} \cos \frac{\theta}{2} \quad \text{and} \quad v = \sqrt{r} \sin \frac{\theta}{2}.$$

Hence

$$\frac{1}{r} v_\theta = \frac{1}{2\sqrt{r}} \cos \frac{\theta}{2} = u_r \quad \text{and} \quad \frac{1}{r} u_\theta = -\frac{1}{2\sqrt{r}} \sin \frac{\theta}{2} = -v_r.$$

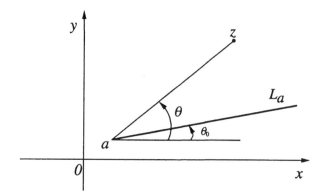

z-plane   Figure 10.29: $D = R^2 - L_a$

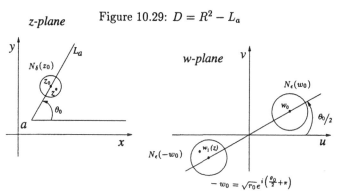

Figure 10.30: $z \in N_\delta(z_0)$, but $w_1(z) \notin N_\epsilon(w_0)$.

(b) We next observe that w_1 is not continuous at any point on the ray $L_a - \{a\}$. For let $z_0 \in [L_a - \{a\}]$, let $r_0 = |z_0 - a|$ and let $w_0 = w_1(z_0)$. Then $|w_0| = \sqrt{r_0} > 0$. Let $0 < \epsilon < |w_0|$. Clearly, $N_\epsilon(w_0) \cap N_\epsilon(-w_0) = \emptyset$. See Figure 10.30. Let δ be **any** positive number. There is some z in $N_\delta(z_0)$ such that $\arg(z - a) < \theta_0 + 2\pi$ but such that $\arg(z - a)$ is sufficiently near $\theta_0 + 2\pi$ that $w_1(z) \in N_\epsilon(-w_0)$. Hence $w_1(z) \notin N_\epsilon(w_0)$. Thus w_1 is not continuous at z_0.

Now let w^* be the single-valued function obtained by taking $k = 1$ in (10.127) where $r = |z - a|$ and $\theta_0 \le \arg(z - a) < \theta_0 + 2\pi$. Then

$$w^* = \sqrt{r}\, e^{i\left(\frac{\theta}{2} + \pi\right)} = -w_1.$$

Since $w^* = -w_1$, it follows that w^* is analytic in D and that w^* is not continuous at any point on the ray $L_a - \{a\}$. Each of these functions (w_1 and w^*) is called a branch in D of the multivalued function given in (10.127).

The ray L_a is called the branch cut (or cut) for w_1 and for w^*. Since θ_0 is arbitrary, we may "cut" R^2 along any ray from a to obtain (analytic) branches of $(z-a)^{\frac{1}{2}}$. [We could cut R^2 along a "curved path" to construct two (analytic) branches of $(z-a)^{\frac{1}{2}}$.]

Definition 10.7.3 Let f be a multivalued function and let D be a domain in R^2.

(a) A **branch in D of f** is a single-valued function g defined on D such that

$$g \text{ is analytic in } D$$

and
$$\text{for each } z \text{ in } D, \quad g(z) \text{ is one of the values of } f(z).$$

(b) The set $\{z : \ g \notin A(z)\}$ is called the **branch cut** (or **cut**) of the branch g.

(c) A **branch point of f** is any point common to the branch cuts for all branches of f. (That is, z_0 is a branch point of f iff there is **no** branch of f which is analytic at z_0.)

It is seen that $(z-a)^{\frac{1}{2}}$ has a branch point at $z=a$.

Example 10.7.4 Let a and b be two distinct complex constants and let f be the multivalued function given by

$$f(z) = [(z-a)(z-b)]^{\frac{1}{2}}.$$

Let L_a and L_b be the rays from a and from b, respectively, which are given by

$$L_a = \{z : \ \theta_0 = \arg(z-a)\} \quad \text{and} \quad L_b = \{z : \ \phi_0 = \arg(z-b)\} \tag{10.130}$$

where $\theta_0 = \text{Arg}(a-b)$ and where $\phi_0 = \theta_0 - \pi$. Find two branches in D of f where

$$D = D_a \cap D_b, \quad D_a = R^2 - L_a, \quad \text{and} \quad D_b = R^2 - L_b. \tag{10.131}$$

Specify the cut for each branch. See Figure 10.31.

Solution. Let w_1 and w_2 be the single-valued functions which are specified in the following array.

$$
\begin{array}{ll}
w_1 = \sqrt{r}\, e^{\frac{i\theta}{2}} & w_2 = \sqrt{s}\, e^{\frac{i\phi}{2}} \\
r = |z-a| & s = |z-b| \\
\theta = \arg(z-a) & \phi = \arg(z-b) \\
\theta \text{ is in } [\theta_0, \theta_0 + 2\pi) & \phi \text{ is in } (\phi_0, \phi_0 + 2\pi] \\
\theta_0 = \text{Arg}(a-b) & \phi_0 = \theta_0 - \pi
\end{array}
\tag{10.132}
$$

Now w_1 is analytic in D_a and w_2 is analytic in D_b by 10.7.2(a). Thus $w_1 w_2$ is analytic in $D_a \cap D_b$. So we let $g = w_1 w_2$. Since g is analytic in D and since $g(z)$ is one of

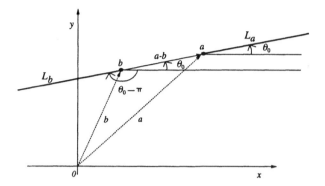

Figure 10.31: $g \in A(D)$, where $g = w_1 w_2$ and $D = R^2 - (L_a \cup L_b)$.

the values of $f(z)$ for each z in D, it follows from Definition 10.7.3 that g is a branch in D of f.

By 10.7.2(b), we know that w_1 is not continuous at any point on the ray L_a except at the point a. Thus w_1 is not analytic at any point on L_a. It is not analytic at a since each neighborhood of a contains points at which w_1 is not continuous. Hence g is not analytic at any point of L_a. For if it were, then $w_1 = \frac{g}{w_2}$ would be analytic at such a point. For we know that w_2 is analytic in L_a and is never zero in L_a.

Similarly, g is not analytic at any point on L_b. Thus $L_a \cup L_b$ is the branch cut for g.

It is clear that $-g$ is another branch in D of f.

Of course, we may cut the plane along any ray from the point a to obtain a branch w_a of $(z - a)^{\frac{1}{2}}$. Likewise, we may cut the plane along any ray from the point b to obtain a branch w_b of $(z - b)^{\frac{1}{2}}$. Then the product $w_a w_b$ is a branch of the multivalued function f given in Example 10.7.4. In Ex. 2 of 10.7.6, we give a branch of f whose cut is the closed line segment from a to b. For use in that exercise, we give another branch of $(z - a)^{\frac{1}{2}}$ and another branch of $(z - b)^{\frac{1}{2}}$ in the following array.

$$
\begin{array}{ll}
w_3 = \sqrt{r}\, e^{\frac{i\theta'}{2}} & w_4 = \sqrt{s}\, e^{\frac{i\phi'}{2}} \\
r = |z - a| & s = |z - b| \\
\theta' = \arg(z - a) & \phi' = \arg(z - b) \qquad (10.133) \\
\theta' \text{ is in } (\theta_0 - \pi, \theta_0 + \pi] & \phi' \text{ is in } [\theta_0, \theta_0 + 2\pi) \\
\theta_0 = \operatorname{Arg}(a - b) & \theta_0 = \operatorname{Arg}(a - b)
\end{array}
$$

In the last mentioned exercise, we shall use the rays L_{a3} and L_{b4} which are given by

$$
L_{a3} = \{z : \ \theta_0 - \pi = \arg(z - a)\} \text{ and } L_{b4} = \{z : \ \theta_0 = \arg(z - b)\}. \qquad (10.134)
$$

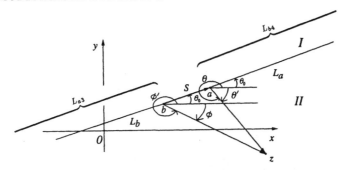

Figure 10.32: $L_{a3} = S \cup L_b$ $L_{b4} = S \cup L_a$

Remark 10.7.5 The number z_0 is a branch point of the multivalued function f if the following condition holds.

In each neighborhood of z_0, there is a circle centered at z_0 such that if w is a value of $f(z)$ which varies continuously as z traverses this circle more than once, then w **must** pass from one branch of f to another.

Exercises 10.7.6

1. In (a)–(b), apply the results of 10.7.4 to obtain two branches of the given multivalued function. Also, specify the branch points of the given multivalued function.

   (a) $f(z) = (z^2 + 1)^{\frac{1}{2}}$
   (b) $f(z) = (1 - z^2)^{\frac{1}{2}}$ **Hint.** $(1 - z^2)^{\frac{1}{2}} = i(z^2 - 1)^{\frac{1}{2}}$

2. Let $h = w_1 \, w_4$ and let $k = w_3 \, w_2$ where w_j is given in (10.132) and (10.133) for $j = 1$ to 4. Using the notation given in (10.130) through (10.134), verify each of the following statements.

   (a) The function w_4 is analytic in $R^2 - L_{b4}$. [Use 10.7.2(a).]
   (b) The function w_1 is analytic in $R^2 - L_{b4}$. [Use 10.7.2(a) and the fact that $L_a \subset L_{b4}$.]
   (c) The function h is analytic in $R^2 - L_{b4}$. [Use (a) and (b).]
   (d) The function k is analytic in $R^2 - L_{a3}$. [Use the pattern established in (a)–(c).]
   (e) Now $h = k$ on $R^2 - L_{a3}$.
      Solution of (e). By (10.132) and (10.133),

      $$h = \sqrt{rs} \, e^{\frac{i}{2}(\theta + \phi')} \quad \text{and} \quad k = \sqrt{rs} \, e^{\frac{i}{2}(\theta' + \phi)}.$$

Now if z is on $L_a - \{a\}$ or on side I of $L_a \cup L_{a3}$ as shown in Figure 10.32, then $\theta = \theta'$ and $\phi = \phi'$. But if z is on the side II in the figure, then $\theta = \theta' + 2\pi$ and $\phi' = \phi + 2\pi$. Thus by the last formulas for h and k, we see that $h = k$ on $R^2 - L_{a3}$.

(f) Hence h is analytic in $L_a - \{a\}$. [Use (d)-(e) and the fact that $(L_a - \{a\}) \subset (R^2 - L_{a3})$.]

(g) Thus h is analytic in $R^2 - S$ where S is the closed line segment from the point b to the point a. [Use (c) and (f).]

(h) If z is any point on $S - \{a, b\}$, then h is not continuous at z. [Now $z \notin L_a$. Thus by 10.7.2(a), the function w_1 is analytic at z and hence is continuous at z. But $z \in L_{b4}$. Thus w_4 is not continuous at z by 10.7.2(b). Hence h is not continuous at z. For if it were, then $w_4 = \frac{h}{w_1}$ would be continuous at z.]

(i) If z is any point on S, then h is not analytic at z. Thus the branch cut for h is the closed segment S. [If $z \in (S - \{a, b\})$, then h is not analytic at z by (h). Each neighborhood of a contains points of $S - \{a, b\}$ where h is not continuous and hence not differentiable. Thus h is not analytic at a. Similiarly, h is not analytic at b.]

3. Let $f(z) = [(z + 2)(z - 3)]^{\frac{1}{2}}$ and let C be a circle centered at $z = 3$. Let $w = \sqrt{|z + 2||z - 3|}\, e^{\frac{i}{2}(\theta + \phi)}$ where $\theta = \arg(z - 3)$ and $\phi = \arg(z + 2)$. Let z traverse C as θ varies from 0 to 2π and ϕ varies continuously from 0. Let w_0 be the value of w at $\theta = \phi = 0$.

(a) Show that if $-2 \in E(C)$, then $w \to -w_0$ as $\theta \to 2\pi$. [Thus w passes from one branch of f to another as z completes the circuit around C if $-2 \in E(C)$.]

(b) Show that if $-2 \in I(C)$, then $w \to w_0$ as $\theta \to 2\pi$. **Hint.** Examine the range of ϕ as θ ranges from 0 to 2π.

(c) Use 10.7.5 to show that -2 and 3 are branch points of f and use 10.7.3(c) to show that -2 and 3 are the only branch points of f. For the last half of (c), we observe that -2 and 3 are the only points common to the cuts for the branch g in 10.7.4 and the branch h in Ex. 2. See Ex. 2(i).

In Ex. 4–12, let $f(z) = \log z$ and for each k in J, let

$$w_k = \text{Log}\,|z| + i(\theta + 2k\pi)$$

where $\theta = \text{Arg}\,z$ is in the interval $(-\pi, \pi)$. Also, let $D = R^2 - L$ where L is the set of all nonpositive reals.

4. (a) Show that w_k is a branch in D of f. [Recall $\frac{dw_k}{dz} = \frac{1}{z}$ for each z in D by 5.4.10.]

(b) Show that for each z in D, we have $w_j(z) \neq w_k(z)$ if $j \neq k$. (The branch w_0 is called the principal branch of $\log z$ and is denoted by **Log z**.)

5. Show that any ray L' from the origin, given by $\theta = \arg z$, is the cut for infinitely many branches of f. Observe that the branches are given by $g_k(z) = \text{Log}\,|z| + i(\theta + 2k\pi)$ where $\theta = \arg z$ is in the interval $(\theta_0, \theta_0 + 2\pi)$. Show that $g_k \in A(D')$, $D' = R^2 - L'$.

6. Use Ex. 5 to show that if $z_0 \neq 0$, then z_0 is not a branch point of f.

7. Use 10.7.5 to show that 0 is a branch point of f.

In Ex. 8–12, draw a figure in the w-plane which represents the image indicated.

8. the image $w_0(D)$ of D under the mapping w_0

9. $w_k(D)$ for $k = 1, -1, 2$

10. $w_0(C)$ where C is the circle given by $|z| = r_0$ where r_0 is a positive constant

11. $w_0(S)$ where $S = \overline{N_r(0)} - L$ for a constant $r > 0$

12. $w_0(H)$ where H is the ray $\theta = \theta_0$

13. Let L_0 be the ray given by $\theta = \theta_0$ and let $D = R^2 - L_0$. For each z in D, let $f(z) = \sqrt{r}\,e^{\frac{i\theta}{2}}$ where $r = |z|$ and where $\theta = \arg z$ is in the interval $(\theta_0, \theta_0 + 2\pi)$. Prove f is analytic in D and $f'(z) = \frac{1}{2f(z)}$ on D. Thus $\frac{d}{dz}\sqrt{z} = \frac{1}{2\sqrt{z}}$ if $\sqrt{z}$ denotes the branch of $z^{\frac{1}{2}}$ given by f. [Use 5.4.9.]

14. Use Ex. 13 and the chain rule to show that $\frac{d}{dz}w_k = \frac{1}{2w_k}$ for $k = 1, 2$ where w_k is given in 10.7.4.

15. Let $g(z) = w_1 w_2$ as in 10.7.4. Show that $g'(z) = \frac{2z-(a+b)}{2w_1 w_2}$ for each z in $R^2 - (L_a \cup L_b)$.

10.8 The Inverse Trigonometric Functions

Remarks 10.8.1 (The Inverse Sine Function). For each z in R^2, there are infinitely many values of w such that $z = \sin w$. To find these values of w (for a given z), we write

$$z = \sin w = \frac{e^{iw} - e^{-iw}}{2i} = \frac{e^{2iw} - 1}{2ie^{iw}}$$

$$= \frac{p^2 - 1}{2ip} \quad \text{where } p = e^{iw}.$$

Thus

$$p^2 - 2izp - 1 = 0.$$

Solving for p,

$$e^{iw} = p = iz \pm \sqrt{1 - z^2} = iz \pm i\sqrt{z^2 - 1},$$

where $\sqrt{z^2 - 1}$ is one of the two values of $(z^2 - 1)^{\frac{1}{2}}$. Thus

$$iw = \log i \left[z \pm \sqrt{z^2 - 1} \right] \quad \text{by 5.4.5.}$$

Now for each z in R^2, there are infinitely many values of w such that

$$\begin{aligned}
w &= -i \log \left[i \left(z \pm \sqrt{z^2 - 1} \right) \right] \\
&= -i \left[\log i + \log \left(z \pm \sqrt{z^2 - 1} \right) \right] \quad \text{by (5.34)} \\
&= -i \log i \mp i \log \left(z + \sqrt{z^2 - 1} \right) \quad \text{by Ex. 6 in 10.8.3.} \quad (10.135)
\end{aligned}$$

Each value of w in (10.135) is a number whose sine is z. [This follows since $e^{\log z} = z$ if $z \neq 0$ by 5.4.5. Note that $z + \sqrt{z^2 - 1} \neq 0$. For suppose $z = -\sqrt{z^2 - 1}$, then $-1 = 0$, a contradiction.] For a fixed number z, each number w given by formula (10.135) is denoted by $\sin^{-1} z$ or $\arcsin z$.

One of the infinitely many branches of w in (10.135) is singled out and called the **principal arcsine of z** and is denoted by $\text{Sin}^{-1}z$. This branch is the following

$$\text{Sin}^{-1}z = \frac{\pi}{2} + i \log \left(z + \sqrt{z^2 - 1} \right) \quad (10.136)$$

where [taking $a = 1$ and $b = -1$ in 10.7.4]

$$\begin{aligned}
\sqrt{z^2 - 1} &= \sqrt{|z^2 - 1|}\, e^{\frac{i}{2}(\theta + \phi)}, \\
\theta &= \arg(z - 1) \text{ is in the interval } [0, 2\pi), \quad \text{and} \\
\phi &= \arg(z + 1) \text{ is in the interval } (-\pi, \pi]. \quad (10.137)
\end{aligned}$$

In 10.8.2, we observe that

$$\text{if } -1 < z = x < 1, \text{ then } -\frac{\pi}{2} < \text{Sin}^{-1}x < \frac{\pi}{2}. \quad (10.138)$$

In Ex. 1 of 10.8.3, we observe that $\text{Sin}^{-1}0 = 0$ and that $\text{Sin}^{-1}(\pm 1) = \pm\frac{\pi}{2}$. Thus if $-1 \leq z = x \leq 1$, then the value of $\text{Arcsin } z$ given in (10.136) agrees with the value of $\text{Arcsin } x$ used in the calculus. In fact, we chose our branch of $(z^2 - 1)^{\frac{1}{2}}$ and the principal branch of the logarithm so that (10.138) will hold.

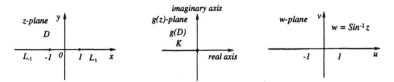

Figure 10.33: $\text{Sin}^{-1}z = \frac{\pi}{2} + i\,\text{Log}\,g(z)$ $D = R^2 - (L_1 \cup L_{-1})$

As in 10.7.4, the branch cut for $\sqrt{(z-1)(z+1)}$, given in (10.137), is $L_1 \cup L_{-1}$ where L_1 is the closed ray $\theta = \arg(z-1) = 0$ and L_{-1} is the closed ray $\phi = \arg(z+1) = \pi$. Using the branch of $(z^2 - 1)^{\frac{1}{2}}$ given in (10.137), we let $g(z) = z + \sqrt{z^2 - 1}$. Then

$$g \text{ is analytic in } D \quad \text{where } D = R^2 - (L_1 \cup L_{-1}). \qquad (10.139)$$

Also, using K for the nonpositive real axis, we notice that

$$\text{Log}\, z \text{ is analytic in } (R^2 - K). \qquad (10.140)$$

Now $g(D) \subset R^2 - K$ by Ex. 2 in 10.8.3. Thus by 2.3.6 and by (10.139) and (10.140),

$$\text{Log}\, g(z) \text{ is analytic in } D. \qquad (10.141)$$

Hence $\text{Sin}^{-1}z$ is analytic in D by (10.136). Thus $L_1 \cup L_{-1}$ is the branch cut for $\text{Sin}^{-1}z$. See Figure 10.33.

Example 10.8.2 Prove (10.138).

Solution. By (10.137),

$$
\begin{aligned}
\sqrt{x^2 - 1} &= \sqrt{|x^2 - 1|}\, e^{\frac{i}{2}(\theta + \phi)} = \sqrt{1 - x^2}\, e^{\frac{i}{2}(\pi + 0)} \\
&= \sqrt{1 - x^2}\, e^{\frac{\pi i}{2}} = i\sqrt{1 - x^2}.
\end{aligned}
$$

Thus by (10.136),

$$
\begin{aligned}
\text{Sin}^{-1}x &= \frac{\pi}{2} + i\,\text{Log}\left(x + i\sqrt{1 - x^2}\right) \\
&= \frac{\pi}{2} + i\left[\text{Log}\,1 + i\text{Arg}\left(x + i\sqrt{1 - x^2}\right)\right] \\
&= \frac{\pi}{2} - \text{Arg}\left(x + i\sqrt{1 - x^2}\right). \qquad (10.142)
\end{aligned}
$$

Now $0 < \text{Arg}\left(x + i\sqrt{1 - x^2}\right) < \pi$ where $-1 < x < 1$. Hence (10.138) follows from (10.142).

Exercises 10.8.3

1. Find each of the following numbers.

   (a) $\text{Sin}^{-1}0$ (b) $\text{Sin}^{-1}1$ (c) $\text{Sin}^{-1}(-1)$

   (d) $\text{Sin}^{-1}\dfrac{1}{2}$ (e) $\text{Sin}^{-1}\left(-\dfrac{1}{\sqrt{2}}\right)$

2. Show that if $z+\sqrt{z^2-1}=t<0$, then z is real and $z\leq -1$. **Hint.** $-1=t^2-2tz$. Thus $z=\frac{1}{2}\left(t+\frac{1}{t}\right)$, which has a maximum (for $t<0$) of -1 at $t=-1$.

3. Let $g(z)=\sqrt{z^2-1}$ be the branch of $(z^2-1)^{\frac{1}{2}}$ given in (10.137). This is the function given in Ex. 15 of 10.7.6 with $a=1$ and $b=-1$. Use that exercise to show that

$$\frac{d}{dz}\sqrt{z^2-1}=\frac{z}{\sqrt{z^2-1}}\qquad\text{for each }z\text{ in }D$$

   where D is given in (10.139).

In Ex. 4–13, unless otherwise indicated, $\sqrt{z^2-1}$ and D will be the same as in Ex. 3.

4. Show that $\dfrac{d}{dz}\,\text{Sin}^{-1}z=\dfrac{1}{\sqrt{1-z^2}}$ for each z in D where $\sqrt{1-z^2}=-i\sqrt{z^2-1}$.

5. (a) Give the value of $\text{Log}\,i$. $\dfrac{\pi i}{2}$

   (b) Recall that the expression for w in (10.135) gives all values of arcsin z for each fixed z in R^2. Simplify the last member of that equation to find all values of arcsin z (using $\frac{\pi i}{2}$ for $\log i$).

$$\arcsin z=\frac{\pi}{2}\pm i\,\log\left(z+\sqrt{z^2-1}\right)$$

   (c) Verify from (5.10) and Ex. 1(b) in 5.2.11 that $\cos\left(\frac{\pi}{2}-w\right)=\sin w$ for each w in R^2.

   (d) Conclude from Part (c) that if w is an arcsin of z, then $\frac{\pi}{2}-w$ is an arccos of z, and conversely. Hence for each fixed z, all values of arccos z are given by

$$\arccos z=\frac{\pi}{2}-\arcsin z.$$

   (e) Use Part (b) and Part (d) to show that all values of arccos z are given by

$$\arccos z=\pm i\log\left(z+\sqrt{z^2-1}\right).$$

(f) Explain why we are not excluding any values of arcsin z in Part (b) when we use $\text{Log}\, i$ instead of $\log i$.

Ans. The additional term represented by $-i \log i$ would be $2k\pi$. But this same term is included in the multivalued function $i \log(z + \sqrt{z^2 - 1})$ of the next term.

6. Show that $\log[z - s(z)] = -\log[z + s(z)]$ where $s(z)$ is a square root of $(z^2 - 1)$. **Hint.** $\log \left(\frac{z - s(z)}{1} \cdot \frac{z + s(z)}{z + s(z)} \right) = \; ?$

7. (a) Solve $z = \cos w = \frac{e^{iw} + e^{-iw}}{2}$ to obtain $w = \arccos z = \cos^{-1} z = -i \log [z \pm s(z)] = \mp i \log[z + s(z)]$ for each z in R^2 where $s(z)$ is given in Ex. 6.

   (b) Reconcile the result in Part (a) with the result in Ex. 5(e). Recall that in these exercises, we use $\sqrt{z^2 - 1}$ to denote the branch of $(z^2 - 1)^{\frac{1}{2}}$ given in (10.137).

8. We shall use $\text{Arccos}\, z$ or $\text{Cos}^{-1}z$ to denote the principal branch of $\arccos z$. In view of Ex. 5(d), it is natural to define the **principal arccosine of z** by

$$\text{Cos}^{-1}z = \frac{\pi}{2} - \text{Sin}^{-1}z \qquad \text{for each } z \text{ in } D$$

where D is given in (10.139). Verify each of the following results.

   (a) The function $\text{Cos}^{-1}z$ is analytic in D. [See the first sentence after (10.141).]

   (b) We have $\dfrac{d}{dz} \text{Arccos}\, z = -\dfrac{1}{\sqrt{1 - z^2}}$ on D where $\sqrt{1 - z^2}$ is given in Ex. 4.

   (c) Also, $\text{Cos}^{-1}z = -i \, \text{Log}\left(z + \sqrt{z^2 - 1}\right)$.

   (d) If $-1 < z = x < 1$, then $0 < \text{Cos}^{-1}x < \pi$. **Hint.** Use (10.138).

9. Solve $z = \tan w = \dfrac{e^{iw} - e^{-iw}}{i(e^{iw} + e^{-iw})}$ for w to show that

$$\tan^{-1} z = \frac{1}{2i} \log \frac{1 + iz}{1 - iz} = \frac{i}{2} \log \frac{i + z}{i - z} \qquad \text{for each } z \neq \pm i.$$

10. The **principal branch** of $\tan^{-1}(z)$ is chosen to be the branch given by

$$\text{Tan}^{-1}z = \frac{i}{2} \, \text{Log} \frac{i + z}{i - z} \qquad \text{for each } z \text{ in } E$$

where $E = R^2 - \{z : \mathcal{R}(z) = 0 \text{ and } |\mathcal{I}(z)| \geq 1\}$.

   (a) Show that $\text{Tan}^{-1}z$ is analytic in E.

   (b) Show that $\frac{d}{dz} \text{Tan}^{-1}z = \frac{1}{z^2 + 1}$ for each z in E.

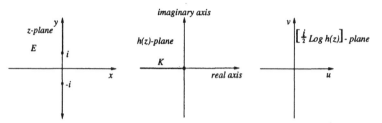

Figure 10.34: $\operatorname{Tan}^{-1} z = \frac{i}{2} \operatorname{Log} \frac{i+z}{i-z}$ for z in $E = R^2 - \{z : \ \mathcal{R}(z) = 0 \text{ and } |\mathcal{I}(z)| \geq 1\}$.

Hint for (a). Let $h(z) = \frac{i+z}{i-z}$. If $h(z) = t \leq 0$, then $z = \left(1 - \frac{2}{t+1}\right) i = yi$. If t is in $(-1, 0]$, then $y \leq -1$. But if $t < -1$, then $y > 1$. Thus $h(E) \subset (R^2 - K)$ where K is the nonpositive real axis. See Figure 10.34.

11. Show that for each z in R^2,

$$\sin^{-1} z = \frac{\pi}{2} + 2k\pi \pm i \operatorname{Log}[z + s(z)] \quad \text{for } k \text{ in } J$$

where $s(z)$ is a square root of $(z^2 - 1)$. [Use (10.135) and Ex. 6. Here, $s(z)$ is not necessarily $\sqrt{z^2 - 1}$ given in (10.137).]

12. Show that for each z in R^2,

$$\cos^{-1} z = 2k\pi \pm i \operatorname{Log}[z + s(z)] \quad \text{for } k \text{ in } J$$

where $s(z)$ is a square root of $(z^2 - 1)$. [Use Ex. 7(a).]

13. Show that for each z in R^2 such that $z \neq \pm i$,

$$\tan^{-1} z = \frac{i}{2} \operatorname{Log} \frac{i + z}{i - z} + k\pi \quad \text{for } k \text{ in } J.$$

14. Let $D_1 = R^2 - \{z : \ \arg(z - 1) = \frac{\pi}{4} \text{ or } \arg(z + 1) = \pi\}$ and let

$$F(z) = \frac{\pi}{2} - i \operatorname{Log}[z + s(z)] \quad \text{for each } z \text{ in } D_1$$

where $s(z)$ is one branch of $(z^2 - 1)^{\frac{1}{2}}$ which is analytic in D_1. [Such a branch exists by 10.7.2(a).] Prove $F \in A(D_1)$. (Thus F is a branch in D_1 of $\sin^{-1} z$ by Ex. 11.) **Hint.** As in Ex. 2, we have $g(D_1) \subset (R^2 - K)$ where $g(z) = z + s(z)$ and K is the nonpositive real axis. See Figure 10.35.

15. Let $D_2 = R^2 - \{z : \ \arg(z - 1) = 0 \text{ or } \arg(z + 1) = \frac{3}{4}\pi\}$ and let

$$G(z) = \frac{\pi}{2} - iL[z + s_1(z)] \quad \text{for each } z \text{ in } D_2$$

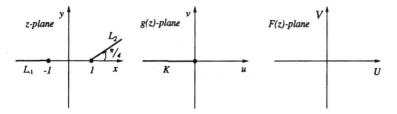

Figure 10.35: F is a branch in D_1 of $\sin^{-1} z$.

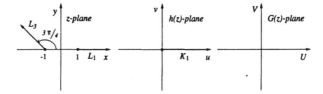

Figure 10.36: $G \in A(D_2)$

where $L(z)$ is the branch of $\log z$ given by

$$L(z) = \operatorname{Log}|z| + i\theta \qquad \text{for } 0 < \theta = \arg z < 2\pi$$

and where $s_1(z)$ is a branch of $(z^2 - 1)^{\frac{1}{2}}$ which is analytic in D_2. Prove $G \in A(D_2)$. **Hint**. If $z \in R^2$ such that $h(z) = z + s_1(z) = t > 0$, then $z \geq 1$. Thus $h(D_2) \subset (R^2 - K_1)$ where K_1 is the non-negative real axis. Also, $L \in A(R^2 - K_1)$. See Figure 10.36.

16. Show that $-\frac{\pi}{2} < \operatorname{Tan}^{-1} x < \frac{\pi}{2}$ for each x in R. **Hint**. From Ex. 10, we have $\operatorname{Tan}^{-1} x = \frac{i}{2}\left[\operatorname{Log} 1 + i\operatorname{Arg}(u + iv)\right]$ where $u = \frac{1-x^2}{x^2+1}$ and $v = -\frac{2x}{x^2+1}$.

Remarks 10.8.4 (a) We observe that the functions F and G in Ex. 14 and 15 of 10.8.3 are branches of $\sin^{-1} z$ [in view of (10.135)] since they are analytic in their domains. Thus we have given three **branches** of $\sin^{-1} z$ including $\operatorname{Sin}^{-1} z$. The question arises for what values of z is there a branch of $\sin^{-1} z$ which is analytic at z? The branch cuts for the three given branches are

$$
\begin{aligned}
L_1 \cup L_{-1} & \quad \text{for } \operatorname{Sin}^{-1} z, \\
L_{-1} \cup L_2 & \quad \text{for } F \text{ in Ex. 14, and} \\
L_3 \cup L_1 & \quad \text{for } G \text{ in Ex. 15}
\end{aligned}
$$

where L_k is shown in the respective figures. Now $z = 1$ and $z = -1$ are the only points common to all three of these branch cuts. Thus if $z \in (R^2 - \{-1, 1\})$, then at least one of the given branches is analytic at z.

(b) For each fixed k in J, let

$$h_k(z) = \frac{\pi}{2} + 2k\pi + i \operatorname{Log} g(z)$$

where $g(z) = z + \sqrt{z^2 - 1}$ and where $\sqrt{z^2 - 1}$ is given by (10.137). By (10.141), we see that $h_k \in A(D)$ where D is given in (10.139). Now $h_k(z)$ is one of the branches in D of $\sin^{-1} z$ by Ex. 11 in 10.8.3. Also by (10.136),

$$h_k(z) = 2k\pi + \operatorname{Sin}^{-1}(z).$$

Thus by Ex. 4 in 10.8.3,

$$h_k'(z) = \frac{1}{\sqrt{1 - z^2}} \qquad \text{for each } z \text{ in } D$$

where $\sqrt{1 - z^2} = -i\sqrt{z^2 - 1}$. But if

$$h_k(z) = \frac{\pi}{2} + 2k\pi - i \operatorname{Log} g(z),$$

then by Ex. 11, it follows that h_k is again a branch of $\sin^{-1} z$, and

$$h_k'(z) = -\frac{1}{\sqrt{1 - z^2}}$$

where it is understood that $\sqrt{1 - z^2}$ is the same square root as the one used previously.

(c) In Ex. 14 of 10.8.3, we may replace $\frac{\pi}{4}$ by any θ_0 in R. Let L_3 be the ray

$$\{z : \arg(z - 1) = \theta_0\}$$

and let $D_3 = R^2 - (L_1 \cup L_3)$. If

$$F_k(z) = \frac{\pi}{2} + 2k\pi \pm i \operatorname{Log}[z + s(z)],$$

where $s(z)$ is a branch in D_3 of $(z^2 - 1)^{\frac{1}{2}}$, then $F_k \in A(D_3)$. Thus in view of Ex. 11, with either choice of sign, F_k is a branch in D_3 of $\sin^{-1} z$.

As in Ex. 13 – 15 of 10.7.6, $s'(z) = \frac{z}{s(z)}$. Thus

$$\begin{aligned}
F_k'(z) &= \pm i \frac{1 + s'(z)}{z + s(z)} = \frac{\pm i}{s(z)} = \frac{\mp 1}{i\, s(z)} \\
&= \frac{1}{\sqrt{1 - z^2}}
\end{aligned}$$

where in this instance, $\sqrt{1 - z^2} = \mp i\, s(z)$.

(d) We have the following general results. If $z \in (R^2 - \{1, -1\})$, then there are branches of $\sin^{-1} z$ and $\cos^{-1} z$ which are analytic at z. Furthermore, for any chosen branch,

$$\frac{d}{dz} \sin^{-1} z = \frac{1}{\sqrt{1 - z^2}} \quad \text{and} \quad \frac{d}{dz} \cos^{-1} z = -\frac{1}{\sqrt{1 - z^2}}$$

for each z in the domain of the branch if the appropriate square root of $(1 - z^2)$ is used. Also, in view of Ex. 10 and 13 of 10.8.3, for each z in E,

$$\frac{d}{dz} \tan^{-1} z = \frac{1}{1 + z^2}.$$

10.9 Inverse Hyperbolic Functions

In view of 5.3.3, we let

$$z = \cosh iw = \cos w.$$

Thus

$$w = \cos^{-1} z \quad \text{and} \quad iw = \cosh^{-1} z.$$

Hence for each z in R^2,

$$\begin{aligned}
\cosh^{-1} z &= i \cos^{-1} z \\
&= i \left[-i \log \left(z + (z^2 - 1)^{\frac{1}{2}} \right) \right] \quad \text{by Ex. 7 in 10.8.3} \\
&= \log \left[z + (z^2 - 1)^{\frac{1}{2}} \right].
\end{aligned} \tag{10.143}$$

Similarly, from 5.3.3, we let

$$z = \sinh iw = i \sin w.$$

Thus

$$w = \sin^{-1}(-iz) \quad \text{and} \quad iw = \sinh^{-1} z.$$

Hence for each z in R^2,

$$\begin{aligned}
\sinh^{-1} z &= i \sin^{-1}(-iz) \\
&= i \left[-i \log i \left(-iz + (-z^2 - 1)^{\frac{1}{2}} \right) \right] \quad \text{by (10.135)} \\
&= \log \left[z + (z^2 + 1)^{\frac{1}{2}} \right].
\end{aligned} \tag{10.144}$$

[Note that $(-1)^{\frac{1}{2}} = \pm i$.]

Exercises 10.9.1

1. Show that $\tanh^{-1} z = \frac{1}{2} \log \frac{1+z}{1-z}$ for each $z \neq \pm 1$. [Use the formula $z = \tanh iw = i \tan w$ and Ex. 9 in 10.8.3.]

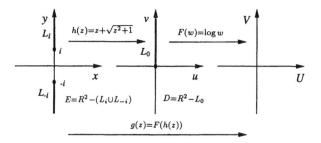

Figure 10.37: If $h(z) \in L_0$ then $z \in L_i$.

2. Show that $\sinh^{-1} z = \frac{1}{\sqrt{1+z^2}}$ for each z in D where $\sinh^{-1} z$ denotes a branch of the multivalued inverse hyperbolic sine function, where $\sqrt{1+z^2}$ denotes the branch of $(1+z^2)^{\frac{1}{2}}$ used in (10.144) to determine the particular chosen branch of the inverse hyperbolic sine function, and where D is the domain of the chosen branch. **Hint.** Differentiate (10.144). See Ex. 11 in 5.4.11 and Ex. 15 in 10.7.6.

   Remark. To obtain a branch of the inverse hyperbolic sine function, we may choose a branch of $(1+z^2)^{\frac{1}{2}}$ and then a branch of $\log z$. We next apply (10.144) to obtain the desired branch of the inverse hyperbolic sine function.

3. Show that $\frac{d}{dz} \cosh^{-1} z = \frac{1}{\sqrt{z^2-1}}$ where $\sqrt{z^2-1}$ is understood to be the branch of $(z^2-1)^{\frac{1}{2}}$ used in (10.143) to determine the chosen branch of the inverse hyperbolic cosine function.

4. Show that $\frac{d}{dz} \tanh^{-1} z = \frac{1}{1-z^2}$ for each $z \neq \pm 1$.

5. Let $\sqrt{z^2+1}$ be a branch in E of $(z^2+1)^{\frac{1}{2}}$ where E is given in Ex. 10 of 10.8.3. Let $g(z) = F\left(z + \sqrt{z^2+1}\right)$ where F is a branch in D of $\log z$, where $D = R^2 - L_0$ and $L_0 = \{z : \ \mathcal{R}(z) = 0 \ \text{and} \ \mathcal{I}(z) \geq 0\}$. Prove g is a branch in E of $\sinh^{-1} z$ and $g'(z) = \frac{1}{\sqrt{1+z^2}}$ for each z in E. **Hint.** Let $h(z) = z + \sqrt{z^2+1}$. To show that $h(E) \subset D$, we observe that (as in Ex. 2 of 10.8.3) if $h(z) = iv$ and $v > 0$, then $z = iy$ where $y = \frac{1}{2}\left(v + \frac{1}{v}\right) \geq 1$. Thus if $h(z) \in L_0$ then $z \in L_i$ where L_i is shown in Figure 10.37. Hence $h(E) \subset D$.

10.10   Branches of log f(z) in a Simply Connected Domain

In Ex. 5 of 10.7.6, we showed the existence of a branch of $\log z$ in special domains; namely, $R^2 - L$ where L is a ray from the origin. Now we shall show the existence of a branch of $\log f(z)$ in any simply connected domain D such that $f \in A(D)$ and such that f has no zero in D.

Theorem 10.10.1 Let $f \in A(D)$ where D is a simply connected domain in R^2. If f has no zero in D, then there exists a branch of $\log f(z)$ in D.

Proof. Now $f' \in A(D)$ by 3.6.10(b) and hence $\frac{f'}{f} \in A(D)$. Let $z_0 \in D$ and let

$$g(z) = \int_{z_0}^{z} \frac{f'(w)}{f(w)}\,dw + \mathrm{Log}\, f(z_0) \quad \text{for each } z \text{ in } D. \tag{10.145}$$

By 3.6.3(a),

$$g \in A(D) \quad \text{and} \quad g' = \frac{f'}{f} \quad \text{on } D. \tag{10.146}$$

Now

$$\begin{aligned} e^{g(z_0)} &= e^{\mathrm{Log}\, f(z_0)} \quad \text{by (10.145)} \\ &= f(z_0) \quad \text{by 5.4.5.} \end{aligned}$$

Hence

$$e^{-g(z_0)} f(z_0) = 1. \tag{10.147}$$

Now on D, we have

$$\begin{aligned} \frac{d}{dz}\left(e^{-g(z)} f(z)\right) &= e^{-g(z)} f'(z) - f(z) e^{-g(z)} g'(z) \\ &= e^{-g(z)} [f'(z) - f(z)g'(z)] \\ &= 0 \quad \text{by (10.146).} \end{aligned}$$

Thus by 2.3.15(a), we have

$$e^{-g(z)} f(z) = C \quad \text{on } D. \tag{10.148}$$

Now by (10.147), we must have $C = 1$. Hence

$$e^{g(z)} = f(z) \quad \text{on } D.$$

Thus $g(z)$ is one of the values of $\log f(z)$ for each z in D. Hence g is a branch in D of $\log f(z)$ by (10.146) and 10.7.3(a).

Corollary 10.10.2 There is a branch of $\log(z - a)$ in any simply connected domain D, not containing $z = a$.

Proof. We take $f(z) = z - a$ in 10.10.1.

Corollary 10.10.3 There is a branch of $(z - a)^{\frac{1}{2}}$ in any simply connected domain D, not containing $z = a$.

Proof. By 10.10.2, let g be a branch in D of $\log(z - a)$. Then g and $\frac{1}{2}g$ are analytic in D. Thus $h(z) = e^{\frac{1}{2}g}$ is analytic in D since e^z is an entire function. But

$$[h(z)]^2 = [e^{\frac{1}{2}g}]^2 = e^g = z - a \quad \text{on } D$$

since $g(z) = \log(z - a)$ on D. Thus $h(z)$ is one of the values of $(z - a)^{\frac{1}{2}}$ for each z in D. Hence h is a branch in D of $(z - a)^{\frac{1}{2}}$ by Definition 10.7.3(a).

Remark 10.10.4 In 10.10.2, we may take $a = 0$ and $D = R^2 - L$ where L is a closed ray from the origin. Thus 10.10.2 implies the existence of branches of $\log z$ in the special domains discussed in Ex. 4 and 5 of 10.7.6.

Exercises 10.10.5

1. Let $f \in A(D)$ where D is a simply connected domain in which f has no zero, and let $n \in (J - \{0, 1, -1\})$. Prove there is a branch in D of $[f(z)]^{\frac{1}{n}}$.

   Solution. By 10.10.1, let g be a branch in D of $\log f(z)$. Then $e^{\frac{g}{n}} \in A(D)$ and $(e^{\frac{g}{n}})^n = f$ on D. Thus $e^{\frac{g}{n}}$ is a branch in D of $f^{\frac{1}{n}}$.

2. Let a and b be two complex constants and let D be any simply connected domain in R^2 which contains neither a nor b. Prove there is a branch in D of $[(z - a)(z - b)]^{\frac{1}{2}}$.

Chapter 11

The Riemann Mapping Theorem

11.1 Sequences of Functions

In this section, we present some standard theorems of complex analysis which we shall need in proving the Riemann Mapping Theorem.

Definition 11.1.1 Let $\{f_n\}$ be a sequence of complex functions defined on a subset D of R^2.

(a) The sequence $\{f_n\}$ **converges (pointwise)** in D to a function f (called the **limit** function) iff for each z in D, the sequence of numbers $\{f_n(z)\}$ converges to the number $f(z)$. We indicate this by writing $f_n \xrightarrow{P\,D} f$.

(b) The sequence $\{f_n\}$ **converges uniformly** to f on D (denoted by $f_n \xrightarrow{U\,D} f$) iff

> for each $\epsilon > 0$, there is an integer M such that
> $|f_n(z) - f(z)| < \epsilon$ for each z in D and for each $n \geq M$.

Remarks 11.1.2 We should recall the definition of the limit of a sequence of numbers, which is given in 4.1.1(b). If we apply that definition to the sequence $\{f_n(z)\}$ of numbers for a fixed value of z, we see that 11.1.1(a) may be restated as indicated here.

Alternative form of 11.1.1(a). The sequence $\{f_n\}$ converges pointwise in D to f if and only if the following criterion is satisfied.

> If z is any **fixed** number in D, then for
> each $\epsilon > 0$, there is an integer $M_{z,\epsilon}$ such
> that $|f_n(z) - f(z)| < \epsilon$ for each $n \geq M_{z,\epsilon}$. $\qquad$ (11.1)

In (11.1), we emphasize that in general the integer $M_{z,\epsilon}$ depends upon **both** z and ϵ. However, if for each $\epsilon > 0$, the integer $M_{z,\epsilon}$ is **independent of z** (that is, the same $M_{z,\epsilon}$ may be used for all values of z in D), then we have uniform convergence on D as defined in 11.1.1(b).

Theorem 11.1.3 (Cauchy Criterion for Uniform Convergence of Sequences). A sequence $\{f_n\}$ of complex **functions** converges uniformly on D to some **function** iff the following criterion is satisfied.

For each $\epsilon > 0$, there is an integer M such that if $m \geq M$

and $n \geq M$, then $|f_m(z) - f_n(z)| < \epsilon$ for each z in D.   (11.2)

Proof. First suppose $f \xrightarrow{U\,D} f$ and let $\epsilon > 0$. Let $M \in N$ such that for each z in D,

$$|f_n(z) - f(z)| < \frac{\epsilon}{2} \quad \text{for each } n \geq M.$$

Then for $m \geq M$ and $n \geq M$ and for each z in D,

$$
\begin{aligned}
|f_m(z) - f_n(z)| &\leq |f_m(z) - f(z)| + |f(z) - f_n(z)| \\
&< \frac{\epsilon}{2} + \frac{\epsilon}{2} = \epsilon.
\end{aligned}
$$

Conversely, suppose (11.2) holds. Then for each z in D, the sequence $\{f_n(z)\}$ is a Cauchy sequence of **numbers** and hence (by 4.5.2) converges to some **number**, which we denote by $f(z)$.

Thus

$$\lim_{n \to \infty} f_n(z) = f(z) \quad \text{for each \textbf{fixed} } z \text{ in } D. \tag{11.3}$$

Let $\epsilon > 0$. By (11.2), we let M be an integer **independent of z** such that

$$|f_n(z) - f_m(z)| < \frac{\epsilon}{2} \quad \text{for each } z \text{ in } D \quad \text{if } m \geq M \text{ and } n \geq M. \tag{11.4}$$

By (11.3), for each **fixed** z in D, we let $P_z \in N$ such that $P_z \geq M$ and such that

$$|f_m(z) - f(z)| < \frac{\epsilon}{2} \quad \text{if } m \geq P_z. \tag{11.5}$$

Now we let z be an arbitrary element of D. If $n \geq M$, then for some $m \geq P_z \geq M$, we have

$$
\begin{aligned}
|f_n(z) - f(z)| &\leq |f_n(z) - f_m(z)| + |f_m(z) - f(z)| \tag{11.6}\\
&< \frac{\epsilon}{2} + \frac{\epsilon}{2} = \epsilon \quad \text{by (11.4) and (11.5).}
\end{aligned}
$$

Thus $f_n \xrightarrow{U\,D} f$.

Theorem 11.1.4 Let $f_n \xrightarrow{U\,D} f$. If each f_n is continuous at the point z_0 in D, then f is continuous at z_0.

Proof. Let $\epsilon > 0$. By 11.1.1(b), we let $M \in N$ such that for each z in D,

$$|f_n(z) - f(z)| < \frac{\epsilon}{3} \quad \text{if } n \geq M. \tag{11.7}$$

Let n be fixed such that $n \geq M$. Since $f_n \in C(z_0)$, we let $\delta > 0$ such that

$$|f_n(z) - f_n(z_0)| < \frac{\epsilon}{3} \quad \text{for each } z \text{ in } D \cap N_\delta(z_0). \tag{11.8}$$

Then for each z in $D \cap N_\delta(z_0)$,

$$\begin{aligned}
|f(z) - f(z_0)| &\leq |f(z) - f_n(z)| + |f_n(z) - f_n(z_0)| + |f_n(z_0) - f(z_0)| \\
&< \frac{\epsilon}{3} + \frac{\epsilon}{3} + \frac{\epsilon}{3} = \epsilon \quad \text{by (11.7) and (11.8)}.
\end{aligned}$$

Thus $f \in C(z_0)$.

Theorem 11.1.5 Let $\{f_n\}$ be a sequence of complex functions each of which is continuous on a contour K in R^2. If $f_n \xrightarrow{U K} f$, then

$$\int_K f_n(z)\,dz \to \int_K f(z)\,dz, \tag{11.9}$$

that is,

$$\lim_{n \to \infty} \int_K f_n(z)\,dz = \int_K \lim_{n \to \infty} f_n(z)\,dz = \int_K f(z)\,dz.$$

Proof. [Each integral in (11.9) exists since $f_n \in C(K)$ by hypothesis and since $f \in C(K)$ by 11.1.4]. Let $\epsilon > 0$. Let $M \in N$ such that if $n \geq M$, then

$$|f_n(z) - f(z)| < \frac{\epsilon}{L} \quad \text{for each } z \text{ on } K \tag{11.10}$$

where L is the length of K. Then for each $n \geq M$,

$$\begin{aligned}
\left| \int_K f_n(z)\,dz - \int_K f(z)\,dz \right| &= \left| \int_K [f_n(z) - f(z)]\,dz \right| \\
&< \frac{\epsilon}{L} \cdot L = \epsilon \quad \text{by (11.10) and 3.4.10(g)}.
\end{aligned}$$

Hence (11.9) is proved.

Theorem 11.1.6 Let $D \subset R^2$ and let $\{g_n\}$ be a sequence of complex functions such that for each z in D, there is a neighborhood N_z of z for which $g_n \xrightarrow{U N_z} g$. If K is a compact subset of D, then $g_n \xrightarrow{U K} g$.

Proof. Since K is compact, let $z_1, z_2, \cdots, z_p$ be points in K such that $\{N_{z_1}, N_{z_2}, \cdots, N_{z_p}\}$ covers K where

$$g_n \xrightarrow{\ U\ N_{z_j}\ } g \quad \text{for } j = 1, 2, \cdots, p.$$

Let $\epsilon > 0$. For each $j \leq p$, let $M_j \in N$ such that

$$\text{if } n \geq M_j, \text{ then } |g_n(z) - g(z)| < \epsilon \text{ on } N_{z_j}.$$

Now let $M = \max\{M_1, M_2, \cdots, M_p\}$. Thus if $n \geq M$, then $|g_n(z) - g(z)| < \epsilon$ on K.

Theorem 11.1.7 For each n in N, let $f_n \in A(D)$ where D is a domain in R^2. If $f_n \xrightarrow{U\ K} f$ for each compact subset K of D, then

(a) $f \in A(D)$ and

(b) $f'_n \xrightarrow{U\ K} f'$ for each compact subset K of D.

Proof of (a). Let $p \in D$ and let $r > 0$ such that $\overline{N_r(p)} \subset D$. By 1.4.15, the closed and bounded set $\overline{N_r(p)}$ is compact. Thus by hypothesis, $f_n \xrightarrow{\ U\ \overline{N_r(p)}\ } f$. Hence $f \in C(\overline{N_r(p)})$ by 11.1.4.

Now let K be any simple closed contour in $N_r(p)$. By Ex. 3 in 3.6.18, the set K is compact. Thus by hypothesis, $f_n \xrightarrow{U\ K} f$. By 11.1.5,

$$\int_K f(z)\, dz = \lim_{n \to \infty} \int_K f_n(z)\, dz = 0 \tag{11.11}$$

[since $\overline{I(K)} \subset N_r(p)$ and hence $\int_K f_n(z)\, dz = 0$ for each $n \in N$ by the Cauchy-Goursat Theorem]. Thus by (11.11) and Morera's Theorem, $f \in A[N_r(p)]$. Thus $f \in A(D)$ since p is an arbitrary point of D.

Proof of (b). Let $p \in D$ and let $r > 0$ such that $\overline{N_{2r}(p)} \subset D$. Let $\epsilon > 0$. Since $f_n \xrightarrow{\ U\ N_{2r}(p)\ } f$, let $M \in N$ such that if $n \geq M$, then

$$|f_n(z) - f(z)| < \frac{r\epsilon}{2} \quad \text{for each } z \text{ in } \overline{N_{2r}(p)}. \tag{11.12}$$

Now if $z \in N_r(p)$, then $\overline{N_r(z)} \subset N_{2r}(p) \subset D$. [For if $t \in \overline{N_r(z)}$, then $|t - z| \leq r$ and $|t - p| \leq |t - z| + |z - p| < r + r = 2r$.] Also $(f_n - f) \in A(\overline{N_r(z)})$ [since f_n and f are analytic in D]. Applying Cauchy's inequalities to $f_n - f$ and using (11.12),

$$|f'_n(z) - f'(z)| \leq \frac{r\frac{\epsilon}{2}}{r} < \epsilon \quad \text{for each } z \text{ in } N_r(p) \text{ and for each } n \geq M.$$

Thus $f'_n \xrightarrow{\ U\ N_r(p)\ } f'$. Hence by 11.1.6, $f'_n \xrightarrow{U\ K} f'$ for each compact subset K of D.

Exercises 11.1.8

1. Prove if $\{a_n\}$ is a bounded sequence of numbers in R^2, then there is a subsequence of $\{a_n\}$ which converges.

2. Prove the Hurwitz Theorem.

   Theorem. (Hurwitz). Let D be a domain in R^2 and let $\{f_n\}$ be a sequence of functions such that each f_n is analytic in D and such that

$$f_n \xrightarrow{U\,K} f \quad \text{for each compact subset } K \text{ of } D.$$

   (a) If no f_n has a zero in D, then either f has no zero in D or $f(x) \equiv 0$ on D.
   (b) If each f_n is one-to-one in D, then f is one-to-one in D or f is constant on D.

3. Give an example to show that in Part (b) of the Hurwitz Theorem, the function f may be constant.
   [**Solution.** Let $f_n(z) = \frac{z}{n}$ and let $D = R^2$.]

4. Give an example to show that in Part (a) of the Hurwitz Theorem, the function f may be identically zero.
   [**Solution.** Let $f_n(z) = \frac{e^z}{n}$ and let $D = R^2$, or let $f_n(z) = \frac{z}{n}$ and $D = R^2 - \{0\}$.]

5. Show by 4.4.1(b) and by 11.1.7 that the (Riemann) zeta function ζ is analytic in H where H is the half-plane $H = \{z : \mathcal{R}(z) > 1\}$. **Note.** We have $\sum_{n=1}^{\infty} f_n \xrightarrow{U\,K} f$, iff $S_n \xrightarrow{U\,K} f$ where $S_n = f_1 + f_2 + \cdots + f_n$. **Hint.** See Ex. 6 in 4.4.11. For each $p > 1$, let $H_p = \{z : \mathcal{R}(z) > p\}$. Show that if K is a compact subset of H, then there is some $q > 1$ such that $K \subset H_q$. To do this, let $\mathcal{C} = \{H_p : p > 1\}$. Now $\mathcal{C}$ is an open cover for K. Since K is compact, let $\{H_{p_1}, H_{p_2}, \cdots, H_{p_n}\}$ be a finite subcover of $\mathcal{C}$ for K. Let $q = \min\{p_1, p_2, \cdots, p_n\}$. Hence $K \subset H_q \subset \overline{H_q} \subset H$.

6. Give the definition of a domain in R^2.

7. Prove if $p \in D$ and D is a domain in R^2, then $D - \{p\}$ is a domain.

8. Give the definition of a compact set in R^2.

9. Prove if K_1 is a compact subset of R^2, then $K = K_1 \cup \{p\}$ is compact where p is any point in R^2.

   Proof of Ex. 1. First, suppose $a_n = c$ for infinitely many values of n. Then we let $n_1 < n_2 < n_3 < \cdots$ such that $a_{n_j} = c$ for each j. Clearly the subsequence $\{a_{n_j}\}$ converges to c.

   Now suppose no point c is repeated infinitely many times. Then the set $\{a_n : n \in N\}$ is an infinite set. This set is bounded by hypothesis. Thus we let a be a limit point of this set by Ex. 3 in 1.4.16. We let $n_1 \in N$ such that $a_{n_1} \in N_1(a)$. By Ex. 4 in 1.4.10, we let $n_2 > n_1$ such that $a_{n_2} \in N_{\frac{1}{2}}(a)$. Continuing, we let $n_j > n_{j-1}$ such that $a_{n_j} \in N_{\frac{1}{j}}(a)$. The subsequence $\{a_{n_j}\}$ clearly converges to a.

Proof of (a) in Ex. 2. Suppose that $f(a) = 0$ for some a in D but that $f \not\equiv 0$ on D. Then a is an isolated zero of f by 6.2.5 and 6.2.6. Hence we let C be a circle centered at a such that $\overline{I(C)} \subset D$ and such that f is never zero on C. Since C is compact, we let $m > 0$ such that $m < |f(z)|$ on C by Ex. 2(c) in 2.4.9. By hypothesis, $f_n \xrightarrow{U C} f$ (since C is compact). Thus we let $M \in N$ such that

$$|f_M(z) - f(z)| < m < |f(z)| \quad \text{on } C.$$

By Rouché's Theorem,

$$f(z) \quad \text{and} \quad f_M(z) = [f_M(z) - f(z)] + f(z)$$

have the same number of zeros in $I(C)$, contradicting the hypothesis on f_M.

Proof of (b) in Ex. 2. Let p be **any** point in D. The set $D_1 = D - \{p\}$ is a domain. Also, if K_1 is a compact subset of D_1, then

$$f_n(z) - f_n(p) \xrightarrow{U K_1} f(z) - f(p). \tag{11.13}$$

To verify, we note that the set $K = K_1 \cup \{p\}$ is a compact subset of D. By hypothesis, $f_n \xrightarrow{U K} f$. Thus we let $M \in N$ such that if $n \geq M$, then $|f_n(z) - f(z)| < \frac{\epsilon}{2}$ for each z in K, including the point p. Now if $n \geq M$, then

$$
\begin{aligned}
|[f_n(z) - f_n(p)] - [f(z) - f(p)]| &\leq |f_n(z) - f(z)| + |f_n(p) - f(p)| \\
&< \frac{\epsilon}{2} + \frac{\epsilon}{2} = \epsilon \quad \text{for each } z \text{ in } K_1.
\end{aligned}
$$

Thus 11.13 is verified.

Now by Part (a), it follows that $f(z) - f(p)$ is either never zero or identically zero on $D_1 = D - \{p\}$. Since p is just any point in D, this means that f is either one-to-one on D or is constant on D.

11.2   Normal Families of Functions

We use $A(D)$ to denote the family of all analytic functions on D.

Definition 11.2.1 Let D be a domain in R^2 and let $\mathcal{F} \subset A(D)$.

(a) The family $\mathcal{F}$ is a **normal family** (in D) iff each sequence of members of $\mathcal{F}$ has a subsequence which converges uniformly on each compact subset of D.

(b) The family $\mathcal{F}$ is **uniformly bounded on each compact subset of D** iff the following condition is satisfied. If K is a compact subset of D, then there is a number M_K such that for each f in $\mathcal{F}$,

$$|f(z)| \leq M_K \quad \text{on } K.$$

(c) Let $K \subset D$. Then $\mathcal{F}$ is **equicontinuous on K** iff the following condition is satisfied.

For each $\epsilon > 0$, there is a number $\delta > 0$ such that
if z and z' are points in K with $|z - z'| < \delta$,
then $|f(z) - f(z')| < \epsilon$ for each f in $\mathcal{F}$.

Lemma 11.2.2 Let $\{z_n\}$ be a sequence in R^2 and let $\{f_n\}$ be a sequence of complex functions such that for each fixed k, the sequence $\{f_n(z_k)\}$ of numbers is bounded. Then there is a subsequence $\{f_{n_j}\}$ of $\{f_n\}$ such that $\{f_{n_j}(z_k)\}$ converges for each k in N.

Proof. Since $\{f_n(z_1)\}$ is a bounded sequence of numbers, it has a convergent subsequence $\{f_{n_j}(z_1)\}$ by Ex. 1 in 11.1.8. This determines a subsequence of functions $\{f_{n_j}\}$ of $\{f_n\}$ which converges at z_1. Likewise the bounded sequence $\{f_{n_j}(z_2)\}$ of numbers has a convergent subsequence $\{f_{p_j}(z_2)\}$. This gives a subsequence $\{f_{p_j}\}$ of $\{f_{n_j}\}$ which converges at z_1 and z_2. Continuing in this way, we obtain successive subsequences which we indicate in the following array.

$$f_{n_1}, f_{n_2}, f_{n_3}, \cdots \text{ (convergent at } z_1)$$
$$f_{p_1}, f_{p_2}, f_{p_3}, \cdots \text{ (convergent at } z_1 \text{ and } z_2)$$
$$f_{q_1}, f_{q_2}, f_{q_3}, \cdots \text{ (convergent at } z_1, z_2, \text{ and } z_3)$$
$$\vdots$$

In this array, each row sequence (after the first) is a subsequence of the previous one, and each such sequence is a subsequence of the original sequence $\{f_n\}$.

Now the "diagonal sequence" $f_{n_1}, f_{p_2}, f_{q_3}, \cdots$ is a subsequence of $\{f_n\}$ which converges at z_k for each k in N.

Lemma 11.2.3 Let D be an open set in R^2. There is a sequence $\{K_n\}$ of compact sets such that

$$\text{if } z \in K_n, \text{ then } N_r(z) \subset K_{n+1} \text{ where } r = \frac{1}{n(n+1)} \text{ and } K_n \subset K_{n+1}^\circ. \quad (11.14)$$

Furthermore,

$$D = \bigcup \{K_n : \ n \in N\}. \quad (11.15)$$

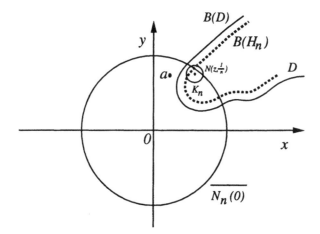

Figure 11.1: $K_n = \overline{N_n(0)} \cap H_n$

Proof. For each n in N, we let

$$K_n = \overline{N_n(0)} \cap H_n \tag{11.16}$$

where

$$H_n = \{z : \ |z - a| \geq \frac{1}{n} \quad \text{for each } a \text{ in } (R^2 - D)\}. \tag{11.17}$$

See Figure 11.1. We notice that H_n may also be written as

$$H_n = R^2 - \bigcup \{N_{\frac{1}{n}}(a) : \ a \in (R^2 - D)\}.$$

Now we see that H_n is closed, since it is the complement of the union of open sets.

Hence K_n is closed since the intersection of two closed sets is closed by Ex. 2(a) in 1.4.20. Now K_n is bounded since $K_n \subset \overline{N_n(0)}$. Thus K_n is compact by 1.4.15.

To prove (11.14), suppose $z \in K_n$ and let $r = \frac{1}{n(n+1)}$. Let $w \in N_r(z)$. Then

$$|w| \leq |w - z| + |z| < \frac{1}{n(n+1)} + n < 1 + n.$$

Thus

$$w \in \overline{N_{n+1}(0)}. \tag{11.18}$$

Also, since $z \in H_n$, we see from (11.17) that for each a in $(R^2 - D)$,

$$|w - a| \geq |z - a| - |z - w| \geq \frac{1}{n} - \frac{1}{n(n+1)} = \frac{1}{n+1}.$$

Thus

$$w \in H_{n+1} \qquad \text{by (11.17)}. \tag{11.19}$$

Hence $w \in K_{n+1}$ by (11.16), (11.18), and (11.19). Thus $N_r(z) \subset K_{n+1}$ and hence the first part of (11.14) is proved since $r = \frac{1}{n(n+1)}$. Now in this paragraph, we have shown if $z \in K_n$, then $N_r(z) \subset K_{n+1}$. This means that $K_n \subset K_{n+1}^\circ$ by 1.4.6(b) and (c). Thus the second part of (11.14) is proved.

To prove (11.15), we first observe that if $z \in H_n$, then z cannot be in $(R^2 - D)$ by (11.17). This means that $H_n \subset D$. Hence each $K_n \subset D$. Thus

$$\bigcup \{K_n : \ n \in N\} \subset D. \tag{11.20}$$

Now we let $z \in D$. Since D is open, we let $s > 0$ such that $N_s(z) \subset D$. Next we let $j \in N$ such that $j > |z|$ and such that $\frac{1}{j} < s$. Then $N_{\frac{1}{j}}(z) \subset N_s(z) \subset D$. Thus $z \in H_j$ by (11.17). Also, $z \in \overline{N_j(0)}$ since $j > |z|$. Hence $z \in K_j$ by (11.16). Therefore,

$$D \subset \bigcup \{K_n : \ n \in N\}. \tag{11.21}$$

Now (11.15) follows from (11.20) and (11.21).

Theorem 11.2.4 Let $\mathcal{F} \subset A(D)$ where D is a domain in R^2. If $\mathcal{F}$ is uniformly bounded on each compact subset of D, then $\mathcal{F}$ is a normal family in D.

Proof. Let $\{K_n\}$ be a sequence of compact sets such that (11.14) and (11.15) hold. For each n in N, let $M_n \in R$ such that for each f in $\mathcal{F}$

$$|f(z)| \le M_n \qquad \text{on } K_n. \tag{11.22}$$

For a fixed n, let z and z' be points in K_n such that $|z - z'| < r_n$ where $0 < r_n < \frac{1}{2n(n+1)}$. Let C be the cco circle with center at z and radius $2r_n < \frac{1}{n(n+1)}$. Then by (11.14),

$$C \subset K_{n+1}. \tag{11.23}$$

By the Cauchy Integral Formula,

$$
\begin{aligned}
|f(z) - f(z')| &= \frac{1}{2\pi} \left| \int_C \frac{f(t)}{t - z} \, dt - \int_C \frac{f(t)}{t - z'} \, dt \right| \\
&= \frac{1}{2\pi} \left| \int_C \frac{f(t)}{(t - z)(t - z')} \, dt \right| |z - z'| \\
&\le \frac{1}{2\pi} \frac{(4\pi r_n) M_{n+1} |z - z'|}{(2r_n)(r_n)} \qquad \text{by (11.22) and (11.23)} \\
&< \frac{M_{n+1}}{r_n} |z - z'|, \tag{11.24}
\end{aligned}
$$

since $|t - z'| \geq |t - z| - |z - z'| > 2r_n - r_n = r_n$. Now (11.24) implies that $\mathcal{F}$ is equicontinuous on K_n. For if $\epsilon > 0$, take $\delta = \min \left\{ r_n, r_n \frac{\epsilon}{M_{n+1}} \right\}$.

Now let $\{f_n\}$ be a sequence of members of $\mathcal{F}$. Let

$$E = \{z : \ z = (x, y) \in D \ \text{ where } x \text{ and } y \text{ are rational} \}.$$

By A.2.11(b), E is countable and hence we may write

$$E = \{z_1, z_2, z_3, \cdots\}.$$

For each fixed k, the set $\{z_k\}$ consisting of the single point z_k is compact. By hypothesis, $\mathcal{F}$ is uniformly bounded on $\{z_k\}$. Thus $\{f_n(z_k)\}$ is bounded for a fixed k. Hence by 11.2.2, let $\{f_{n_j}\}$ be a subsequence of $\{f_n\}$ such that

$$\{f_{n_j}\} \text{ converges at each } z \text{ in } E. \tag{11.25}$$

To show that $\{f_{n_j}\}$ converges uniformly on a fixed $\mathcal{F}_n$, let $\epsilon > 0$. Since $\mathcal{F}$ is equicontinuous on K_n, let $\delta > 0$ such that if $f \in \mathcal{F}$ and if z and z' are in K_n with $|z - z'| < \delta$, then

$$|f(z) - f(z')| < \frac{\epsilon}{3}. \tag{11.26}$$

Now $\mathcal{C} = \{N_\delta(z) : \ z \in E\}$ is an open cover for K_n. [For let $p \in K_n$. There exists an r such that $0 < r < \delta$ and such that $N_r(p) \subset D$. But $N_r(p)$ surely contains a point $q = (x, y)$ of E, that is, $q \in D$ and x and y are rational. Thus $|p - q| < r < \delta$, and hence $p \in N_\delta(q)$.] Since K_n is compact, let $\{N_\delta(z_j) : \ j = 1, 2, \cdots, m\}$ be a finite subcover of $\mathcal{C}$ for K_n. Thus

$$K_n \subset N_\delta(z_1) \cup N_\delta(z_2) \cup \cdots \cup N_\delta(z_m). \tag{11.27}$$

By (11.25) and 4.5.2, we see that $\{f_{n_j}(z_k)\}$ is a Cauchy sequence for each $k = 1, 2, \cdots, m$. Thus we obtain an integer M such that for each $k = 1$ to m,

$$|f_{n_j}(z_k) - f_{n_r}(z_k)| < \frac{\epsilon}{3} \quad \text{if } j \geq M \text{ and } r \geq M. \tag{11.28}$$

Now let $z \in K_n$ and by (11.27), let $z \in N_\delta(z_t)$ for some integer t such that $1 \leq t \leq m$. Thus if $j \geq M$ and $r \geq M$, then

$$\begin{aligned}
|f_{n_j}(z) - f_{n_r}(z)| &\leq |f_{n_j}(z) - f_{n_j}(z_t)| + |f_{n_j}(z_t) - f_{n_r}(z_t)| \\
&\quad + |f_{n_r}(z_t) - f_{n_r}(z)| \\
&< \frac{\epsilon}{3} + \frac{\epsilon}{3} + \frac{\epsilon}{3} = \epsilon \quad \text{by (11.26) and (11.28)}.
\end{aligned}$$

Now M was obtained **before** z was chosen (that is, M is independent of z in K_n). Thus by the Cauchy condition for uniform convergence of a sequence of functions, $\{f_{n_j}\}$ converges uniformly on K_n where n is arbitrary.

Now let K be **any** compact subset of D. We shall show that

$$K \subset K_s \quad \text{for some } s \text{ in } N. \tag{11.29}$$

Since $\{f_{n_j}\}$ converges uniformly on K_s, it certainly converges uniformly on K. Thus $\mathcal{F}$ is a normal family. To prove (11.29), we observe that

$$
\begin{aligned}
K \subset D &= (K_1 \cup K_2 \cup K_3 \cup \cdots) \quad \text{by (11.15)} \\
&= K_1^\circ \cup K_2^\circ \cup K_3^\circ \cup \cdots \quad \text{by the second part of (11.14)} \quad (11.30)
\end{aligned}
$$

(since each term of the first union is contained in some term of the second union and vice versa). By (11.30), we see that $\{K_n^\circ : \; n \in N\}$ is an open cover for K. Since K is compact, there is an s such that

$$K \subset K_1^\circ \cup K_2^\circ \cup \cdots \cup K_s^\circ \subset K_s.$$

11.3   The Riemann Mapping Theorem

In this section, we let D be a simply connected domain in R^2 such that $D \neq R^2$, and we let

$$\mathcal{F} = \{f : \; f \in A(D), \; f \text{ is one-to-one on } D \text{ and } f(D) \subset U\}$$

where U is the open unit disk $\{z : \; |z| < 1\}$.

Lemma 11.3.1 The set $\mathcal{F}$ is not empty. That is, there exists a one-to-one analytic mapping of D into U.

Proof. Let

$$a \in (R^2 - D). \tag{11.31}$$

By 10.10.3, we let

$$f \text{ be a branch in } D \text{ of } (z-a)^{\frac{1}{2}}. \tag{11.32}$$

Now f is one-to-one on D. For if $f(z_1) = f(z_2)$, then

$$z_1 - a = [f(z_1)]^2 = [f(z_2)]^2 = z_2 - a, \text{ and hence } z_1 = z_2.$$

Thus

$$f \text{ is one-to-one on } D. \tag{11.33}$$

Hence $f(D)$ is open by the Open Mapping Theorem, 8.2.14. Now let $h(w) = -w$ for each $w \in f(D)$ and let

$$E = h[f(D)].$$

Then by the Open Mapping Theorem,

$$E \text{ is open.} \tag{11.34}$$

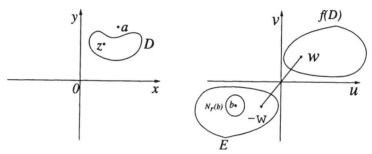

Figure 11.2: $E = h[f(D)]$

See Figure 11.2.

Let z_1 and z_2 be points in D such that $z_1 \neq z_2$. Then $f(z_1) \neq -f(z_2)$. [For if $f(z_1) = -f(z_2)$, then squaring both sides, we obtain $z_1 - a = z_2 - a$ which implies $z_1 = z_2$.] This means [since $0 \notin f(D)$ by (11.31) and (11.32)] that

$$E \cap f(D) = \emptyset. \tag{11.35}$$

Let $b \in E$. By (11.34), let $r > 0$ such that $N_r(b) \subset E$. Hence $N_r(b) \cap f(D) = \emptyset$ by (11.35). Thus

$$|f(z) - b| \geq r \quad \text{for each } z \text{ in } D. \tag{11.36}$$

Now let

$$g(z) = \frac{1}{2} \frac{r}{f(z) - b} \quad \text{for each } z \text{ in } D.$$

Then $g \in A(D)$ and g is one-to-one on D by (11.33). Also by (11.36),

$$|g(z)| \leq \frac{1}{2} < 1.$$

Hence $g(D) \subset U$. Thus $g \in \mathcal{F}$.

Remarks 11.3.2 Let $a \in U$. We shall use T_a to denote the bilinear transformation

$$T_a(z) = \frac{z - a}{1 - \bar{a}z} = w. \tag{11.37}$$

By 10.3.8, we see that T_a is a one-to-one analytic mapping of U **onto** U. Thus

$$\text{if } f \in \mathcal{F}, \text{ then } (T_a \circ f) \in \mathcal{F}. \tag{11.38}$$

See 2.3.6. By (11.37),

$$a \text{ is the only zero of } T_a. \tag{11.39}$$

Solving (11.37) for z, we obtain

$$T_a^{-1} = T_{-a}. \tag{11.40}$$

Lemma 11.3.3 If $f \in \mathcal{F}$ such that $f(D) \neq U$, then for any z_0 in D, there is some g in $\mathcal{F}$ such that

$$|g'(z_0)| > |f'(z_0)|. \tag{11.41}$$

Proof. Let $z_0 \in D$, and let

$$f_1 = T_c \circ f \quad \text{where } c = f(z_0) \in U. \tag{11.42}$$

Then $f_1 \in \mathcal{F}$ by (11.38), and

$$f_1(z_0) = T_c(f(z_0)) = 0. \tag{11.43}$$

Now $f_1(D) \neq U$, since $f(D) \neq U$ and since T_c is a one-to-one mapping of U onto U. Thus let

$$a \in [U - f_1(D)]. \tag{11.44}$$

Since $f_1 \in \mathcal{F}$,

$$(T_a \circ f_1) \in \mathcal{F} \quad \text{by (11.38)}$$

and

$$T_a \circ f_1 \text{ has no zero in } D \quad \text{by (11.39) and (11.44)}.$$

Thus by Ex. 1 in 10.10.5, we let

$$h \text{ be a branch in } D \text{ of } [(T_a \circ f_1)(z)]^{\frac{1}{2}}. \tag{11.45}$$

Then

$$h \in \mathcal{F} \tag{11.46}$$

since $T_a(f_1(z)) \in U$ implies $\sqrt{T_a(f_1(z))} \in U$.

Finally, we let

$$g = T_b \circ h \quad \text{where } b = h(z_0). \tag{11.47}$$

Then $g \in \mathcal{F}$ by (11.38), (11.46), and (11.47), and we shall prove that (11.41) holds.

By (11.47),

$$h = T_b^{-1} \circ g. \tag{11.48}$$

Squaring both sides in (11.48) and using (11.45) and (11.40),

$$\begin{aligned} T_a \circ f_1 &= h^2 = (T_b^{-1} \circ g)^2 \\ &= s \circ (T_{-b} \circ g) \quad \text{where } s(w) = w^2. \end{aligned} \tag{11.49}$$

By (11.49),

$$f_1 = (T_{-a} \circ s \circ T_{-b}) \circ g = H \circ g \tag{11.50}$$

where

$$H = T_{-a} \circ s \circ T_{-b}. \tag{11.51}$$

Using the chain rule in (11.50), we have

$$
\begin{aligned}
f_1'(z_0) &= H'(g(z_0))\,g'(z_0) \\
&= H'(0)\,g'(z_0)
\end{aligned}
\tag{11.52}
$$

since by (11.47),

$$
g(z_0) = T_b\,(h(z_0)) = T_b(b) = 0 \quad \text{by (11.37).}
\tag{11.53}
$$

We shall prove

$$
H(0) = 0, \quad |H'(0)| < 1 \quad \text{and} \quad g'(z_0) \neq 0.
$$

Now

$$
\begin{aligned}
0 &= f_1(z_0) \quad \text{by (11.43)} \\
&= H(g(z_0)) = H(0) \quad \text{by (11.50) and (11.53).}
\end{aligned}
$$

Since T_{-b} is a one-to-one mapping of U onto U and s is not one-to-one on U, it follows that H is not one-to-one on U. Thus H is **not** of the form $H(z) = kz$ where $k \in R^2$. Hence by Schwarz's Lemma (6.2.13), since $H(0) = 0$ and $H(U) \subset U$,

$$
|H'(0)| < 1.
\tag{11.54}
$$

Now $g \in A(D)$ and g is one-to-one on D. Thus by 8.2.13,

$$
g'(z_0) \neq 0.
\tag{11.55}
$$

By (11.52), (11.54), and (11.55),

$$
|g'(z_0)| > |f_1'(z_0)|.
\tag{11.56}
$$

By (11.42),

$$
\begin{aligned}
|f_1'(z_0)| &= |T_c'(c)f'(z_0)| \\[2mm]
&= \left| \frac{1}{1 - |c|^2} f'(z_0) \right| \\[2mm]
&\geq |f'(z_0)| \quad \text{since } \frac{1}{1 - |c|^2} \geq 1.
\end{aligned}
\tag{11.57}
$$

Now (11.41) follows from (11.56) and (11.57).

Lemma 11.3.4 For a fixed z_0 in D, let

$$
B = \mathrm{lub}\,\{|f'(z_0)| : \ f \in \mathcal{F}\}.
\tag{11.58}
$$

Then B is finite, and there is some member g of $\mathcal{F}$ such that

$$
|g'(z_0)| = B.
\tag{11.59}
$$

Proof. By 11.3.1, the set $\mathcal{F}$ is not empty. Thus whether B is finite or infinite, there exists a sequence $\{f_n\}$ in $\mathcal{F}$ such that

$$|f_n'(z_0)| \to B \quad \text{as } n \to \infty. \tag{11.60}$$

[For if B is finite, then for each n in N, we let $f_n \in \mathcal{F}$ such that $|f_n'(z_0)| > B - \frac{1}{n}$. But if $B = +\infty$, then we let $f_n \in \mathcal{F}$ such that $|f_n'(z_0)| > n$. In either case, $|f_n'(z_0)| \to B$ as $n \to \infty$.] Now $\mathcal{F}$ is uniformly bounded on compact subsets in D [since $|f(z)| < 1$ on D for each f in $\mathcal{F}$]. By 11.2.4, the set $\mathcal{F}$ is a normal family in D. Hence there exists a subsequence $\{g_n\}$ of $\{f_n\}$ such that if K is a compact subset of D,

$$g_n \xrightarrow{U\,K} g, \tag{11.61}$$

where g is defined on D. By (11.61) and 11.1.7(a),

$$g \in A(D). \tag{11.62}$$

By 11.1.7(b),

$$g_n'(z_0) \to g'(z_0) \quad \text{as } n \to \infty. \tag{11.63}$$

Hence

$$|g'(z_0)| = B \quad \text{by (11.60) and (11.63).}$$

Thus B is finite.

Now g_n is one-to-one on D (since $g_n \in \mathcal{F}$). Thus g is one-to-one on D or g is constant by (11.61) and Ex. 2(b) in 11.1.8. But g is not constant since $|g'(z_0)| = B \neq 0$. (We know that $B \neq 0$ since f is one-to-one on D for each f in $\mathcal{F}$ and thus $f'(z_0) \neq 0$ by 8.2.13.) Thus

$$g \text{ is one-to-one on } D. \tag{11.64}$$

Finally, $g_n(D) \subset U$. Thus $g(D) \subset \overline{U}$ by (11.61). But by the Open Mapping Theorem and (11.64), we know that $g(D)$ is an open subset of $\overline{U}$. Hence

$$g(D) \subset U. \tag{11.65}$$

Now $g \in \mathcal{F}$ by (11.62), (11.64), and (11.65).

Theorem 11.3.5 (Riemann Mapping Theorem). If D is a simply connected domain in R^2 such that $D \neq R^2$, then there exists a one-to-one analytic mapping of D **onto** the open unit disk U.

Proof. Let $z_0 \in D$ and let

$$\mathcal{F} = \{f : \ f \in A(D), \ f \text{ is one-to-one on } D \ \text{and} \ f(D) \subset U\}.$$

By 11.3.4, let $g \in \mathcal{F}$ such that

$$|g'(z_0)| = \text{lub}\,\{|f'(z_0)| : \ f \in \mathcal{F}\}. \tag{11.66}$$

Then g maps D **onto** U by 11.3.3. [For if $g(D)$ were not equal to U, then (by 11.3.3) there would be some member h of $\mathcal{F}$ such that $|h'(z_0)| > |g'(z_0)|$, contrary to (11.66).]

Exercises 11.3.6

1. State why there is no analytic mapping of R^2 onto the unit disk. (See Liouville's Theorem, 3.6.14.)

2. Let D_1 and D_2 be simply connected domains in R^2 such that neither of them is R^2. Prove there is a one-to-one analytic mapping of D_1 onto D_2.

3. Let D be a simply connected domain in R^2 such that $D \neq R^2$ and let $z_0 \in D$. Prove there is a one-to-one analytic mapping g of D onto U such that $g(z_0) = 0$ and such that $g'(z_0) > 0$.

4. Prove that for a fixed z_0 in D, the function g in Ex. 3 is unique.

Proof of Ex. 2. By 11.3.5 for $k = 1, 2$, let g_k be a one-to-one analytic mapping of D_k onto U. Then g_2' is never zero in D_2 by 8.2.13. Also, g_2^{-1} is analytic in U by 8.2.16. Thus $g_2^{-1} \circ g_1$ is a one-to-one analytic mapping of D_1 onto D_2.

Proof of Ex. 3. By 11.3.5, let f be a one-to-one analytic mapping of D onto U. Let

$$h = T_{f(z_0)} \circ f$$

where T_a is given in (11.37). Then

$$h(z) = \frac{f(z) - f(z_0)}{1 - \overline{f(z_0)}\, f(z)}.$$

Now h is a one-to-one analytic mapping of D onto U by 11.3.2 and

$$h(z_0) = 0. \qquad (11.67)$$

By 8.2.13,

$$h'(z_0) = |h'(z_0)|\, e^{i\alpha} \neq 0 \qquad (11.68)$$

where $\alpha = \operatorname{Arg} h'(z_0)$. Finally, let

$$g(z) = e^{-i\alpha}\, h(z) \qquad \text{for each } z \text{ in } D. \qquad (11.69)$$

Then $g(z_0) = 0$ by (11.67), and g is h followed by a rotation of U, since $|e^{-i\alpha}| = 1$. Thus g is a one-to-one analytic mapping of D onto U. By (11.69),

$$\begin{aligned}
g'(z_0) &= e^{-i\alpha}\, h'(z_0) \\
&= |h'(z_0)| > 0 \quad \text{by (11.68)}.
\end{aligned}$$

Proof of Ex. 4. Let f be **any** one-to-one analytic mapping of D onto U such that $f(z_0) = 0$ and $f'(z_0) > 0$. We must show that $f = g$. Let

$$h = f \circ g^{-1} \quad \text{on } U. \qquad (11.70)$$

Then, h is a one-to-one analytic mapping of U onto U by 8.2.16. Also, by (11.70) and the hypothesis that $f(z_0) = g(z_0) = 0$, we have

$$h(0) = 0. \tag{11.71}$$

Thus h satisfies the hypotheses of 6.2.13 (Schwarz's Lemma). Now let $z \in D$ and let

$$w = g(z) \quad \text{so that } z = g^{-1}(w). \tag{11.72}$$

Then

$$\begin{aligned}
|f(z)| &= |f(g^{-1}(w))| \quad \text{by (11.72)} \\
&= |h(w)| \quad \text{by (11.70)} \\
&\leq |w| \quad \text{by 6.2.13} \\
&= |g(z)| \quad \text{by (11.72)}.
\end{aligned} \tag{11.73}$$

Similarly by interchanging f and g,

$$|g(z)| \leq |f(z)| \quad \text{for each } z \text{ in } D. \tag{11.74}$$

Thus by (11.73) and (11.74),

$$|f(z)| = |g(z)| \quad \text{for each } z \text{ in } D. \tag{11.75}$$

By (11.70), (11.71), and the chain rule,

$$\begin{aligned}
h'(0) &= f'(z_0)\frac{1}{g'(z_0)} \quad \text{by 8.2.16} \\
&> 0 \quad \text{since } f'(z_0) > 0 \text{ and } g'(z_0) > 0 \quad \text{by hypothesis.}
\end{aligned} \tag{11.76}$$

Now for each w in U,

$$\begin{aligned}
|h(w)| &= |f(g^{-1}(w))| \quad \text{by (11.70)} \\
&= |g(g^{-1}(w))| \quad \text{by (11.75)} \\
&= |w|.
\end{aligned} \tag{11.77}$$

Thus by Schwarz's Lemma, for each w in U,

$$h(w) = aw \quad \text{where } |a| = 1. \tag{11.78}$$

By (11.76) and (11.78),

$$0 < h'(0) = a.$$

Hence by (11.78),

$$h(w) = w \quad \text{for each } w \text{ in } U.$$

Thus $h = e$ where e is the identity mapping of U onto U. By (11.70), this means that $f \circ g^{-1} = e$. Hence $f = e \circ g = g$. Since f was just any function with the prescribed properties in Ex. 3, it follows that g is unique.

Remark 11.3.7 Let D and D_1 be bounded simply connected domains such that $B = B(D)$ and $B_1 = B(D_1)$ are simple closed curves. It can be shown that each one-to-one analytic mapping of D onto D_1 can be extended to a homeomorphism of $\overline{D}$ onto $\overline{D}_1$.

A mapping f is a **homeomorphism** of X onto Y iff

f is a one-to-one mapping of X onto Y such that

f is continuous on X and such that f^{-1} is continuous on Y.

Appendix A

A.1 The System of Real Numbers

We are acquainted with the set of natural numbers; namely,

$$N = \{x : \ x \text{ is a natural number}\} = \{1, 2, 3, \cdots\},$$

the set of numbers used in counting. It was not until the year 1889 that G. Peano gave an axiomatic development of N. Starting with just five axioms, known as **Peano's axioms** (or **postulates**), he proved the familiar properties of N.

Definition A.1.1 An **algebraic system** is a set S together with some operations and relations on S. The real number system $(R, +, \cdot, <)$ is an example of an algebraic system. Another such example is the complex number system $(R^2, +, \cdot, |\,|)$. In these two systems, we notice that with each ordered pair a and b of numbers, the operation $+$ associates exactly one number denoted by $a + b$. Thus in the system of complex numbers, we recognize that the operation $+$ is in fact a function whose domain is $R^2 \times R^2$ and whose range is R^2 (where $R^2 \times R^2$ is the set of all ordered pairs of members of R^2).

The operations and relations mentioned in A.1.1 are usually binary operations and binary relations. But what is a binary operation in general? If S is an abstract set, then a **binary operation on S** is a function whose domain is $S \times S$ and whose range is contained in S (where $S \times S$ is the set of all ordered pairs of members of S).

A particular example of a binary relation on the set S is obtained if there is a rule such that for each ordered pair a and b of members of S, this rule determines whether or not a is related to b. We may then let H be the set of all ordered pairs (a, b) such that a is related to b. If this set H of ordered pairs is known, then the relation is completely determined. Thus to obtain a precise mathematical definition of a binary relation on a set S, we adopt the following definition.

A **binary relation on S** is a subset of $S \times S$.

We often use symbols such as $+$, $\cdot$, and $\circ$ to denote binary operations and we may use symbols $<$, $\approx$, $\geq$, $=$, etc. to denote binary relations. In general, if an algebraic system consists of the set S together with the binary operations $\cdot$ and $\circ$ and the binary relation $\approx$, then we denote this system by

$$(S, \cdot, \circ, \approx).$$

[We should notice that in the system of complex numbers, the thing denoted by $|\ |$ is neither a relation nor a binary operation. However, we may refer to it as a "monary operation," since it is a function which associates with each number z in R^2 exactly one number $|z|$ in R, which is a subset of R^2.]

Now from the system $(N, +, \cdot, \leq)$ of natural numbers, the system $(J, +, \cdot, \leq)$ of integers is developed where

$$J = \{x : x \text{ is an integer}\} = \{\cdots, -3, -2, -1, 0, 1, 2, 3, \cdots\}.$$

From the system of integers, the system $(Q, +, \cdot, \leq)$ of rational numbers is developed where

$$Q = \left\{\frac{x}{y} : x \in J, y \in J, \text{ and } y \neq 0\right\}.$$

From the system of rational numbers, the system $(R, +, \cdot, \leq)$ of real numbers is developed. All of these number systems are based upon the Peano postulates and the definitions of the operations and relations involved.

Definition A.1.2 A **field** is an algebraic system $(F, +, \cdot)$ where F is a set of at least two elements and where $+$ and $\cdot$ are binary operations on F such that the following conditions are satisfied.

(a) If a, b, and c are members of F, then the following properties hold.

$a + b = b + a \quad a \cdot b = b \cdot a \quad$ (commutative laws)

$$\left.\begin{array}{l} (a + b) + c = a + (b + c) \\ (a \cdot b) \cdot c = a \cdot (b \cdot c) \end{array}\right\} \quad \text{(associative laws)}$$

$a \cdot (b + c) = a \cdot b + a \cdot c \quad$ (distributive law)

(b) There are additive and multiplicative identities in F, denoted by 0 and 1, such that

$$0 + x = x \quad \text{and} \quad 1 \cdot x = x \quad \text{for each } x \text{ in } F.$$

(c) For each x in F, there is an additive inverse $-x$ in F such that

$$-x + x = 0$$

and if $x \neq 0$, there is a multiplicative inverse x^{-1} in F such that

$$x \cdot x^{-1} = 1.$$

We shall write ab for $a \cdot b$ and $a - b$ for $a + (-b)$.

Definition A.1.3 An **ordered field** is an algebraic system $(F, +, \cdot, >)$ in which $(F, +, \cdot)$ is a field and in which $>$ is a binary relation on F such that if a and b are any members of F, then the following conditions are satisfied.

(a) Exactly one of the relations

$$0 > a, \quad a = 0, \quad \text{or} \quad a > 0$$

must hold.

(b) If $a > 0$ and $b > 0$, then $a + b > 0$ and $ab > 0$.

(c) $a > b$ iff $a - b > 0$.

Notation A.1.4 (a) If $(F, +, \cdot, >)$ is an ordered field, we often refer to the field F without explicitly mentioning the operations and relations since they are clearly understood to be a part of the system. Also, if we wish to stress the fact that the field is ordered, we may refer to the ordered field $(F, >)$ without mentioning the operations.

(b) When $>$ is an order relation, we may write

$$x < y \quad \text{iff} \quad y > x,$$
$$x \geq y \quad \text{iff} \quad x > y \text{ or } x = y, \quad \text{and}$$
$$x \leq y \quad \text{iff} \quad x < y \text{ or } x = y.$$

Definition A.1.5 Let $E \subset F$ and let $b \in F$ where $(F, >)$ is an ordered field. Then b is an **upper bound** of E iff $x \leq b$ for each x in E. Also, b is the **least upper bound** of E iff

b is an upper bound of E and

$b \leq c$ for each upper bound c of E.

We denote the least upper bound of E by lub E.

Definition A.1.6 An ordered field $(F, >)$ is **complete** iff each nonempty subset of F with an upper bound in F has a least upper bound in F.

Remarks A.1.7 When the real number system is developed, it is proved that $(R, +, \cdot, >)$ is a complete ordered field. Furthermore, any complete ordered field is essentially the real number system. Thus we shall define the **real number system** as a complete ordered field.

Definition A.1.8 The **real number system** $(R, +, \cdot, >)$ is a system which satisfies the conditions in A.1.2, A.1.3, and A.1.6.

Since the completeness property of R is extremely important in analysis, we state it specifically.

Completeness Property for R A.1.9   Each nonempty subset of R with an upper bound in R has a least upper bound in R.

In the development of the real number system, after the set of positive integers is identified with the set N of natural numbers and other similar identifications are made, we have

$$N \subset J \subset Q \subset R.$$

A number x is called an **irrational number** iff $x \in (R - Q)$. The following theorems A.1.10(a) – (d), stated without proofs, show the existence of many irrational numbers. Speaking intuitively, the irrational numbers are more numerous than the rational numbers.

Theorem A.1.10 **(a)** If p is a prime, then $\sqrt{p} \in (R - Q)$.

(b) The values of the trigonometric functions are irrational at nonzero rational values of the arguments.

(c) The numbers π and e are irrational.

(d) If $r \in Q$ and $r \neq 0$, then e^r is irrational. Also, if $r \in Q$, $r > 0$ and $r \neq 1$, then $\text{Log}\, r$ is irrational.

Definition A.1.11 **(a)** Let $E \subset F$ where $(F, >)$ is an ordered field. The set E is **dense in F** iff the following condition is satisfied. If a and b are members of F and $a < b$, then there is some c in E such that $a < c < b$. (Such an element c is said to be between a and b.)

(b) To indicate that F is dense in F, we simply say "F is **dense**."

Theorem A.1.12 Let $(F, >)$ be an ordered field. Then each of the following relations hold.
 (a) $1 > 0$ (b) $2 > 1$ where 2 denotes $1 + 1$ (c) $2^{-1} > 0$

Proof of (a). By A.1.3(a) and Ex. 3(f) in A.1.17, we have $1 < 0$ or $1 > 0$. Suppose $1 < 0$. Then $-1 > 0$ by Ex. 10 in A.1.17. So by Ex. 5(d) in A.1.17 and A.1.3(b), we have $1 = (-1)(-1) > 0$, a contradiction to our assumption. Thus $1 > 0$.

Proof of (b). Now $2 - 1 = (1 + 1) - 1 = 1 + (1 - 1) = 1 > 0$. Thus $2 > 1$ by A.1.3(c).

Proof of (c). By (a) and (b), we have $0 < 1 < 2$. Thus $0 < 2^{-1}$ by Ex. 17 in A.1.17.

Theorem A.1.13 If $(F, >)$ is an ordered field, then F is dense.

Proof. Let a and b be members of F such that $a < b$. Then by Ex. 8 in A.1.17,

$$2a = (1 + 1)a = a + a < a + b < b + b = 2b.$$

Thus by Ex. 16 in A.1.17 and A.1.12(c),

$$a = 2^{-1}(2a) < 2^{-1}(a + b) < 2^{-1}(2b) = b.$$

Hence $c = \frac{a+b}{2}$ is a member of F such that $a < c < b$.

Using Ex. 11 and 12 in A.1.17, we obtain the following corollaries from A.1.13.

Corollary A.1.14 Between any two members of an ordered field F, there are infinitely many elements of F.

Corollary A.1.15 Between any two real numbers, there are infinitely many real numbers.

Proof. This follows from A.1.14 since R is an ordered field by A.1.7.

Definition A.1.16 Let $(F, >)$ be an ordered field. Let $a \in F$. Then

$$|a| = \left\{ \begin{array}{ll} a & \text{if } a \geq 0 \\ -a & \text{if } a < 0. \end{array} \right.$$

Exercises A.1.17

1. Let $F = \{0, 1\}$ and let $0 + 0 = 1 + 1 = 0$, $0 + 1 = 1 + 0 = 1$, $0 \cdot 0 = 0 \cdot 1 = 1 \cdot 0 = 0$ and $1 \cdot 1 = 1$. Prove $(F, +, \cdot)$ is a field but there is no order relation which makes F an ordered field.

2. Let $S = \{0, 1, 2\}$. Let $+$ and $\cdot$ be binary operations on S given by the following tables.

$+$	0	1	2
0	0	1	2
1	1	0	0
2	2	0	1

$\cdot$	0	1	2
0	0	0	0
1	0	1	2
2	0	2	1

   Show that $(S, +, \cdot)$ is a field, but there is no order relation which makes S an ordered field.

3. Let F be a field and let a, b, and x be members of F. Prove each of the following properties.

(a) There is only one additive identity in F. **Hint.** $0 = 0 + 0' = 0'$

(b) **Cancellation Law.** If $a + x = b + x$, then $a = b$.
   Hint. $a = a + 0 = a + (x - x) = (a + x) - x$.
   Alternative Hint. If $a + x = b + x$, then $a + x + (-x) = b + x + (-x)$.

(c) There is only one additive inverse of x.
      Proof. Suppose $a + x = 0 = b + x$. Then $a = b = -x$ by Part (b).

(d) $0x = 0$ **Hint.** We have $0x + 0x = (0 + 0)x = 0x = 0 + 0x$. Use Part (b).

(e) There is only one multiplicative identity in F. **Hint.** $1 = 1 \cdot 1' = 1'$

(f) The additive identity is not equal to the multiplicative identity. **Hint.** By Part (a) and A.1.2, let $b \in F$ such that $b \neq 0$. If $0 = 1$, then $0 = 0b = 1b = b$, a contradiction.

4. Prove that in a field, if $ab = 0$, then $a = 0$ or $b = 0$. **Hint.** Assume $a \neq 0$ and show that b must be 0. Multiply both sides of $ab = 0$ by a^{-1} and use Ex. 3(d).

5. Let a and b be elements of a field F. Prove each of the following results.

   (a) $-(-a) = a$.   **Proof.** Now $a + (-a) = 0$. Thus a is an additive inverse of $-a$. Use Ex. 3(c).

   (b) $a(-b) = -(ab)$.   **Proof.** $0 = a \cdot 0 = a[b + (-b)] = ab + a(-b)$

   (c) $(-a)b = -(ab)$.   **Hint.** Multiplication is commutative.

   (d) $(-a)(-b) = ab$.   [This follows from (b), (c), and (a).]

6. Let $S = \{0, 1, 2, 3\}$. If $+$ and $\cdot$ are given by the following tables, is the algebraic system $(S, +, \cdot)$ a field? **Hint.** See Ex. 4 and check the second table for $2 \cdot 2$.

+	0	1	2	3
0	0	1	2	3
1	1	2	3	0
2	2	3	0	1
3	3	0	1	2

·	0	1	2	3
0	0	0	0	0
1	0	1	2	3
2	0	2	0	2
3	0	3	2	1

7. Prove if a and b are members of a field, then $-(a + b) = -a - b$.

In Ex. 8–19, let $(F, >)$ be an ordered field, and let a, b, and c be elements of F. Prove the stated results.

8. If $a > b$, then $a + c > b + c$.   Notice $(a + c) - (b + c) = a - b$ and use A.1.3(c).

9. If $a > 0$, then $-a < 0$.   By Ex. 8, we have $0 = a - a > 0 - a = -a$.

10. If $a < 0$, then $-a > 0$.   Use Ex. 8 or A.1.3(c).

11. Exactly one of the conditions $a > b$, $a = b$, or $a < b$ holds.

12. **The Transitive Law.** If $a > b$ and $b > c$, then $a > c$.

13. If $a > 0$ and $b < 0$, then $ab < 0$.

14. If $a < 0$ and $b < 0$, then $ab > 0$.

15. If $a \neq 0$, then $aa > 0$.

16. If $a < b$ and $c > 0$, then $ac < bc$.

17. (a) If $a > 0$, then $a^{-1} > 0$. **Hint.** Use A.1.3(a), Ex. 3(d), A.1.12(a), and Ex. 13.

    (b) If $0 < a < b$, then $0 < b^{-1} < a^{-1}$.

18. $|ab| = |a||b|$

19. **Triangle Inequality.** $|a + b| \leq |a| + |b|$

$$\text{Hint. } |a + b|^2 = (a + b)^2 = |a|^2 + 2ab + |b|^2$$
$$\leq |a|^2 + 2|a||b| + |b|^2 = (|a| + |b|)^2$$

20. Prove if F is a complete ordered field, then F must satisfy the following property, which is called the **Archimedean Property.**

    If $a \in F$, $b \in F$, and $a > 0$, then $na > b$ for some positive integer n where na denotes the sum $a + a + a + \cdots + a$ (n terms).

    Proof. We suppose $a > 0$ and b is an upper bound of the set $E = \{ka : k \in N\}$. Since F is complete, we let $L = \text{lub } E$. Then $ka \leq L$ for each k in N. Thus for each integer $k > 1$,

    $$(k - 1)a = ka - a \leq L - a \quad \text{by Ex. 8.}$$

    Now $-a < 0$ by Ex. 9. Hence $L - a < L + 0 = L$ by Ex. 8. Now the last displayed inequality means that $L - a$ is an upper bound of E, contrary to our choice of L. Thus we must reject our assumption that b is an upper bound of E. This means that $na > b$ for some natural number n.

21. Let c and d be real numbers such that $0 < c < d$. Prove there is some rational number r such that $c < r < d$. **Hint.** By Ex. 20, with $a = d - c$ and $b = 1$, let $q \in N$ such that $1 < q(d - c)$. Then $qc + 1 < qd$. By Ex. 20, with $a = 1$ and $b = qc$, we have $qc < m$ for some m in N. Let p be the least member of N such that $qc < p$. Then $p - 1 \leq qc$. Thus $p \leq qc + 1 < qd$. Hence $qc < p < qd$. Now $\frac{1}{q} > 0$ by Ex. 17. Finally, use Ex. 16.

22. Prove that between any two real numbers, there are infinitely many rational numbers and infinitely many irrational numbers. **Hint.** Let a and b be real numbers such that $0 < a < b$. By Ex. 21, let c and d be rational numbers such that $a < c < d < b$. Now for each prime p, let

$$x_p = c + \frac{d - c}{\sqrt{p}}.$$

23. Let $E \subset F$ and let $b \in F$ where $(F, >)$ is an ordered field. Then b is called a **lower bound** of E iff $b \leq x$ for each x in E. Also, b is called the **greatest lower bound** of E (glb E) iff

    b is a lower bound of E, and

    $b \geq c$ if c is any lower bound of E.

Prove that in a complete ordered field, each nonempty set with a lower bound has a greatest lower bound. (Thus each nonempty set E of real numbers with a lower bound has a greatest lower bound.) **Hint.** Let $H = \{x : x$ is a lower bound of $E\}$. Then $H \neq \emptyset$ by hypothesis. Each element of E is an upper bound of H. Let $L = \mathrm{lub}\, H$. To show that L is a lower bound of E, suppose it is not and reach a contradiction. Then show that $L = \mathrm{glb}\, E$ by the method of contradiction.

A.2   Countable Sets

If $k \in N$, we use N_k to denote the set $\{1, 2, 3, \cdots, k\}$.

Definitions A.2.1 (a) A set X is **countable** iff $X = \emptyset$ (the empty or null set) or there is a mapping of N onto X. A set is **uncountable** iff it is not countable.

(b) A set X is **finite** iff $X = \emptyset$ or there is a mapping of N_k onto X for some k in N. A set is **infinite** iff it is not finite.

Remarks A.2.2 (a) Roughly speaking, A.2.1(a) asserts that X is countable iff the number of elements of X is not greater than the number of elements of N.

(b) Clearly, N is countable. For let f be the identity mapping given by $f(x) = x$ for each x in N. Then f maps N onto N. Thus N is countable by A.2.1(a).

(c) Let $B \subset X$. If X is countable, then B is countable. For if $B \neq \emptyset$, let $b_0 \in B$ and let f be a mapping of N onto X. Now for each x in N, let

$$g(x) = \begin{cases} f(x) & \text{if } f(x) \in B \\ b_0 & \text{if } f(x) \notin B. \end{cases}$$

Then g maps N onto B. Thus B is countable by A.2.1(a).

(d) Let h map X onto B. If X is countable, then B is countable. For if $B \neq \emptyset$, let f be a mapping of N onto X. The composite mapping, given by $g(x) = h(f(x))$, maps N onto B.

(e) Each finite set is countable. This follows from (b), (c), and (d). [Let B be a nonempty finite set. Let h map N_k onto B for some k in N. By (b) and (c), the set N_k is countable. Thus B is countable by (d).]

If S is a set, it is customary to use $S \times S$ to denote the set of all ordered pairs of members of S. Thus

$$S \times S = \{(x,y) : \ x \in S \text{ and } y \in S\}.$$

Theorem A.2.3 The set $N \times N$ is countable.

Proof. Let $A = \{x : \ x = 2^m 3^n \text{ where } m \in N \text{ and } n \in N\}$. Then $A \subset N$ and hence by A.2.2(b) and (c),

$$A \text{ is countable}. \tag{A.1}$$

Now for each x in A, let $h(x) = (m,n)$ where (m,n) is the unique member of $N \times N$ such that $x = 2^m 3^n$. (Note that $2^m 3^n = 2^p 3^q$ implies $m = p$ and $n = q$.) Thus h is a well defined mapping of A onto $N \times N$. To see that h maps A **onto** $N \times N$, let $(r,s) \in N \times N$. Then $x = 2^r 3^s \in A$ and $h(x) = (r,s)$. Thus $N \times N$ is countable by (A.1) and A.2.2(d).

Remark A.2.4 A nonempty set E is countable iff E is the range of a sequence, that is, $E = \{x_1, x_2, x_3, \cdots\}$. (Of course, if E is not infinite, some member is repeated infinitely many times.)

Theorem A.2.5 The union of a countable collection of countable sets is countable.

Proof. Let $\mathcal{C}$ be a countable collection of sets, each of which is countable. If $\bigcup \mathcal{C} = \emptyset$, then this union is countable by A.2.1(a). Suppose then that $\bigcup \mathcal{C} \neq \emptyset$, and let $\mathcal{B} = \{A : \ A \in \mathcal{C} \text{ and } A \neq \emptyset\}$. Then

$$\bigcup \mathcal{B} = \bigcup \mathcal{C}. \tag{A.2}$$

By A.2.4, we may write $\mathcal{B} = \{A_1, A_2, A_3, \cdots\}$ where each a_k may be written as indicated in the following display.

$$
\begin{aligned}
A_1 &= \{x_{11}, x_{12}, x_{13}, \cdots, x_{1n}, \cdots\} \\
A_2 &= \{x_{21}, x_{22}, x_{23}, \cdots, x_{2n}, \cdots\} \\
&\vdots \\
A_m &= \{x_{m1}, x_{m2}, x_{m3}, \cdots, x_{mn}, \cdots\} \\
&\vdots
\end{aligned}
$$

Now for each (m,n) in $N \times N$, let $h(m,n) = x_{mn}$. Clearly, h maps $N \times N$ onto $\bigcup \mathcal{B}$. Thus $\bigcup \mathcal{B}$ is countable by A.2.3 and A.2.2(d). Hence $\bigcup \mathcal{C}$ is countable by (A.2).

Corollary A.2.6 (a) The set J of all integers is countable.

(b) The set Q of all rational numbers is countable.

Proof of (a). Let $h(x) = -x$ for each x in N. Then $h(N) = N_-$ where N_- denotes the set of all negative integers. Thus N_- is countable by A.2.1(a). Now $J = N \cup \{0\} \cup N_-$. Thus J is countable by A.2.5.

Proof of (b). For each fixed k in N, we let $E_k = \left\{ \frac{m}{k} : m \in J \right\}$. Then each E_k is countable by Part (a) and A.2.2(d). [For a fixed k, we let $h_k(m) = \frac{m}{k}$ for each m in J.] Also, the collection $\mathcal{C} = \{E_1, E_2, E_3, \cdots\}$ is countable by A.2.4. But $Q = \bigcup \mathcal{C}$. Hence Q is countable by A.2.5.

Remarks A.2.7 We observe that a real number may or may not have two different decimal representations. For example, $4.57000 \cdots$ and $4.56999 \cdots$ represent the same number. However, the number represented by $3.75242424 \cdots$ has only the given decimal representation. It is important to observe that if a number has two decimal representations, then one repeats the digit 0 from some place onward and the other repeats the digit 9 from some place onward.

Theorem A.2.8 The set R of all real numbers is uncountable.

Proof. Let f be **any** mapping of N into R. We shall show that f is not **onto** R. To do this, we construct a real number r which is not in the range of f. Let each number $f(n)$ have a definite decimal representation, and let

$$r = 8.d_1 d_2 d_3 \cdots$$

where

$$d_n = \begin{cases} 2 & \text{if the } n\text{th decimal place in our representation of } f(n) \text{ is } 6 \\ 6 & \text{otherwise.} \end{cases}$$

Hence r differs from $f(n)$ in the nth decimal place for each n in N. Since the digits 9 and 0 do not appear in the decimal representation of r, it follows from A.2.7 that the decimal representation of r is unique. Thus for each n in N, we see that $r \neq f(n)$ and hence f is not onto R. Since f is an arbitrary mapping of N into R, it follows that there is no mapping of N **onto** R. Thus R is uncountable.

Remark A.2.9 In A.2.10, we give a short proof that Q is countable which uses only A.2.1 and the Fundamental Theorem of Arithmetic.

Alternate Proof that Q is Countable A.2.10 For each x in N, let

$$f(x) = \begin{cases} \dfrac{m}{n} & \text{if } x = 2^m 3^n \text{ where } m \in N \text{ and } n \in N \\[2mm] -\dfrac{m}{n} & \text{if } x = 5^m 7^n \text{ where } m \in N \text{ and } n \in N \\[2mm] 0 & \text{otherwise.} \end{cases}$$

Clearly, f maps N **onto** Q. Thus Q is countable by A.2.1. [For a given x, we know that $f(x)$ is uniquely determined by the Fundamental Theorem of Arithmetic, which states that x has a unique factorization into primes except for order.]

Theorem A.2.11 Let Q be the set of all rational numbers.

(a) The set $Q \times Q$ is countable.

(b) Any subset of $Q \times Q$ is countable.

Proof. By A.2.6(b), let f be a mapping of N onto Q. For each (m, n) in $N \times N$, let $g(m, n) = [f(m), f(n)]$. Clearly, g maps $N \times N$ onto $Q \times Q$. Thus $Q \times Q$ is countable by A.2.2(d) and A.2.3.

Now Part (b) follows from Part (a) and A.2.2(c).

By definition, a real number x is called an **irrational number** iff x is not rational. Thus our final theorem follows from A.2.8, A.2.6(b), and A.2.5.

Theorem A.2.12 The set of all irrational numbers is uncountable.

Index